Frontiers of Semiconductor Lasers

Frontiers of Semiconductor Lasers

Editors

Yongyi Chen
Li Qin

MDPI • Basel • Beijing • Wuhan • Barcelona • Belgrade • Manchester • Tokyo • Cluj • Tianjin

Editors
Yongyi Chen
Chinese Academy of Sciences
Changchun, China

Li Qin
Chinese Academy of Sciences
Changchun, China

Editorial Office
MDPI
St. Alban-Anlage 66
4052 Basel, Switzerland

This is a reprint of articles from the Special Issue published online in the open access journal *Crystals* (ISSN 2073-4352) (available at: https://www.mdpi.com/journal/crystals/special_issues/semicondutor_lasers).

For citation purposes, cite each article independently as indicated on the article page online and as indicated below:

LastName, A.A.; LastName, B.B.; LastName, C.C. Article Title. *Journal Name* **Year**, *Volume Number*, Page Range.

ISBN 978-3-0365-6940-6 (Hbk)
ISBN 978-3-0365-6941-3 (PDF)

© 2023 by the authors. Articles in this book are Open Access and distributed under the Creative Commons Attribution (CC BY) license, which allows users to download, copy and build upon published articles, as long as the author and publisher are properly credited, which ensures maximum dissemination and a wider impact of our publications.

The book as a whole is distributed by MDPI under the terms and conditions of the Creative Commons license CC BY-NC-ND.

Contents

About the Editors vii

Preface to "Frontiers of Semiconductor Lasers" ix

Shunhua Wu, Te Li, Zhenfu Wang, Lang Chen, Jiachen Zhang, Junyue Zhang, Jiachen Liu, et al.
Study of Temperature Effects on the Design of Active Region for 808 nm High-Power Semiconductor Laser
Reprinted from: *Crystals* **2023**, *13*, 85, doi:10.3390/cryst13010085 1

Qiaoxia Gong, Mengxin Zhang, Chaonan Lin, Xun Yang, Xihong Fu, Fengying Ma, Yongsheng Hu, et al.
Analysis of Thermal Effects in Kilowatt High Power Diamond Raman Lasers
Reprinted from: *Crystals* **2022**, *12*, 1824, doi:10.3390/cryst12121824 17

Jinliang Han, Jun Zhang, Xiaonan Shan, Yawei Zhang, Hangyu Peng, Li Qin and Lijun Wang
Tunable, High-Power, Narrow-Linewidth Diode Laser for Potassium Alkali Metal Vapor Laser Pumping
Reprinted from: *Crystals* **2022**, *12*, 1675, doi:10.3390/cryst12111675 33

Zhuo Zhang, Jianwei Zhang, Yuxiang Gong, Yinli Zhou, Xing Zhang, Chao Chen, Hao Wu, et al.
Long-Distance High-Power Wireless Optical Energy Transmission Based on VECSELs
Reprinted from: *Crystals* **2022**, *12*, 1475, doi:10.3390/cryst12101475 43

Yuhang Ma, Keke Ding, Long Wei, Xuan Li, Junce Shi, Zaijin Li, Yi Qu, et al.
Research on Mid-Infrared External Cavity Quantum Cascade Lasers and Applications
Reprinted from: *Crystals* **2022**, *12*, 1564, doi:10.3390/cryst12111564 55

Nan Zhang, Quanxin Na, Qijie Xie and Siqi Jia
Development of Solution-Processed Perovskite Semiconductors Lasers
Reprinted from: *Crystals* **2022**, *12*, 1274, doi:10.3390/cryst12091274 71

Yanxin Shen, Xinpeng Fu, Cong Yao, Wenyuan Li, Yubin Wang, Xinrui Zhao, Xihong Fu, et al.
Optical Crystals for 1.3 μm All-Solid-State Passively Q-Switched Laser
Reprinted from: *Crystals* **2022**, *12*, 1060, doi:10.3390/cryst12081060 93

Bin Wang, Yugang Zeng, Yue Song, Ye Wang, Lei Liang, Li Qin, Jianwei Zhang, et al.
Principles of Selective Area Epitaxy and Applications in III–V Semiconductor Lasers Using MOCVD: A Review
Reprinted from: *Crystals* **2022**, *12*, 1011, doi:10.3390/cryst12071011 111

Shen Niu, Yue Song, Ligong Zhang, Yongyi Chen, Lei Liang, Ye Wang, Li Qin, et al.
Research Progress of Monolithic Integrated DFB Laser Arrays for Optical Communication
Reprinted from: *Crystals* **2022**, *12*, 1006, doi:10.3390/cryst12071006 139

Keke Ding, Yuhang Ma, Long Wei, Xuan Li, Junce Shi, Zaijin Li, Yi Qu, et al.
Research on Narrow Linewidth External Cavity Semiconductor Lasers
Reprinted from: *Crystals* **2022**, *12*, 956, doi:10.3390/cryst12070956 161

Yue Song, Zhiyong Lv, Jiaming Bai, Shen Niu, Zibo Wu, Li Qin, Yongyi Chen, et al.
Processes of the Reliability and Degradation Mechanism of High-Power Semiconductor Lasers
Reprinted from: *Crystals* **2022**, *12*, 765, doi:10.3390/cryst12060765 173

Xuan Li, Junce Shi, Long Wei, Keke Ding, Yuhang Ma, Zaijin Li, Lin Li, et al.
Research on Silicon-Substrate-Integrated Widely Tunable, Narrow Linewidth External Cavity Lasers
Reprinted from: *Crystals* **2022**, *12*, 674, doi:10.3390/cryst12050674 . . . . **199**

Yongyi Chen and Li Qin
Editorial for Special Issue "Frontiers of Semiconductor Lasers"
Reprinted from: *Crystals* **2023**, *13*, 349, doi:10.3390/cryst13020349 . . . . **217**

About the Editors

Yongyi Chen

Yongyi Chen received his B.S. degree from Nanjing University, Nanjing, China in 2008 and his Ph.D. from Changchun Institute of Optics, fine Mechanics and Physics (CIOMP), Chinese Academy of Sciences, Changchun, China, in 2013. Since 2016, he has been an Assistant Professor in CIOMP. He became a professor in 2022. He is the co-author of two books, more than 70 articles, and more than 20 inventions. His research interests include simulation and fabrication of semiconductor lasers and applications, nano-scale gratings and photonic crystals, and semiconductor materials. Mr. Chen's awards include first prize of the Science and Technology of Jilin Province in 2015. And has obtained the Outstanding Youth Funds of Jilin Province in 2023.

Li Qin

Li Qin received her B.S. degree from Jilin University, Changchun, China in 1993 and her Ph.D. degree in Jilin University, Changchun, China in 1999. From 2000 to 2002, she conducted her postdoctoral research in the Changchun Institute of Optics and Mechanics. Since 2002, she has been engaged in the research of semiconductor lasers in the Changchun Institute of Optics and Mechanics. In 2008, she became a professor. She is the co-author of two books, more than 100 articles, and has a track record of more than 30 inventions. Her research interests include simulation and fabrication of semiconductor lasers and applications.

Preface to "Frontiers of Semiconductor Lasers"

Semiconductor lasers are now in every aspect of our lives. Thanks to the fast development of semiconductor lasers, our lives have greatly changed. We are now having faster communication speed owning to the semiconductor lasers in optical communication systems. We are now using more and more semiconductor lasers in medical treatment, in laser displays, as well as in industry as pumping sources, and so on. In this Special Issue of Crystals, we have gathered twelve peer-reviewed papers that shed light on recent advances in the field of semiconductor lasers and their applications.

Yongyi Chen and Li Qin
Editors

Article

Study of Temperature Effects on the Design of Active Region for 808 nm High-Power Semiconductor Laser

Shunhua Wu [1,2], Te Li [1,*], Zhenfu Wang [1], Lang Chen [1], Jiachen Zhang [1], Junyue Zhang [1,2], Jiachen Liu [1,2], Yeqi Zhang [1,2] and Liting Deng [1,2]

[1] State Key Laboratory of Transient Optics and Photonics, Xi'an Institute of Optics and Precision Mechanics, Chinese Academy of Sciences, Xi'an 710119, China
[2] University of Chinese Academy of Sciences, Beijing 100049, China
* Correspondence: lite@opt.ac.cn

Abstract: High-power, broad-area, semiconductor lasers are attractive sources for material processing, aerospace, and laser pumping. The design of the active region is crucial to achieve the required high power and electro-optical conversion efficiency, since the temperature significantly affects the performance of the quantum well, including the internal quantum efficiency and mode gain. In this work, the temperature effects on the active region of a 808 nm high-power semiconductor laser were investigated theoretically and experimentally. The simulations were performed with a Quasi-3D model, which involved complete steady-state semiconductor and carrier confinement efficiency combined with a new mathematical method. The critical aluminum content of the quantum barrier was proposed and the relationship between temperature and various loss sources was disclosed in the temperature range of 213 to 333 K, which provides a reliable reference for the design of epitaxial structures of high-power semiconductor lasers in different operating conditions. Subsequently, the optimized epitaxial structure was determined and used to fabricate standard laser bar chips with a cavity length of 2 mm. The experimental electro-optical conversion efficiency of 71% was demonstrated with a slope efficiency of 1.34 W/A and an injection current of 600 A at the heatsink temperature of 223 K. A record high electro-optical conversion efficiency of 73.5% was reached at the injection current of 400 A, while the carrier confinement efficiency was as high as 98%.

Keywords: semiconductor laser; temperature effects; carrier confinement; internal quantum efficiency

1. Introduction

High-power semiconductor lasers have various excellent characteristics, including high output power and electro-optical conversion efficiency, compact structure, high reliability, long operating lifetime, and simple electric driving conditions, and have thus already been widely applied in material processing, the medical field, communication, aerospace, laser pumping, and so on [1]. In some specific fields, such as communication and aerospace, semiconductor lasers are required to adapt to harsh working conditions, especially drastic temperatures, which will cause the device performance to deteriorate sharply or even fail. Therefore, the temperature characteristic of high-power semiconductor lasers has always been a research hotspot. This paper mainly studies the performance of semiconductor lasers in the temperature range of 213 to 333 K, which is a common requirement for industrial applications [2]. Generally, temperature has significant effects on the properties of semiconductor materials, including mobility [3], energy band structure [4], carrier concentration, and refractive index [5], as well as dynamic processes, including drift-diffusion equations, current injection [6], current distribution [7], and gain-absorption [8], making the temperature analysis of semiconductor lasers rather complicated. From the perspective of the vertical epitaxial structure, temperature affects the active region more significantly than other non-active layers, including internal parameters such as mode gain, injected carrier concentration, carrier leakage, and internal optical absorption loss, which ultimately affect the threshold current, slope efficiency, voltage, and resistance [9]. For this reason, this paper mainly focuses on the design of the active region and analyzes differences in performance in the active region at different temperatures.

There is extensive literature on many aspects of epitaxial design from the perspective of temperature. In 2017, Y. F. Song reported 808 nm semiconductor laser arrays with a 1.5 mm cavity length and conduction cooling package. When the temperature dropped from 298 to 233 K, the electro-optical conversion efficiency increased from 56.7% to 66.8% and the carrier leakage ratio dropped from 16.6% to 3.1%, indicating that the significant reduction in carrier leakage loss was the main reason leading to the increased electro-optical conversion efficiency at low temperature [10]. In 2015, C. Frevert elaborated on the power and voltage characteristics of 9xx nm GaAs-based semiconductor lasers in the temperature range of 208 to 298 K, showing the ratio of quantum barrier height to temperature, $\Delta E/(k_B T)$, had a remarkable effect on the differential internal quantum efficiency. These results showed that the carrier leakage was significantly affected by temperature when $\Delta E/(k_B T)$ was less than seven [11]. In 2017, K. H. Hasler conducted low-temperature research on 9xx nm GaAs-based high-power semiconductor lasers and analyzed the quantum barrier and waveguide at 200 and 300 K for AlGaAs materials with different aluminum (Al) contents. The results indicated that the increase in the mode gain of the active region and the decrease in the accumulation of free electrons in the P-waveguide layer at a low temperature of 200 K led to a decrease in the threshold current and an increase in the slope efficiency, thereby increasing the power and electro-optical conversion efficiency [12]. In 2019, M. P. Wang studied the output power, electro-optical conversion efficiency, and spectral variation of high-power semiconductor lasers in the temperature range of 213 to 273 K. The results showed that the energy distribution of injected carriers became narrower at low temperatures, so the leakage of carriers was reduced. Coupled with the decrease in the transparent carrier concentration and internal optical loss, the threshold current eventually decreased as the temperature dropped [13].

Although temperature characteristics research of high-power semiconductor lasers is relatively intensive, few studies have been able to match theory and experiment perfectly. For one thing, studies only utilized experimental methods to obtain output data and qualitatively describe related internal physical quantities. However, lasers designed for a specific temperature scope are not always suitable for all other temperature ranges. Therefore, the variable temperature-dependent experimental test with a fixed structure did not reflect the best performance of the device, reducing the practicality of the experimental data. For another, the related theoretical analysis of internal quantum efficiency, η_i, and material gain, g_0, is less reported. There is no distinct explanation describing how carrier leakage loss is affected by temperature and barrier height, or how the gain of the active region changes with temperature and materials, which is inconvenient to the design of the active region at a specific temperature.

In this work, the epitaxial structure of the 808 nm GaAs-based semiconductor laser was optimized in detail. The active region consisted of a commonly used InAlGaAs/AlGaAs strained quantum well, and the other epitaxial layers were based on an asymmetric, wide waveguide structure [14]. The trends of η_i and g_0 were theoretically investigated in the scope of the active region. To simplify the calculation of η_i, a mathematical model to calculate the specific amount of carrier leakage in quantum wells was uniquely proposed and its approximate expression and application scope were derived, which was simple, time-saving, and accurate. Together with simulation tools, this model can make accurate judgments on the output performance of semiconductor lasers at different temperatures.

In Section 2, the mathematical model for internal quantum efficiency and mode gain was derived, and then temperature dependence was preliminarily analyzed. Section 3 is devoted first to the introduction of the simulation tools and then to the summary of the most relevant physical effects of the model. In Section 4, the mathematical model of Section 2 was embedded in the simulation tools of Section 3 to analyze the temperature effects. In Section 5, the simulated results are compared with the experimental data to verify the correctness of analysis in Section 4. The paper ends with a conclusion in Section 6.

2. Theory

The core output performance of a semiconductor laser is the output power, P_{out}, and its empirical equation is [15]:

$$P_{out} = \eta_{slop}(I - I_{th}) = \frac{hc}{q\lambda}\eta_i\frac{\alpha_m}{\alpha_i + \alpha_m}exp\left(\frac{-\Delta T}{T_1}\right)\left[I - WL\frac{J_{tr}}{\eta_i}exp\left(\frac{\alpha_i + \alpha_m}{\Gamma g_0}\right)exp\left(\frac{\Delta T}{T_0}\right)\right] \quad (1)$$

where

η_{slop} is the slope efficiency;
I_{th} is the threshold current;
h is the Planck's constant;
c is the speed of light in vacuum;
q is the amount of elementary charge;
λ is the lasing wavelength;
η_i is the internal quantum efficiency;
α_m is the mirror loss, and α_i is the internal optical loss;
ΔT is the temperature rise of the active region relative to the heatsink;
T_0 and T_1 are the characteristic temperatures that depict the temperature sensitivity of threshold current and slope efficiency, respectively;
L is the length of the resonant cavity, and W is the width of the device electrode;
J_{tr} is the transparent current density;
Γg_0 is the mode gain, which is the product of the optical confinement factor Γ in the quantum well and the material gain g_0.

According to Equation (1), it is essential to reduce the threshold current and increase the slope efficiency as much as possible to increase the output power. Generally, the state of the active region has the most significant influence on these two parameters. Therefore, the relationships among the internal quantum efficiency η_i, the material gain g_0 and the temperature T are derived as follows based on the theory of semiconductor lasers.

The internal quantum efficiency is defined as the ratio of the number of photons generated in the active region to the electron-hole pairs injected from the electrode. Due to the existence of impurity defects in the active region, the heterojunction interface state, the carrier leakage in the quantum well, etc., the electron-hole pairs injected into the active region cannot produce 100% radiative recombination, so η_i is always less than 1. According to the reasons for the loss in carrier utilization, the internal quantum efficiency can be divided into three parts, as follows:

$$\eta_i = \eta_{inj} \cdot \eta_{con} \cdot \eta_{rad} \quad (2)$$

where η_{inj} is the ratio of the carriers injected into the active area to those injected from the electrode, which is assumed as 1 in this paper [6].

η_{con} is defined as the proportion of carriers injected into the active region that is effectively confined in the quantum well. This part of loss is mainly caused by the insufficient height of the quantum barrier and the excessively high temperature of the active region, causing the carriers to cross over the barrier into the waveguide layer. Therefore, it is critical to study the barrier height of the active region at different ambient temperatures to improve internal quantum efficiency.

η_{rad} is the ratio of the number of photons generated by effective radiation recombination to the number of carriers confined in the quantum well [16].

Next, the new carrier confinement efficiency model is derived. The electron concentration in the energy range from the bottom of the conduction band E_c in the quantum well to any higher energy level E' is:

$$n(E') = \int_{E_c}^{E'} g_c(E)f(E)dE \qquad (3)$$

Similarly, the hole concentration in the energy range from the bottom of the valence band E_v in the quantum well to any lower energy level E' is:

$$p(E') = \int_{E_v}^{E'} g_v(E)[1-f(E)]dE \qquad (4)$$

where $g_c(E)$ and $g_v(E)$ are the state density of the conduction band and valence band, respectively. $f(E)$ is the Fermi-Dirac distribution function. The leakage of carriers in the quantum well originates from the part over the quantum barrier height [17]. Therefore, η_{con} is approximately equal to the ratio of the carriers confined in the potential well ΔE_c and ΔE_v to the carrier concentration over the whole energy band:

$$\eta_{con} = \frac{n(E_{c,barrier})+p(E_{v,barrier})}{n(\infty)+p(-\infty)} = \frac{\int_{E_{c,well}}^{E_{c,barrier}} g_c(E)f(E)dE + \int_{E_{v,well}}^{E_{v,barrier}} g_v(E)[1-f(E)]dE}{\int_{E_{c,well}}^{\infty} g_c(E)f(E)dE + \int_{E_{v,well}}^{-\infty} g_v(E)[1-f(E)]dE}$$
$$\approx \frac{e^{-x1}-e^{-x2}}{e^{-x1}} \qquad (5)$$

where $E_{c,barrier}$ and $E_{v,barrier}$ are the bottom of the conduction band and valence band in the quantum barrier, respectively, and the position of the Fermi energy level is crucial to the accuracy of the model, which is extrapolated by the simulation tools.

The first-order approximation of the model was derived when ignoring the hole confinement and the higher subband in the quantum well, as well as approximating the Fermi distribution function as a Boltzmann distribution. This is shown in the second line of Equation (5), where $x1 = \frac{E_{i1}+E_c-E_F}{k_BT}$, $x2 = \frac{\Delta E_c+E_c-E_F}{k_BT}$. E_{i1} is the energy difference between the first electron subband and the conduction band bottom of the quantum well, ΔE_c is the energy difference between the conduction band bottom of the quantum well and the quantum barrier, and k_B is the Boltzmann constant.

It can be deduced that the carrier confinement efficiency is mainly affected by the barrier height (ΔE_c and ΔE_v), the active region temperature T, and the injected current density J (affecting the position of the Fermi energy level). As for the AlGaAs quantum barrier, the larger the Al content, the higher the ΔE_c and ΔE_v, and thus, the better the carriers are confined. However, high Al content will cause the resistance to increase, resulting in lower electro-optical conversion efficiency. Thus, for different operating temperatures, choosing a suitable Al content for the barrier layers can maximize the power and electro-optical conversion efficiency of the semiconductor laser.

In terms of the material gain in the quantum well, the gain spectrum equation is as follows [15]:

$$g(\hbar\omega) = \sum_{n,m} g_{max}[f_c^n(E_t = \hbar\omega - E_{hm}^{en}) - f_v^m(E_t = \hbar\omega - E_{hm}^{en})]H(\hbar\omega - E_{hm}^{en}) \qquad (6)$$

where g_{max} is the peak gain and $H(\hbar\omega - E_{hm}^{en})$ represents the unit step function. The occupation probability of electrons in the nth conduction subband and the mth hole subband are shown as follows, respectively [15],

$$f_c^n(E_t = \hbar\omega - E_{hm}^{en}) = \frac{1}{1+e^{[E_{en}+(m_r^*/m_e^*)(\hbar\omega - E_{hm}^{en})-F_c]/k_BT}} \qquad (7)$$

$$f_v^m(E_t = \hbar\omega - E_{hm}^{en}) = \frac{1}{1+e^{[E_{hm}-(m_r^*/m_h^*)(\hbar\omega - E_{hm}^{en})-F_v]/k_BT}} \qquad (8)$$

When $f_c^n > f_v^m$, population inversion is achieved, and the net gain will be generated. According to Equation (6), the active region temperature T mainly affects the energy distribution states f_c^n and f_v^m of the injected carriers, thereby changing the gain peak. Simultaneously, the temperature also causes the quantum well subband transition energy E_{hm}^{en} to change and then shifts the peak wavelength.

3. Simulation Model and Epitaxial Parameters

The simulation tools we applied were exploited to analyze the performance of different types of semiconductor devices, such as F-P lasers [18], tapered semiconductor optical amplifiers (SOA) [19], VCSELs [20], and LEDs [21], where it has demonstrated predictive capabilities. In brief, the program self-consistently solves the complete steady-state electrical, optical, and thermal equations for customers. The simulators included Maxwell's wave equation solver for the normal modes in the waveguide; the 3D electrical solver of Poisson and continuity equations; the energy band structure solver with strain effect, quantum effect, and band-mixing effect; and the 3D thermal solver of heat-flow equation. Table 1 summarizes the basic physical effects included in the simulations and their dependence on temperature, carrier concentration, and wavelength.

Table 1. Main physical effects included in the simulations.

Physical Effects	Notes
Temperature dependence of energy band structure	Adopting k-p theory
Carriers capture and escape process in the quantum well	Defined by electron and hole capture times
Temperature dependence of electron and hole mobility	Affected by temperature, carrier concentration and applied electric field
Main non-radiative recombination	Including Auger recombination and Shockley-Hall-Read (SHR) recombination
Temperature dependence of refractive index	Affected by temperature, carrier concentration and wavelength
Free carrier absorption	Proportional to carrier concentration and wave intensity
Temperature dependence of material gain	Affected by temperature, carrier concentration and spectrum
Temperature dependence of carrier concentration	Defined by Fermi-Dirac distribution, applying Poisson and continuity equations
Local heat sources	Not included, treated as constant temperature and no thermal gradients

The elaborate epitaxial structure we applied to analyze the temperature effects was based on our original epitaxial structure, as shown in Table 2. The quantum well thickness was selected to be 8 nm with a compressed strain of approximately 1%, and the quantum barrier thickness was fixed to 50 nm. The Al content of the barrier ranged from 0.1 to 0.35, which needed to be optimized through the temperature analysis. The simulated temperature ranged from 213 to 363 K, which was divided into 16 groups with the same gap. The chip of cm-bar contained 44 emitting units, each with a 170 μm electrode width, a 2 mm cavity length, and a front and rear reflectivity of 3% and 91.5%, respectively. Only the single-emitting unit of laser bars needed to be simulated due to the consistency of the epitaxial structure.

Table 2. Optimized epitaxial structure [22].

Description	Materials	Thickness (μm)	Dopant	Doping Concentration (cm^{-3})
P-clap	GaAs	0.2	C	$3 \times 10^{19} \rightarrow 1 \times 10^{20}$
P-cladding	$Al_{0.4}Ga_{0.6}As$-$Al_{0.5}Ga_{0.5}As$	0.5	C	$2 \times 10^{18} \rightarrow 4.5 \times 10^{18}$
P-waveguide	$Al_xGa_{1-x}As$-$Al_{0.4}Ga_{0.6}As$	0.8	C	$5 \times 10^{16} \rightarrow 2 \times 10^{18}$
Quantum Barrier	$Al_xGa_{1-x}As$	0.05	Undoped	None
Quantum Well	$In_{0.14}Al_{0.14}Ga_{0.72}As$	0.008	Undoped	None
Quantum Barrier	$Al_xGa_{1-x}As$	0.05	Undoped	None
N-waveguide	$Al_{0.35}Ga_{0.65}As$-$Al_xGa_{1-x}As$	1.2	Si	$2 \times 10^{17} \rightarrow 5 \times 10^{16}$
N-cladding	$Al_{0.35}Ga_{0.65}As$	1.5	Si	$2 \times 10^{18} \rightarrow 2 \times 10^{17}$
N-buffer	GaAs	0.5	Si	2×10^{18}
N-substrate	GaAs	150	Si	2×10^{18}

4. Simulation Results and Discussion

4.1. Temperature Effects in the Quantum Well

The temperature effects on the carrier leakage process and mode gain in the active region were first simulated and analyzed. Figure 1a shows the distribution of carrier concentration with energy level, which was calculated according to the model in Section 2, where the quantum barrier was $Al_{0.25}Ga_{0.75}As$ and the temperature of the active region was 223 K. The position of the curve jump was the location of the quantum well sub-energy level, and the near-exponential decay was a distinctive feature of the Fermi-Dirac distribution. A high carrier confinement efficiency of 98% was obtained by calculating the proportion of the curve integral in the red bar area (i.e., the part below the quantum barrier height) to the total range. Figure 1b shows the variation in carrier confinement efficiency with the Al content of the quantum barrier at different temperatures, where the carrier confinement efficiency decreased exponentially as the Al content decreased from 0.35 to 0.1. The 96% carrier confinement efficiency was considered as an acceptable value in this design, as shown in Table 3. Thus, the operating temperature was a significant factor in the selection of the quantum barrier material. Since the Al content of 0.25 in the quantum barrier ensured 98% carrier confinement efficiency at 223 K, the followed simulations were based on the $In_{0.14}Al_{0.14}Ga_{0.72}As/Al_{0.25}Ga_{0.75}As$ quantum well.

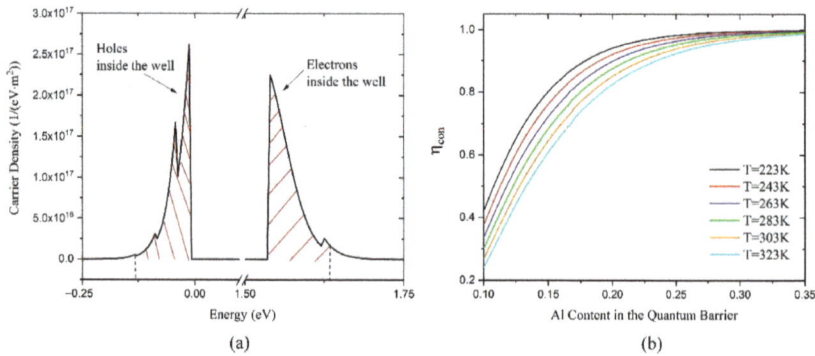

Figure 1. Simulated carrier confinement efficiency model: (**a**) the carrier concentration distribution with energy level for $In_{0.14}Al_{0.14}Ga_{0.72}As/Al_{0.25}Ga_{0.75}As$ quantum well at 223 K (**b**) carrier confinement efficiency vs. Al content of the quantum barrier at different temperatures.

Table 3. Al content of quantum barrier that satisfies $\eta_{con} = 96\%$.

T (K)	223	243	263	283	303	323
Al Content	0.216	0.230	0.246	0.26	0.274	0.288

The gain spectrum curves are shown in Figure 2. When the carrier concentration in the quantum well was 0.5×10^{18} cm^{-3}, meaning the injection current was lower than the threshold current, the material gain was negative. When the carrier concentration increased to 5×10^{18} cm^{-3}, the material gain at the lasing wavelength was positive and lasing occurred. In addition, as the temperature rose, the peak material gain gradually decreased, and red shifts of the lasing wavelength were observed. Figure 2b is drawn with the peak gain as the Y-axis to better illustrate the relationship between material gain, temperature, and carrier concentration. As the temperature rose from 223 to 323 K, the slope of material gain gradually decreased, and the transparent carrier density gradually increased from 0.98×10^{18} to 1.54×10^{18} cm^{-3}, indicating the weaker gain capability of the quantum well.

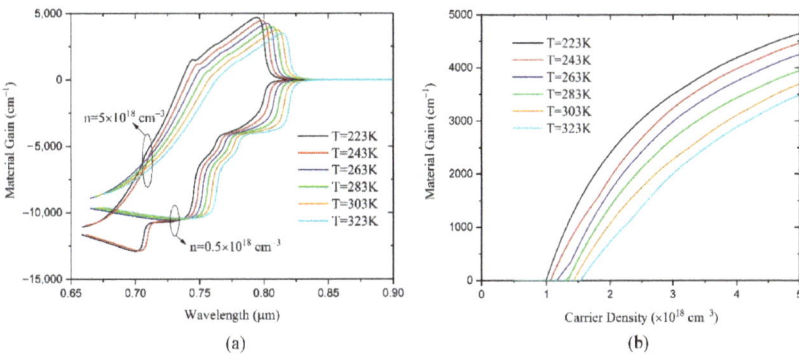

Figure 2. Temperature effects on material gain: (**a**) gain spectrum curves at different temperatures and injected carrier concentrations, (**b**) peak gain vs. injected carrier concentration at different temperatures.

4.2. Temperature Effects in the Whole Epitaxial Direction

The carrier leakage in the quantum well was enhanced and the material gain decreased as temperature rose, which brought about a series of catastrophic chain reactions. Therefore, the trends in energy band structure, carrier concentration, electric resistivity, and free carrier absorption intensity with temperature were analyzed along the epitaxial direction, as shown in Figures 3 and 4. According to the threshold gain condition of semiconductor lasers [15]:

$$\Gamma g_{th}(T, n) = \alpha_i + \alpha_m \qquad (9)$$

The model gain must compensate for the optical loss in the resonant cavity, thus the carrier concentration injected into the quantum well must increase to compensate for the decrease in gain capability of the active region, which was consistent with the increase in carrier concentration and electron Fermi energy level illustrated in Figure 3.

The electric resistivity and free carrier absorption intensity are important indicators that affect the voltage and output power of the device. These two values will also be affected as the temperature changes the distribution of carrier concentration, as shown in Figure 4. The free carrier absorption intensity $\alpha(x)$ is defined as follows [23],

$$\alpha(x) = \frac{I(x)}{S}[\sigma_n \cdot n(x) + \sigma_p \cdot p(x)]$$
$$S = \int_{-\infty}^{+\infty} I(x)dx \qquad (10)$$

where $I(x)$, $n(x)$, and $p(x)$ are the wave intensity, electron concentration, and hole concentration along the epitaxial direction, respectively. σ_n and σ_p are the absorption coefficients of free electrons and holes, respectively.

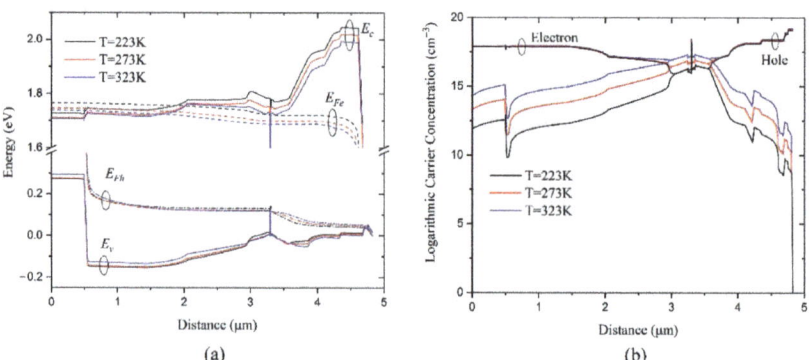

Figure 3. Temperature effects in the epitaxial direction at an injection current of 10 A (**a**) energy band structure (**b**) carrier concentration (the position of 0 μm is located at the start position of the N-buffer layer, while the ordinate of Figure 3a has a break range from 0.4 to 1.6 and the ordinate of Figure 3b is logarithmically transformed by 10).

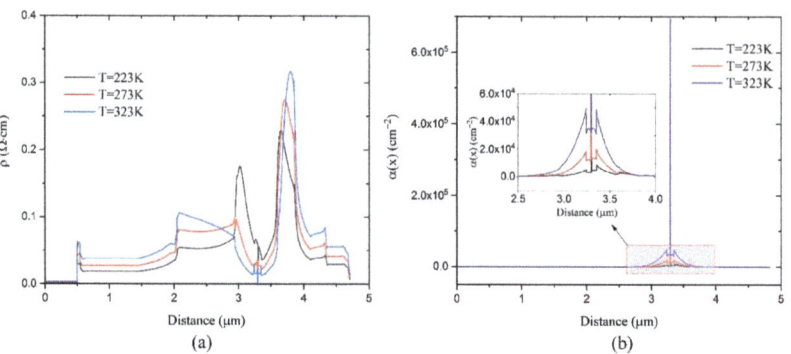

Figure 4. Temperature effects in the epitaxial direction at an injection current of 10 A (**a**) electrical resistivity, (**b**) free carrier absorption intensity.

On the one hand, as the temperature increased, the resistivity of the local area near quantum well decreased, while the value of both N-cladding and P-cladding increased significantly, resulting in a rise of bulk resistance from 25.5 mΩ at 223 K to 36.9 mΩ at 323 K. On the other hand, the closer to the quantum well, the higher the free carrier absorption intensity and the higher contribution to the internal optical loss. The value of free carrier absorption intensity in the quantum well was approximately 50 times that of the waveguide at 223 K, mainly resulting from the high carrier concentration and wave intensity in the quantum well.

4.3. Temperature Effects on the Slop Efficiency

According to Equations (1) and (2), the carrier confinement efficiency needs to be combined with the radiation recombination efficiency and internal loss to determine the slop efficiency. Therefore, in order to compare the simulated and experimental data, the temperature effects on the radiation recombination efficiency and the internal loss were simulated in this section.

Figure 5 illustrates the position of the electron and hole Fermi energy levels in the quantum well at various temperatures. The electron Fermi energy level was approximately linear with temperature when the injection current was fixed at 10 A, and the fitted result was $E_c - E_{Fe} = -0.0192 - 1.828 \times 10^{-4} T$ (eV). The hole Fermi energy level was approximately parabolic with temperature, and the fitted result was $E_v - E_{Fh} = 0.02362 - 1.03878 \times 10^{-4} T + 1.78482 \times 10^{-7} T^2$ (eV).

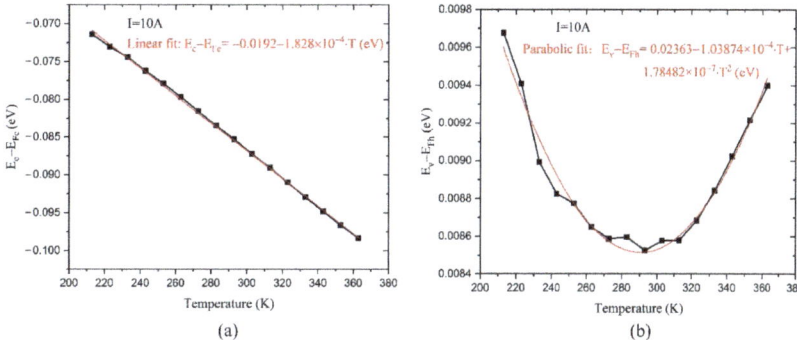

Figure 5. Temperature effects on Fermi energy level at an injection current of 10 A (**a**) electron Fermi energy level, (**b**) hole Fermi energy level.

The carrier concentration in the quantum well was clamped over the threshold; however, this clamped state was shifted by the temperature, as shown in Figure 6, which implied a nonlinear relationship between the threshold carrier concentration and temperature. The threshold carrier concentration increased exponentially from 1.60×10^{18} to 2.75×10^{18} cm^{-3} when the temperature rose from 223 to 323 K.

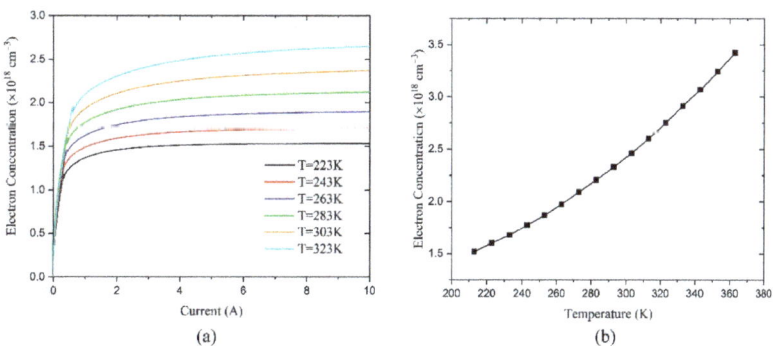

Figure 6. Temperature effects on electron concentration at the center of quantum well (**a**) electron concentration vs. injection current at different temperatures (**b**) electron concentration vs. temperature at an injection current of 10 A.

Figure 7 shows the trend in radiation recombination efficiency with injection current and temperature. The radiation recombination efficiency is derived from the formula:

$$\eta_{rad} = \frac{R_{spon} + R_{stim}}{R_{spon} + R_{stim} + R_{auger} + R_{SHR}} \quad (11)$$

where R_{spon}, R_{stim}, R_{auger}, and R_{SHR} represent the spontaneous radiation recombination rate, stimulated radiation recombination rate, Auger non-radiation recombination rate, and SHR non-radiation recombination rate, respectively. Each of them is affected by the carrier concentration in the quantum well. Therefore, under the threshold, the radiation recombination efficiency increased rapidly with the injection current before becoming stable. Additionally, it was exponentially reduced from 94.9% at 223 K to 92.6% at 323 K at an injection current of 10 A. This finding meant that as the temperature rose, the proportion of non-radiative recombination increased, and this part of the lost energy would eventually become the local thermal source in the chip to further reduce the carrier confinement efficiency and material gain.

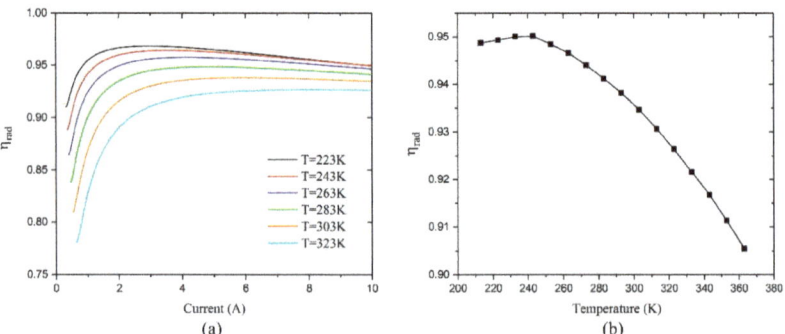

Figure 7. Temperature effects on radiation recombination efficiency (**a**) radiation recombination efficiency vs. injection current at different temperatures (**b**) radiation recombination efficiency vs. temperature at an injection current of 10 A.

Figure 8 shows the trend in internal optical loss accompanied by current and temperature. The internal loss is derived from the formula [24]:

$$\alpha_i = \int_{-\infty}^{+\infty} \alpha(x)dx + \alpha_{scat} \quad (12)$$

where α_{scat} represents the scattering loss and $\alpha(x)$ is the free carrier absorption intensity defined in Equation (10). It can be estimated that the internal optical loss did not change significantly with the injection current after reaching the threshold current. The internal optical loss increased exponentially from 0.57 to 1.67 cm^{-1} as the temperature rose from 223 to 323 K. According to Equation (9), the threshold material gain has to increase since the optical confinement factor of the quantum well hardly changed with temperature. In summary, the material gain ability gradually weakened as the temperature increased, so the carrier concentration in the quantum well needed to increase, resulting in a corresponding increase in internal optical loss. Finally, increasing the carrier concentration in the quantum well is needed to achieve the new threshold gain condition, forming a vicious circle and causing the carrier concentration in the quantum well to increase exponentially. This is an important reason for the rapid decline in the performance of semiconductor lasers at high temperatures.

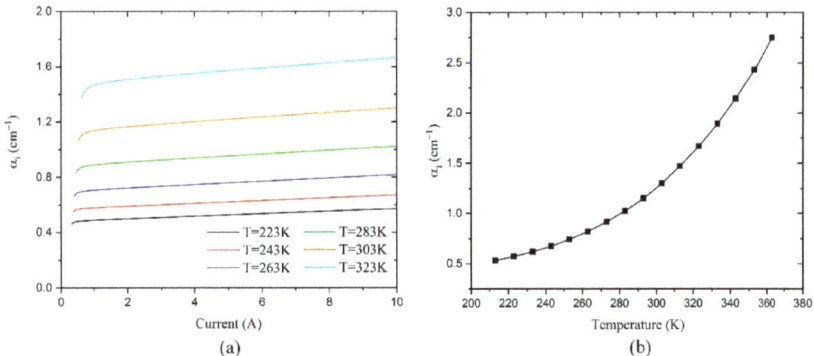

Figure 8. Temperature effects on internal optical loss (**a**) internal optical loss vs. injection current at different temperatures, (**b**) internal optical loss vs. temperature at an injection current of 10 A.

5. Contrast of Simulated and Experimental Data

In order to achieve a carrier confinement efficiency as high as 98% at 223 K, the Al content of the quantum barrier in Table 2 was optimized to be 0.25. The semiconductor laser with the optimized epitaxial structure was fabricated using a standard process. The laser bars with 44 emitting units and 2 mm cavity length were tested at heatsink temperatures ranging from 213 to 333 K in the quasi-continuous-wave (QCW) mode of 250 μs pulse and 200 Hz frequency.

Figure 9 shows the output power, injection current, and applied voltage (L-I-U) characteristics of the laser bar at 223 K compared with simulated values, which were deduced from the single-emitting simulation multiplier. According to the experimental results, the output power of the bar reached 799 W when the injection current was 600 A, while the electro-optical conversion efficiency was 71.2%. The device reached the maximum electro-optical conversion efficiency of 73.5% when the injection current was 400 A. The simulated and experimental results were highly consistent, indicating that the theoretical model we used in Sections 2 and 4 accurately anticipated the performance of high-power semiconductor lasers.

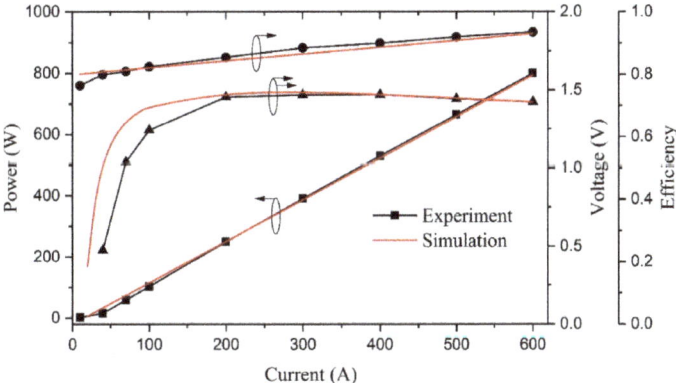

Figure 9. Contrast of simulated and experimental L-I-U curves of laser bars at 223 K.

The output power-current (L-I) curves ranging from 213 to 333 K are shown in Figure 10. The L-I curves showed an obvious linear relationship below a heatsink temperature of 273 K, with the slope efficiency rising from 1.25 W/A at 263 K to 1.34 W/A at 223 K. Additionally, the temperature effects of rising power were saturated below 213 K when the carrier leakage was almost negligible. The threshold current was increasing rapidly above a heatsink temperature of 293 K, from 41.7 A at 293 K to 70.5 A at 333 K, and the slope efficiency decreased significantly, from 1.09 W/A at 293 K to 0.81 W/A at 333 K. The output power dropped from 808 W at 213 K to 311 W at 333 K when the injection current was 600 A, with a difference of 497 W, meaning the device was only suitable for low temperatures.

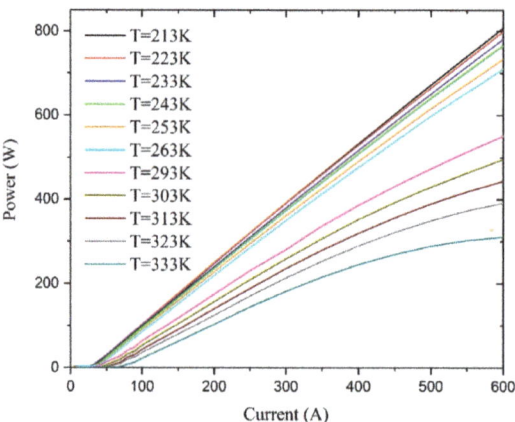

Figure 10. The experimental L-I curves of laser bars from 213 to 333 K.

The temperature of the active region was deduced using the method proposed in reference [25], and the detailed data is illustrated in Table 4. When the injection current was 600 A, the temperature of the active region was on average 30 K higher than the heatsink temperature. The trend in the threshold current and slope efficiency with the corrected temperature of the active region are illustrated in Figure 11. Both the variation in the threshold current and slope efficiency with temperature showed exponential forms, which were closely related to the exponential decay of the carrier confinement efficiency and material gain analyzed in Section 4.1. The simulated and experimental values were highly consistent in terms of temperature trends, indicating the reasonableness of the temperature effect analysis in Section 4. In addition, the simulation results in Section 4 were obtained by embedding the mathematical model in Section 2 into the simulation tools in Section 3, which further confirmed the accuracy of the newly proposed carrier confinement efficiency calculation model.

Table 4. The temperature of the active region at different operating conditions.

T (K) Heatsink	λ (nm) I = 600 A	Waste Heat Power (W)	T (K) Active Region	R_{th} (K/W)
213	797.58	27.74	240.1	0.97
223	799.77	27.40	249.2	0.95
233	802.37	28.86	259.9	0.93
243	804.90	29.69	270.3	0.92
253	807.26	30.80	280.1	0.87
263	810.09	32.26	291.8	0.89
293	818.11	45.27	324.9	0.70
303	819.16	49.12	329.2	0.53
313	823.06	53.27	345.3	0.60
323	825.30	56.99	354.5	0.55
333	827.99	62.95	365.6	0.52

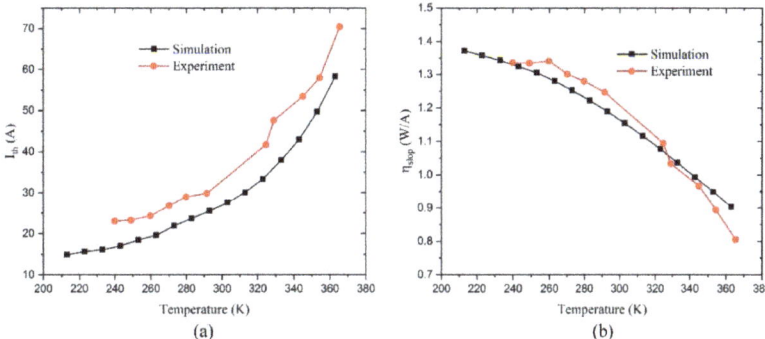

Figure 11. Comparison of simulated and experimental data of: (**a**) the threshold current, (**b**) the slope efficiency.

As for the absolute difference between the simulated and experimental values of the threshold current, this value mainly comes from the calculation deviation of the non-radiative recombination because the default non-radiative recombination parameters in the software material library were applied to calculate the radiation recombination efficiency. The non-radiative recombination rate is not only material dependent but also closely related to the quality of the epitaxial growth process, so it is usually roughly estimated. The fitting for the threshold current will be advanced in our future work to help better predict the device output characteristics.

6. Conclusions

The relationships among internal quantum efficiency, η_i, model gain, Γg_0, and temperature T were analyzed theoretically and experimentally from the perspective of the active region of high-power semiconductor lasers. Based on the results obtained from different temperatures and injection currents, the following design guidelines were derived: (i) the carrier confinement efficiency will drop sharply when the barrier height is lower than a certain critical value, meaning optimization is necessary for the active region at different operating temperatures; (ii) the material gain of the active region decreases quickly as temperature rises, resulting in exponential soaring of carrier concentration and internal optical loss. The optimized design of the quantum well ensured that the carrier confinement efficiency reached 98% at 223 K and the carrier concentration in the quantum well was as low as 1.6×10^{18} cm^{-3}, which further achieved a high radiation recombination efficiency of 95% and low internal optical loss of 0.57 cm^{-1} together with the optimized doping profile, achieving a high slope efficiency of 1.34 W/A. The output power of laser bars reached 799 W at 600 A injection current, while the electro-optical conversion efficiency reached 71%. The maximum electro-optical conversion efficiency of 73.5% was reached at the injection current of 400 A.

Author Contributions: Conceptualization, S.W. and T.L.; software, methodology, writing—original draft preparation, S.W.; writing—review and editing, T.L. and Z.W.; visualization, L.C. and J.Z. (Jiachen Zhang); investigation, J.Z. (Junyue Zhang) and J.L.; data curation, Y.Z. and L.D. All authors have read and agreed to the published version of the manuscript.

Funding: This work was supported by the National Natural Science Foundation of China (Grant No.61504167), the Natural Science Foundation of Shannxi Province, China (Grant Nos. 2019ZY-CXPT-03-05, 2018JM6010, and 2015JQ6263), and the Talent Project of Science and Technology Department of Shannxi Province (No.2017KJXX-72).

Data Availability Statement: The Data is not applicable due to confidentiality.

Acknowledgments: This work was performed within the project financially supported by the State Key Laboratory of Transient Optics and Photonics, Xi'an Institute of Optics and Precision Mechanics, Chinese Academy of Sciences. Additionally, we thank Xia from the Crosslight Company for his constructive guidance on the technical discussion and Yu for the provision of the experimental equipment. The simulation work was mainly supported by the Crosslight Company.

Conflicts of Interest: The authors declare that they have no conflict of interest.

References

1. Ma, X.Y.; Wang, J.; Liu, S.P. Present situation of investigations and applications in high power semiconductor lasers. *Infrared Laser Eng.* **2008**, *37*, 189–194.
2. Zhang, Y.; Hao, S.; Zhang, J.; Guo, C.; Wang, W. Study on the Application of Laboratory Environmental Test Methods for Military Materiel-GJB 150A—2009. *China Personal Protective Equipment* **2017**, *03*, 28–31. [CrossRef]
3. Sotoodeh, M.; Khalid, A.H.; Rezazadeh, A.A. Empirical low-field mobility model for III–V compounds applicable in device simulation codes. *J. Appl. Phys.* **2000**, *87*, 2890–2900. [CrossRef]
4. Vurgaftman, I.; Meyer, J.Á.; Ram-Mohan, L.Á. Band parameters for III–V compound semiconductors and their alloys. *J. Appl. Phys.* **2001**, *89*, 5815–5875. [CrossRef]
5. Ho, C.H.; Li, J.H.; Lin, Y.S. Optical characterization of a GaAs/In0.5(AlxGa1-x)0.5P/GaAs heterostructure cavity by piezoreflectance spectroscopy. *Opt. Express* **2007**, *15*, 13886–13893. [CrossRef] [PubMed]
6. Tansu, N.; Mawst, L.J. Current injection efficiency of InGaAsN quantum-well lasers. *J. Appl. Phys.* **2005**, *97*, 054502. [CrossRef]
7. Laikhtman, B.; Gourevitch, A.; Donetsky, D.; Westerfeld, D.; Belenky, G. Current spread and overheating of high power laser bars. *J. Appl. Phys.* **2004**, *95*, 3880–3889. [CrossRef]
8. Arafin, S.; Bachmann, A.; Vizbaras, K.; Hangauer, A.; Gustavsson, J.; Bengtsson, J.; Larsson, A.; Amann, M.C. Comprehensive analysis of electrically-pumped GaSb-based VCSELs. *Opt. Express* **2011**, *19*, 17267–17282. [CrossRef] [PubMed]
9. Ryu, H.Y.; Ha, K.H. Effect of active-layer structures on temperature characteristics of InGaN blue laser diodes. *Opt. Express* **2008**, *16*, 10849–10857. [CrossRef] [PubMed]
10. Song, Y.F.; Wang, Z.F.; Li, T.; Yang, G.W. Efficiency analysis of 808 nm laser diode array under different operating temperatures. *Acta Phys. Sin.* **2017**, *66*, 104202. [CrossRef]
11. Frevert, C.; Crump, P.; Bugge, F.; Knigge, S.; Erbert, G. The impact of low Al-content waveguides on power and efficiency of 9xx nm diode lasers between 200 and 300 K. *Semicond. Sci. Technol.* **2016**, *31*, 025003. [CrossRef]
12. Hasler, K.H.; Frevert, C.; Crump, P.; Erbert, G.; Wenzel, H. Numerical study of high-power semiconductor lasers for operation at sub-zero temperatures. *Semicond. Sci. Technol.* **2017**, *32*, 045004. [CrossRef]
13. Wang, M.P.; Zhang, P.; Nie, Z.Q.; Liu, H.; Sun, Y.B.; Wu, D.H.; Zhao, Y.L. Analysis of Cryogenic Characteristics of High Power Semiconductor Lasers. *Acta Phtonica Sin.* **2019**, *48*, 0914002. [CrossRef]
14. Wang, Z.; Li, T.; Yang, G.; Song, Y. High power, high efficiency continuous-wave 808 nm laser diode arrays. *Opt. Laser Technol.* **2017**, *97*, 297–301. [CrossRef]
15. Chuang, S.L. Fundamentals of Semiconductor Lasers. In *Physics of Photonic Devices*, 2nd ed.; Jia, D.F., Wang, Z.Y., Sang, M., Yang, T.X., Eds.; Publishing House of Electronics Industry: Beijing, China, 2013; Volume 10, pp. 268–315.
16. Zou, Y.; Osinski, J.S.; Grodzinski, P.; Dapkus, P.D.; Rideout, W.C.; Sharfin, W.F.; Schlafer, J.; Crawford, F.D. Experimental study of Auger recombination, gain, and temperature sensitivity of 1.5 μm compressively strained semiconductor lasers. *IEEE J. Quantum Elect.* **1993**, *29*, 1565–1575. [CrossRef]
17. Romero, B.; Arias, J.; Esquivias, I.; Cada, M. Simple model for calculating the ratio of the carrier capture and escape times in quantum-well lasers. *Appl. Phys. Lett.* **2000**, *76*, 1504–1506. [CrossRef]
18. Zbroszczyk, M.; Bugajski, M. Design optimization of InGaAlAs/GaAs single and double quantum well lasers emitting at 808 nm. In Proceedings of the Conference on Physics and Simulation of Optoelectronic Devices XII, San Jose, CA, USA, 26 January 2004; pp. 446–453.
19. Tijero, J.M.G.; Borruel, L.; Vilera, M.; Consoli, A.; Esquivias, I. Simulation and geometrical design of multi-section tapered semiconductor optical amplifiers at 1.57 μm. In Proceedings of the Semiconductor Lasers and Laser Dynamics VI, Brussels, Belgium, 2 May 2014; p. 91342A.
20. Calciati, M.; Debernardi, P.; Goano, M.; Bertazzi, F. Towards a comprehensive 3D VCSEL model: Electrical simulations with PICS3D. In Proceedings of the Fotonica AEIT Italian Conference on Photonics Technologies, Turin, Italy, 6–8 May 2015.
21. Xia, C.S.; Li, Z.S.; Crosslight, Y.S. Effect of last barrier on efficiency improvement of blue InGaN/GaN light-emitting diodes. In Proceedings of the 13th International Conference on Numerical Simulation of Optoelectronic Devices, Vancouver, BC, Canada, 19–22 August 2013.
22. Wu, S.H.; Li, T.; Wang, D.; Yu, X.C.; Wang, Z.F.; Liu, G.J. Optimization of the Epitaxial Structure of Low-Loss 885nm High-Power Laser Diodes. In Proceedings of the Sixteenth National Conference on Laser Technology and Optoelectronics, Shanghai, China, 31 August 2021; p. 53.
23. Bulashevich, K.A.; Mymrin, V.F.; Karpov, S.Y.; Demidov, D.M.; Ter-Martirosyan, A.L. Effect of free-carrier absorption on performance of 808 nm AlGaAs-based high-power laser diodes. *Semicond. Sci. Technol.* **2007**, *22*, 502–510. [CrossRef]

24. Pikhtin, N.A.; Slipchenko, S.O.; Sokolova, Z.N.; Tarasov, I.S. Internal optical loss in semiconductor lasers. *Semiconductors* **2004**, *38*, 360–367. [CrossRef]
25. Johnson, L.A.; Teh, A. *Measuring High Power Laser Diode Junction Temperature and Package Thermal Impedance*; Application Note #30; ILX Ligthwave Corporation: Bozeman, MT, USA, 2008.

Disclaimer/Publisher's Note: The statements, opinions and data contained in all publications are solely those of the individual author(s) and contributor(s) and not of MDPI and/or the editor(s). MDPI and/or the editor(s) disclaim responsibility for any injury to people or property resulting from any ideas, methods, instructions or products referred to in the content.

Article

Analysis of Thermal Effects in Kilowatt High Power Diamond Raman Lasers

Qiaoxia Gong [1], Mengxin Zhang [1], Chaonan Lin [1], Xun Yang [1], Xihong Fu [2], Fengying Ma [1], Yongsheng Hu [1,*], Lin Dong [1,3] and Chongxin Shan [1,3,*]

[1] Key Laboratory of Materials Physics of Ministry of Education, School of Physics and Microelectronics, Zhengzhou University, Zhengzhou 450001, China
[2] State Key Laboratory of Luminescence and Applications, Changchun Institute of Optics Fine Mechanics and Physics, Chinese Academy of Sciences, Changchun 130033, China
[3] Henan Key Laboratory of Diamond Optoelectronic Materials and Devices, School of Physics and Microelectronics, Zhengzhou University, Zhengzhou 450052, China
* Correspondence: huyongsheng@zzu.edu.cn (Y.H.); cxshan@zzu.edu.cn (C.S.)

Abstract: Chemical vapor deposition (CVD) diamond crystal is considered as an ideal material platform for Raman lasers with both high power and good beam quality due to its excellent Raman and thermal characteristics. With the continuous development of CVD diamond crystal growth technology, diamond Raman lasers (DRLs) have shown significant advantages in achieving wavelength expansion with both high beam quality and high-power operation. However, with the output power of DRLs reaching the kilowatt level, the adverse effect of the thermal impact on the beam quality is progressively worsening. Aiming to enunciate the underlying restrictions of the thermal effects for high-power DRLs (e.g., recently reported 1.2 kW), we here establish a thermal-structural coupling model, based on which the influence of the pump power, cavity structure, and crystal size have been systematically studied. The results show that a symmetrical concentric cavity has less thermal impact on the device than an asymmetrical concentric cavity. Under the ideal heat dissipation condition, the highest temperature rise in the diamond crystal is 23.4 K for an output power of ~2.8 kW. The transient simulation further shows that the heating and cooling process of DRLs is almost unaffected by the pump power, and the times to reach a steady state are only 1.5 ms and 2.5 ms, respectively. In addition, it is also found that increasing the curvature radius of the cavity mirror, the length and width of the crystal, or decreasing the thickness of the crystal is beneficial to alleviating the thermal impact of the device. The findings of this work provide some helpful insights into the design of the cavity structure and heat dissipation system of DRLs, which might facilitate their future development towards a higher power.

Keywords: diamond; thermal effect; high power; thermal-structural coupling model

1. Introduction

High-power lasers with high beam quality have extensive applications in fields such as industrial processing, eye-safe lidar, laser medicine, military countermeasures, laser satellite communications, space exploration, etc. [1–7]. Raman lasers have received much attention in developing high-beam quality and high-power operation lasers because they can achieve beam purification by using unique automatic phase-matching characteristics [8]. Diamond crystal is considered an ideal material for Raman lasers with both high-power and good beam quality due to its high Raman gain coefficient (1332.3 cm^{-1}), low Raman linewidth (1.5 cm^{-1}), low thermal expansion coefficient (1.1 × 10^{-6} K^{-1}), high damage threshold, and thermal conductivity (2200 Wm^{-1}K^{-1}) that is 2–3 orders of magnitude higher than other crystals materials [8–11].

With continual developments in the chemical vapor deposition (CVD) manufacturing process of large-size single-crystal diamonds, diamond Raman lasers (DRLs) with both

high power and high beam quality have made remarkable progress in the last decade or so [8,11–24]. In 2014, the Mildren's group at Macquarie University reported a first-order Stokes light output with a power of 108 W. This is the first time that the output power of DRLs has reached the 100-watt scale with an optical-optical conversion efficiency of 34% and a beam quality factor of $M^2 < 1.1$ [25]. In 2018, they obtained a second-order Stokes light output of 302 W with an optical-optical conversion efficiency of 36% and a beam quality factor of $M^2 = 1.1$ [26]. In 2019, they further got the stokes light output of the kilowatt level (1.2 kW) with an optical-optical conversion efficiency of 53% and a beam quality factor of $M^2 = 1.25$ [27]. Despite these significant advances, it can be seen that the beam quality of the laser begins to deteriorate as the output power of DRLs reaches the kilowatt level, indicating that the adverse effect of the thermal impact on the beam quality is progressively increasing.

Part of the energy for the pump light is converted into heat in the crystal, causing a temperature rise that is the root cause of the thermal effects [28–33]. The uneven temperature distribution inside the laser crystal will cause uneven expansion inside the medium and generate thermal stress to deform the crystal, resulting in a decrease in the conversion efficiency of the pumped light inside the crystal. It is generally believed that the configuration of the laser resonant cavity, the power and beam distribution of the pump light, the size of the laser crystal, as well as the heat dissipation method and the structure of the heat dissipation system are important factors affecting the thermal effect of the laser crystal [34–39].

Up to now, there has been a lack of research on the thermal effects of DRLs because the excellent thermal conductivity of diamond allows the neglect of its thermal effects for low input power. However, the thermal effects of the diamond play a key role in limiting the output power and beam quality of the laser when the output power of DRLs reaches the hundred-watt or even kilowatt level [6,40–42]. In 2021, Bai et al. studied the thermal effects of DRLs by using a simpler point source as a heat source model and found that the temperature rise of the device reached 70 K at an output power of 302 W [40]; recently, the group further found that the temperature rise of the device was only 7.15 K at an output power of 132 W by using the heat source model with crystal internal heat transfer, which is in better agreement with the experimental results [41]. These results provide a more significant contribution to the preliminary understanding of the thermal effects of DRLs. However, we note that there is a lack of research on the thermal effects of high-power DRLs for the kW level and a lack of research and understanding of the conformational relationship between cavity structure and thermal effects.

In this paper, aiming to discuss the fundamental limits of the thermal effects for high-power devices (e.g., 1.2 kW as recently reported [27]), we here establish a thermal-structural coupling model, based on which we first studied the relationship between device thermal effect and pump power; then, the influence of cavity structure, including cavity type, cavity mirror radius of curvature, and diamond size is further discussed. The results show that: (1) a symmetrical concentric cavity has less thermal impact on the device than an asymmetrical concentric cavity; (2) the maximum temperature rise in the crystal is ~23.4 K for DRLs with a symmetric concentric cavity at an output power of ~2.8 kW under the ideal heat dissipation condition; meanwhile, the warming and cooling processes of the device are fast, the heating and cooling processes of DRLs are almost unaffected by the pump power, and the times to reach a steady state are only ~1.5 ms and ~2.5 ms, respectively; (3) increasing the radius of curvature of the cavity mirror, increasing the length and width of the diamond crystal, or reducing the thickness of the crystal are all beneficial to the improvement of the thermal effect of the device. The results of this paper can help to deepen the understanding of the constitutive relationship of thermal effects for DRLs with high power and might provide some insights into the design of cavity structures and heat dissipation systems.

2. System Structure and Heat Source Model

According to the literature [27], we first built the cavity structure of DRLs as shown in Figure 1a, where the diamond crystals with dimensions of 1.2 mm × 4 mm × 8.6 mm are placed in a nearly concentric cavity consisting of concave mirrors with radii of curvature of 150 mm and 92 mm, respectively. The DRLs are pumped by a 1064 nm Nd: YAG laser, the device achieves a maximum output power of 1.2 kW, and an optical-to-optical conversion efficiency of 52.2% for a pump power of 2.3 kW. The system uses single-side water cooling, where the diamond is placed on a copper heat sink with dimensions of 3 mm × 8.6 mm × 8.6 mm (Figure 1b). Most of the heat inside the crystal is removed through a temperature control device (298 K) and water circulation inside it, while the rest of the heat is removed through heat exchange with air on the remaining five sides of the crystal. Here, we consider the ideal heat dissipation case, i.e., a constant temperature of 298 K at the bottom of the diamond crystal (also the upper surface of the copper heat sink).

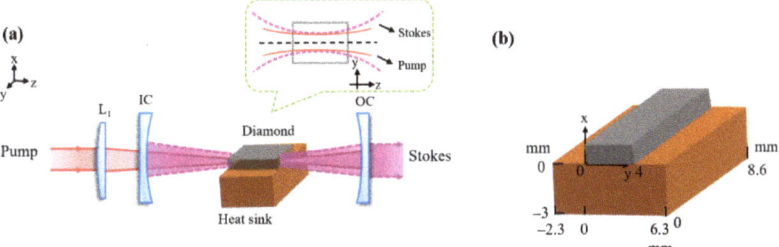

Figure 1. (a) Schematic of DRLs used for the simulation. (b) Model of the diamond crystal and copper heat sink used for the simulation.

2.1. Thermal Model of Diamond Crystal

In this work, we use a Gaussian heat source model that is closer to the actual pump beam characteristics than the point heat source model, which is widely accepted in solid-state laser heat dissipation studies [28,31,36,43]. The pump light is assumed to be approximated as a Gaussian beam, which is focused through the lens to the center of the diamond crystal. Meanwhile, assuming that the pump beam waist radius is equal to the Stokes beam waist radius due to the consideration of the requirement of good mode matching [26,40,44], the heat source function can then be expressed as:

$$Q_v(x,y,z) = \frac{2\xi P_{pump}}{\pi \omega_p^2 l} exp\left[-2\frac{(x-\frac{a}{2})^2+(y-\frac{b}{2})^2}{\omega_p^2}\right] exp(-\alpha z) \quad (1)$$
$$0 \leq x \leq a,\ 0 \leq y \leq b,\ 0 \leq z \leq l$$

$$\xi = 1 - \eta \quad (2)$$

$$\omega_p = \omega_0 \sqrt{1 + \left(\frac{(z-\frac{l}{2})}{z_p}\right)^2} \quad (3)$$

$$z_p = \frac{\pi n \omega_0^2}{M^2 \lambda_p} \quad (4)$$

where a, b, and l are the thickness, width, and length of the diamond crystal, respectively, P_{pump} is the pump power of the laser, z_p is the Rayleigh length, ω_0 is beam radius of the pump light, ω_p is the beam waist radius of the pump light at any position within the crystal, as shown in Table 1, α is the absorption coefficient of the diamond crystal, ξ is the thermal conversion coefficient, $\eta = \lambda_p/\lambda_s$ is the quantum conversion efficiency, λ_p, λ_s are the wavelengths of the pump light and Stokes light, respectively, and n is the refractive index of the diamond crystal.

Table 1. Parameters used in simulation.

	Parameters	Value
Diamond	Thermal conductivity K	2000 Wm^{-1}K^{-1}
	Coefficient of thermal expansion α_T	1.1×10^{-6} K^{-1}
	Absorption coefficient α	0.375 m^{-1}
	Thermal coefficient of the refractive index d_n/d_t	1.5×10^{-5} K^{-1}
	Thermal conversion coefficient ζ	0.142
	Density	3510 kgm^{-3}
	Crystal refractive index n	2.39 @1 μm
	Crystal size	1.2 mm × 4 mm × 8.6 mm
	Young's modulus	1100 GPa
	Poisson's ratio ν	0.069
Copper	Size	3 mm × 8.6 mm × 8.6 mm
	Thermal conductivity	385 Wm^{-1}K^{-1}
DRLs	Pump wavelength λ_p	1064 nm
	Quality factor of pumping beam M_p^2	15
	Stokes wavelength λ_s	1240 nm
Water	Constant temperature	298 K (25 °C)
Air	Ambient temperature T_0	298 K (25 °C)

Assuming that the thermal conductivities of diamond crystals in the x, y, and z directions are k_x, k_y, and k_z, respectively, the three-dimensional heat conduction equation is [29–31]:

$$k_x \frac{\partial^2 T(x,y,z)}{\partial x^2} + k_y \frac{\partial^2 T(x,y,z)}{\partial y^2} + k_z \frac{\partial^2 T(x,y,z)}{\partial z^2} + Q_v(x,y,z) = 0 \quad (5)$$
$$0 \leq x \leq a,\ 0 \leq y \leq b,\ 0 \leq z \leq l$$

where $T(x, y, z)$ is the temperature at any point position within the crystal. At the beginning of heating, the initial moment conditions are set assuming that the diamond crystal and the copper heat sink are in equilibrium as follows:

$$T(x, y, z, 0) = T_0 \quad (6)$$

T_0 is the external ambient temperature, where the natural convection of air and the radiation of the thermal environment between the diamond and the copper heat sink and the surrounding environment are taken into account during the heating process.

2.2. Thermo-Elasticity Model of Diamond Crystal

When the pump light hits the crystal, part of the energy is absorbed, and the rest of the power is converted into heat unevenly distributed inside the crystal, which leads to uneven temperature distribution, resulting in uneven thermal expansion in the crystal. The magnitude of thermal stresses and thermal strains caused by the uneven temperature variations in the crystal can be solved by coupled thermal-stress analysis, i.e., the thermal analysis of the diamond crystal is first performed, and the mechanical analysis is performed based on the results of the thermal analysis simulation, where the position and stress state of the crystal is determined by a set of thermoelastic equations, including geometric, physical, and equilibrium differential equations [31,40,41], as follows:

Firstly, the relationship between strain and displacement can be described by the geometric equation, which is as follows:

$$\begin{cases} \varepsilon_x = \frac{\partial u_x}{\partial x}, \varepsilon_y = \frac{\partial u_y}{\partial y}, \varepsilon_z = \frac{\partial u_z}{\partial z} \\ \varepsilon_{xy} = \frac{\gamma_{xy}}{2} = \frac{1}{2}\left(\frac{\partial u_x}{\partial y} + \frac{\partial u_y}{\partial x}\right) \\ \varepsilon_{yz} = \frac{\gamma_{yz}}{2} = \frac{1}{2}\left(\frac{\partial u_y}{\partial z} + \frac{\partial u_z}{\partial y}\right) \\ \varepsilon_{zx} = \frac{\gamma_{zx}}{2} = \frac{1}{2}\left(\frac{\partial u_z}{\partial x} + \frac{\partial u_x}{\partial z}\right) \end{cases} \quad (7)$$

where u_x, u_y, u_z are the displacement components of the crystal in the x, y, and z directions, respectively, ε_x, ε_y, and ε_z are the line strain components of the crystal in the x, y, and z directions, respectively, and γ_{xy}, γ_{yz}, and γ_{zx} are the shear strain components of the crystal in the three planes, respectively.

Secondly, the relationship between stress and strain satisfies Hooke's law and can be described by the physical equation:

$$\begin{cases} \varepsilon_x = \frac{\partial u_x}{\partial x} = \frac{1}{E}\left[\sigma_x - \mu(\sigma_y + \sigma_z)\right] + \alpha_T \Delta T \\ \varepsilon_y = \frac{\partial u_y}{\partial y} = \frac{1}{E}\left[\sigma_y - \mu(\sigma_z + \sigma_x)\right] + \alpha_T \Delta T \\ \varepsilon_z = \frac{\partial u_z}{\partial z} = \frac{1}{E}\left[\sigma_z - \mu(\sigma_x + \sigma_y)\right] + \alpha_T \Delta T \end{cases} \quad (8)$$

$$\gamma_{xy} = \frac{\tau_{xy}}{G}, \gamma_{yz} = \frac{\tau_{yz}}{G}, \gamma_{zx} = \frac{\tau_{zx}}{G} \quad (9)$$

σ_x, σ_y, σ_z are the stress components of the crystal in the x, y, and z directions, respectively, μ is the Poisson's ratio, E is the tensile modulus of the crystal, α_T is the coefficient of the thermal expansion of the crystal, ΔT is the temperature change, and $G = \frac{E}{2(1+\mu)}$, is the shear modulus of elasticity.

The crystal satisfies the hydrostatic equilibrium condition when a diamond crystal is in equilibrium:

$$\Sigma X = \Sigma Y = \Sigma Z = \Sigma M_x = \Sigma M_y = \Sigma M_z = 0 \quad (10)$$

Using the equilibrium condition for the moments $\Sigma M_x = \Sigma M_y = \Sigma M_z = 0$, we can obtain:

$$\begin{cases} \tau_{xy} = \tau_{yx} \\ \tau_{xz} = \tau_{zx} \\ \tau_{yz} = \tau_{zy} \end{cases} \quad (11)$$

Using the displacement equilibrium condition $\Sigma X = \Sigma Y = \Sigma Z = 0$, we can obtain the equilibrium differential equations for the forces in the x, y, and z directions of the crystal, respectively, as:

$$\begin{cases} \frac{\partial \sigma_x}{\partial x} + \frac{\partial \tau_{yx}}{\partial y} + \frac{\partial \tau_{zx}}{\partial z} + F_x = 0 \\ \frac{\partial \tau_{xy}}{\partial x} + \frac{\partial \sigma_y}{\partial y} + \frac{\partial \tau_{zy}}{\partial z} + F_y = 0 \\ \frac{\partial \tau_{xz}}{\partial x} + \frac{\partial \tau_{yz}}{\partial y} + \frac{\partial \sigma_z}{\partial z} + F_z = 0 \end{cases} \quad (12)$$

where, F_x, F_y, F_z are the external force components acting on the diamond crystal in x, y, and z directions, respectively, and $F_x = F_y = F_z = 0$ since the surface of the diamond crystal is freely bounded. Bringing Equations (8) and (9) into Equation (12), the equilibrium differential equation expressed in terms of displacement components can be obtained.

2.3. Thermal Lensing Strength of Diamond Crystal

A large heat load is accumulated in diamond crystal due to the attenuation of the optical phonons generated by Raman and the absorption of impurities or defects in the crystal by pump light and Stokes light from DRLs. The heat generated affects the conversion efficiency and beam quality of DRLs through the thermal lensing effect, thermally induced stress birefringence, and thermally induced stress fracture, where the thermal lensing effect is one of the critical parameters in the study of thermal effects. To evaluate the effect of thermal lensing inside the crystal, the magnitude of the thermal lensing intensity within the crystal can be calculated. Assuming a uniform thermal accumulation along the length of the crystal as P_h, the thermal lens intensity is mainly influenced by the thermally induced radial perturbation of the refractive index, end-face bulging, and photo-elastic effects, which can be expressed as the equation of the intensity as [6,45,46]:

$$f^{-1} \approx \frac{P_h}{2\pi \omega_p^2 K} \cdot \left(d_n/d_t + (n-1)(\nu+1)\alpha_T + n^3 \alpha_T C_{r,\varphi} \right) \qquad (13)$$

$$P_h = \xi P_{dep} + \alpha_p P_{pump} + \alpha_s P_s / (1 - R_s) \qquad (14)$$

As shown in Table 1, where K is the thermal conductivity of the diamond, d_n/d_t is the thermo-optical coefficient of the refractive index, n is the refractive index of the diamond crystal, ν is the Poisson's ratio, α_T is the coefficient of thermal expansion, $C_{r,\varphi}$ is the photoelasticity coefficient, where P_{dep} is the heat deposited (per unit time) due to the production of Raman phonons and absorption of the Stokes beam, and P_{pump} is the pump power of the laser. In this work, we assumed that $P_{dep} = P_{pump}$, which is more convenient for our calculations. P_s is the output power of Stokes light; α_p and α_s are the absorption coefficients of the diamond crystal for pump light and Stokes light, respectively. Since the wavelength difference between pump light and Stokes light is insignificant, it is assumed that $\alpha_p = \alpha_s$ (0.375 m^{-1} @1 µm), and R_s is the reflectance of the output coupling mirror to Stokes light.

For diamond crystals, since the thermo-optical effect is several orders of magnitude higher than the photo-elastic effect [24], under the first-order approximation, only the thermo-refractive index change is usually calculated, while neglecting the effect of the last two terms in Equation (13), so the thermal lensing intensity of the crystal can be simplified as:

$$f^{-1} \approx \frac{P_h}{2\pi \omega_p^2 K} \cdot d_n/d_t \qquad (15)$$

Thus, the thermal lensing intensity of the crystal is proportional to the pump power; the higher the pump power, the more serious the thermal lensing effect; inversely proportional to the square of the pump spot radius, the larger the pump spot, the smaller the thermal lensing effect.

3. Results and Discussion
3.1. The Effect of Pump Power on Thermal Effect
3.1.1. Temperature Distribution at Different Pumping Power Levels

Firstly, we examine the steady-state temperature distribution inside the diamond crystal for different pump power levels. Due to the optical-to-optical conversion efficiency being 52.2% [27], when pump power levels are 800 W, 2.3 kW, 3.8 kW, and 5.3 kW, respectively, the corresponding output power levels would be about 0.42 kW, 1.2 kW, 2.0 kW, and 2.8 kW. As shown in Figure 2, the temperature variation of the diamond crystal is mainly concentrated in the pumped region, i.e., near the central crystal axis. With the pumping power rising from 800 W to 5.3 kW, the maximum temperature rise inside the crystal gradually increases from 3.2 K to 23.4 K. In the transverse direction (y-direction), the temperature distribution inside the crystal is symmetrical, and it gradually decreases from the center to both sides of the face. Finally, both sides of the surface and ambient temperature are almost the same.

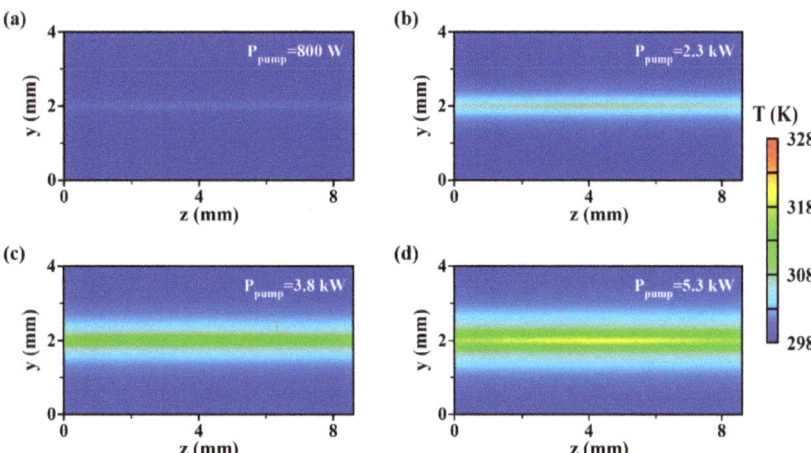

Figure 2. Cross-sectional temperature distribution of diamond crystal at pumping power (**a**) 800 W (**b**) 2.3 kW (**c**) 3.8 kW, and (**d**) 5.3 kW, respectively.

Figure 3 further gives the temperature variations in three directions inside the diamond for different pumping power levels. It can be seen from Figure 3a that the temperature is highest at the center of the crystal on the lateral side [31,41], and the temperature difference between the center temperature and the lateral edge increases gradually from 3.2 K to 21.5 K as the pumping power increases. From Figure 3b, it can be seen that the temperature also decreases gradually from the center of the crystal to the two end faces in the crystal axis direction, with the lowest temperature at the two ends, which is because the pump beam waist position coincides with the center of the diamond crystal so that the pump energy density is the largest, and at the same time, both ends of the crystal are in contact with the air, so the heat dissipation effect is better than the center of the crystal, which is consistent with the findings in the Ref. [47]. With the pumping power increasing from 800 W to 5.3 kW, the temperature gradient between the center of the crystal and the two end faces increases from 0.6 K to 4.0 K. Figure 3c shows the temperature distribution along the x-direction (thickness direction) at the central face of the crystal. It can be seen that the highest temperature also occurs at the central position of the crystal, which is 301.5 K, 308.1 K, 314.8 K, and 321.4 K at the pump power of 800 W, 2.3 kW, 3.8 kW, and 5.3 kW, respectively. The temperature decreases gradually from the center of the crystal to the upper and lower surfaces. Since the bottom surface of the diamond crystal is in contact with the copper heat sink and is fixed at the ideal 298 K, its heat dissipation effect is better than that of the upper surface. The temperature decreases rapidly from the center of the crystal to 298 K, while the temperature of the upper surface increases with the increase in pumping power.

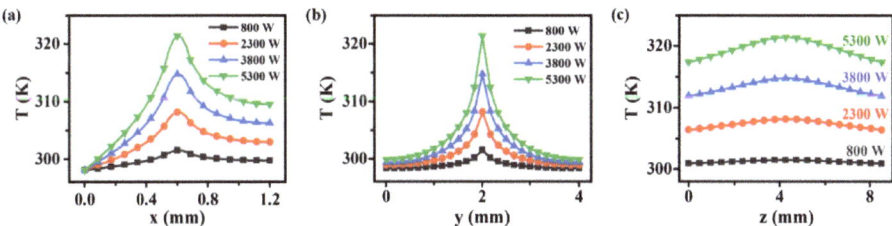

Figure 3. Temperature distribution of (**a**) x-axis intercept in crystal center plane, (**b**) y-axis intercept in crystal center plane, and (**c**) crystal axis at different pumping power levels.

It should be noted that the temperature of the bottom surface of the diamond crystal (or the top of the copper heat sink) is assumed to be constant at 298 K in this work, which is an ideal state, while if the temperature at the bottom of the copper heat sink is fixed at 298 K, the maximum temperature inside the diamond crystal will reach 355.7 K at a pump power of 2.3 kW. The corresponding maximum temperature rise will be significantly increased from 10.1 K to 57.7 K, which is even much higher than the ideal case with a pump power of 5.3 kW. This shows that it is crucial for DRLs to rationally design the heat dissipation structure of the diamond crystal.

We further investigated the warming process of the diamond crystal in the case of pulsed pumping and the cooling process after stopping pumping. The results are shown in Figure 4. T_c and T_s are the temperatures of the diamond in the volume center and the upper surface center, respectively, $T_c - T_s$ is their temperature difference, and the time for the temperature gradient to reach 99% of the steady-state value is defined as the constant thermal time [25,40]. It can be seen from Figure 4a that the volume center temperature of the diamond crystal is higher and rises faster than the upper surface center temperature at the beginning stage after the pump pulse action, which is because the pump energy is concentrated at the center of the crystal. The upper surface center temperature starts to rise from ~0.01 ms. It stabilizes at about 1.5 m, during which the temperature of the volume center is always higher than the upper surface temperature of the crystal, which is consistent with the results of the steady-state temperature distribution within the diamond. It is worth noting that the thermal constant time is almost independent of the pumping power because the thermal equilibrium time is mainly determined by the pumping beam area and the basic thermal properties of the crystal (specific heat capacity and thermal conductivity) [25].

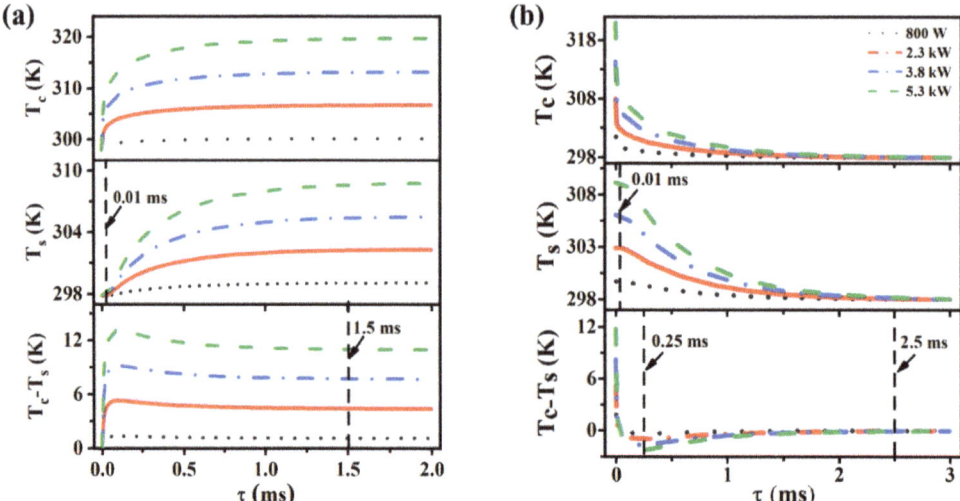

Figure 4. Transient temperature variation of diamond crystal at different pumping power levels during (**a**) heating and (**b**) cooling.

Figure 4b shows the results of the cooling process for the diamond crystal after the pumping stopped. Within 0.01 ms of removing the heat source, the center temperature of the upper surface of the diamond crystal remains almost unchanged. In contrast, the temperature of the volume center drops rapidly to be close to that of the upper surface. After that, the center temperature of the upper surface of the diamond crystal starts to decrease slowly. The hysteresis of the cooling is mainly caused by the different heat dissipation conditions on the upper and lower surfaces, which can also be seen from the variation of the temperature difference $T_c - T_s$: after the heat source is removed, the temperature difference $T_c - T_s$ decreases rapidly, and becomes negative after ~0.01 ms, reaching the maximum value at ~0.25 ms, and then slowly converges to the ambient temperature (298 K). The thermal effect reaches equilibrium after the pumping stops after ~2.5 ms. Similarly, the time required to reach equilibrium finally is almost unaffected for different pump power levels, despite the slightly different range of temperature variations. Based on this result, we can speculate that kW-level Raman laser emission without heat accumulation can be achieved at repetition frequencies of ~250 Hz for the diamond crystals cooled on one side in this paper.

3.1.2. Thermal Stress Distribution under Different Pumping Power

Figure 5 shows the distribution of thermal stresses inside the diamond at different pumping power levels. It can be seen that the stress is highest at the center of the crystal and lowest at the center of both end faces, and the maximum thermal stress inside the crystal increases from 20.7 MPa to 137.4 MPa as the pumping power increases from 800 W to 5.3 kW. It is noteworthy that this distribution and trend of thermal stress is similar to the results of the temperature distribution inside the diamond (Figure 2), which is mainly because the higher thermal expansion at higher temperature locations within the crystal requires more stress generation to avoid thermal deformation. Also, due to the diamond having a smaller thermal expansion coefficient, this thermal stress is significantly lower than other crystals at the same pumping power level, such as Tm: YAP and Tm: LuAG et al. [31,36].

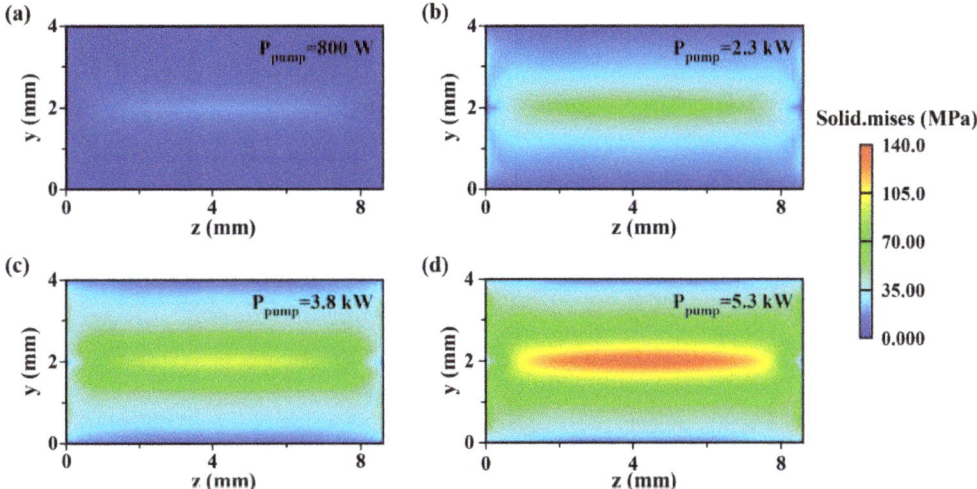

Figure 5. Cross-sectional thermal stress distribution of diamond crystal at different output power levels, (**a**) 800 W, (**b**) 2.3 kW, (**c**) 3.8 kW, and (**d**) 5.3 kW.

3.1.3. Heat Deformation Distribution under Different Pumping Power

According to Equation (8), the relationship between deformation and stress satisfies Hooke's law, i.e., thermal stress tends to prevent thermal deformation, which also means that the distribution of thermal deformation will be opposite to the distribution of thermal stress [41]. The corresponding results are given in Figure 6, where it can be seen that the deformation at the center of the crystal is the smallest, and the deformation at both ends is the largest for different pumping power levels. With the pumping power increasing from 800 W to 5.3 kW, the maximum deformation in the crystal increases from 0.02 μm to 0.13 μm. Nevertheless, this result is significantly lower than other crystal materials [31,36].

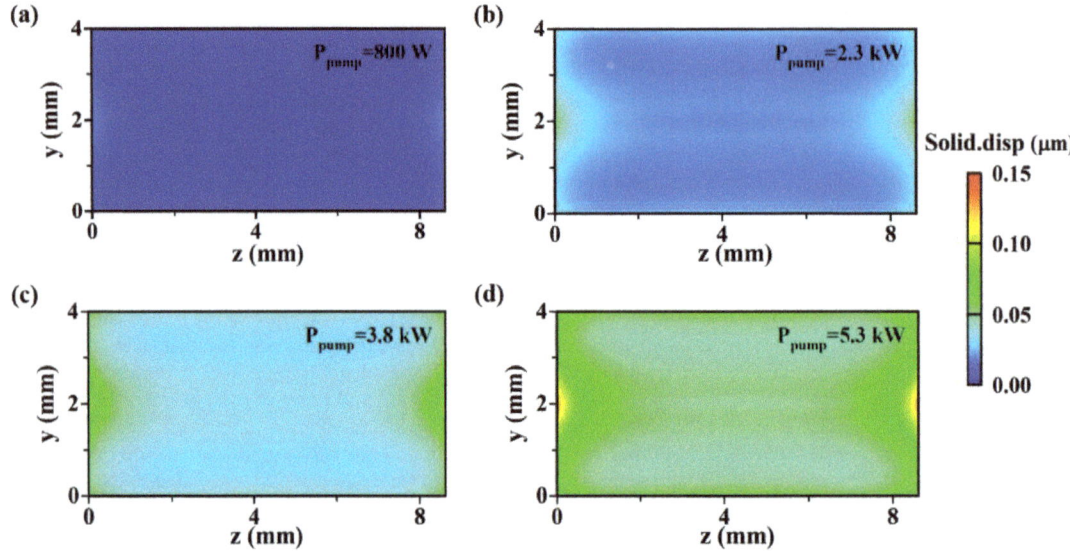

Figure 6. Cross-sectional thermal deformation distribution of diamond crystal at different output power levels, (**a**) 800 W, (**b**) 2.3 kW, (**c**) 3.8 kW, and (**d**) 5.3 kW.

3.2. Effect of Resonant Cavity Configuration on the Thermal Effects

To analyze the factors influencing the thermal effect more comprehensively, we further investigated the thermal impact of the devices at different cavity lengths and cavity curvature radii on the basis of the literature of [27] (pumping power of 2.3 kW). Considering the cavity lengths of 100 mm, 150 mm, 200 mm, and 250 mm, respectively, Figure 7a–d is the temperature distribution of the crystal axes at different cavity radii of curvature. It can be seen that the symmetric concentric cavity ($L = R1 + R2$) structures all exhibit the lowest crystal temperature for different cavity lengths, and the longer the cavity length (i.e., the larger the radius of curvature), the lower the temperature. The minimum temperatures for the four cavity lengths are 309.5 K, 308.7 K, 308.3 K, and 308.0 K, respectively. The above temperature distribution pattern is mainly caused by the variation of pump light energy density: the increase in cavity length and the radius of curvature of the cavity mirror causes the increase in the beam waist radius of the first-order Stokes light, which leads to the rise in the pump light beam waist radius (a requirement for good mode matching) so that the pump energy density decreases, and the temperature inside the crystal decreases. It is easy to calculate from the ABCD matrix to obtain the beam waist radii of Stokes light at four cavity lengths; they have slight variations, which are 66 µm, 73 µm, 78 µm, and 83 µm, respectively; therefore, the difference in minimum temperature is also tiny, only ~1.5 K.

Figure 7e,f summarizes the relationship between the maximum thermal stress and thermal deformation of the concentric cavity DRLs with the radius of cavity curvature for different cavity lengths, respectively. It can be seen that the maximum thermal stresses all gradually decrease with the increase in the radius of curvature of the cavity mirror for all four cavity lengths, but the overall effect is weak and eventually tends to stabilize at ~60 MPa. The maximum thermal deformations under different conditions (Figure 7f) are independent of the cavity length and radius of cavity curvature, which all remain around ~0.057 µm.

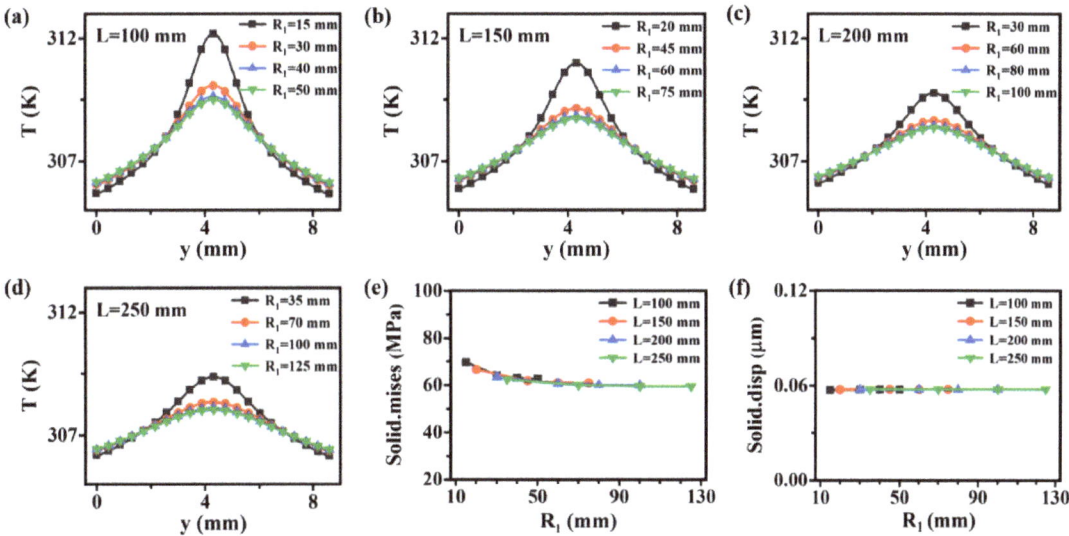

Figure 7. Cross-sectional temperature distribution of diamond crystal with concentric cavity of different radii of curvature at different cavity lengths, (**a**) 100 mm, (**b**) 150 mm, (**c**) 200 mm, and (**d**) 250 mm. Distribution of (**e**) maximum thermal stress, and (**f**) maximum thermal deformation inside crystals with a symmetrical concentric cavity at different radii of curvature.

3.3. Effect of Diamond Crystal Size on the Thermal Effect

Considering that the application of larger-size single-crystal diamonds will become possible in the future, we further discussed the effect of the diamond crystal size on thermal effects. Figure 8a shows the variation of the maximum temperature inside the diamond crystal (crystal cross-section is 1.2 mm × 4 mm) for different lengths. It can be seen that the maximum temperature inside the crystal gradually decreases from 310.1 K to 306.3 K as the crystal length increases from 7 mm to 11 mm. This is because the increase in crystal length will make the conversion of pump light to Stokes light more adequate, thus improving the utilization of pump light and reducing heat accumulation. Figure 8b shows the variation of maximum thermal stress with crystal length. It can be seen that the effect of crystal length on the maximum thermal stresses inside the crystal is negligible; they all remain around 60 MPa. In contrast, the ultimate thermal deformation gradually decreases from 0.07 μm to 0.04 μm as the crystal length increases (Figure 8c). Figure 8d shows the relationship between the maximum temperature inside the diamond crystal and the width and thickness of the crystal for a fixed length of 8.6 mm. It can be seen that the maximum temperature increases with the increase in the crystal thickness when the width of the crystal is certain; this is because the bottom of the diamond is fixed at 298 K, and increasing the thickness of the diamond is equivalent to increasing the distance between the heat source and the heat sink, which indicates that the thermal effect of DRLs increases with the increase in the crystal thickness; it is consistent with the findings of the literature [48] regarding the effect of the thickness of the Ti: Sa crystal on the thermal effect of the laser system; meanwhile, the maximum temperature decreases with increasing the width of the crystal when the thickness of the crystal is fixed, which is because expanding the width is equivalent to increasing the contact area between the diamond and the heat sink. Figure 8e shows the corresponding maximum thermal stresses. It can be seen that the maximum thermal stress of the device decreases slightly with increasing crystal width, and the effect of crystal thickness on the maximum thermal stress is negligible. Figure 8f shows the maximum thermal deformation as a function of crystal size. It can be seen that it tends to follow the same trend as the temperature change, where the thermal deformation decreases as the crystal width increases, and the smaller the crystal thickness is, the smaller the thermal deformation is.

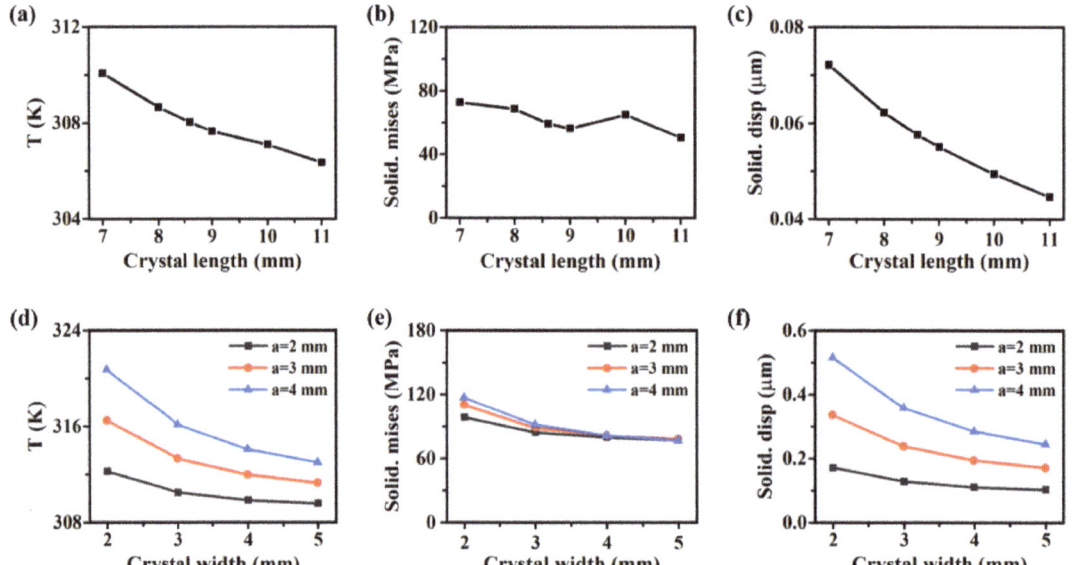

Figure 8. Distribution of (**a**) maximum temperature, (**b**) maximum thermal stress, and (**c**) maximum thermal deformation inside diamond crystals of different lengths. Distribution of (**d**) maximum temperature, (**e**) maximum thermal stress, and (**f**) maximum thermal deformation inside the diamond crystal for different crystal widths and thicknesses.

3.4. Thermal Lensing Effect Analysis

Finally, we discuss the variation of the thermal lens strength of DRLs with the pump power for different cavity radii of curvature (symmetric concentric cavity) and crystal lengths. It can be seen from Figure 9a that the thermal lens strength decreases with increasing the cavity radius of curvature when the pumping power is constant because an increase in the radius of curvature of the cavity mirror will cause an increase in the radius of the first-order Stokes beam waist and a consequent increase in the pump beam waist, which will lead to a decrease in the temperature rise in the crystal and, consequently, a decrease in the refractive index change. It can also be seen from Figure 9a that the thermal lens strength increases significantly with the increase in the pumping power, indicating that the rise of the pumping power causes the thermal accumulation of the crystal to increase rapidly, which in turn leads to the rapid change of the refractive index. However, with the pump power increasing, the thermal lens strength increases more flatly for structures with a large cavity radius of curvature. When the pump power increases to 5300 W and the cavity radii of curvature are 50 mm, 75 mm, 100 mm, and 125 mm, respectively, the thermal lens strengths are 163.6 m^{-1}, 133.7 m^{-1}, 117.1 m^{-1}, and 103.5 m^{-1}, which are similar to the thermal lens strength of diamond crystal in the literature [6,45], while significantly lower than those of other crystals such as Nd: GdVO$_4$ and Tm: YLF [34,35]. There is a similar conclusion for different crystal lengths (Figure 9b), where the thermal lens strength decreases with increasing pump power and crystal length.

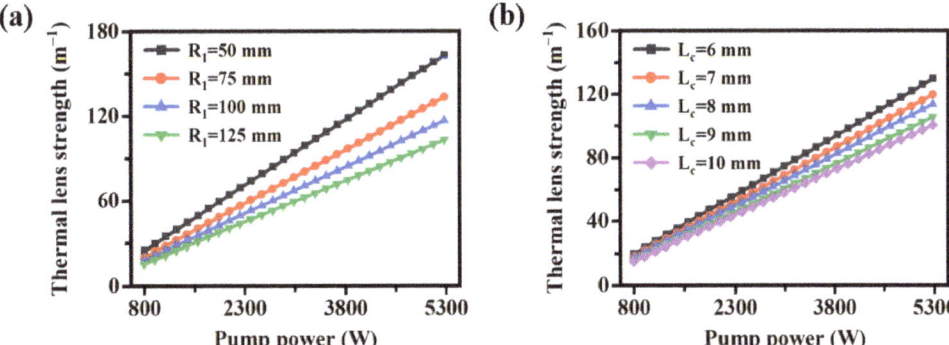

Figure 9. Relationship between thermal lens intensity and pump power at different (**a**) radii of curvature of cavity mirrors, and (**b**) crystal lengths.

4. Conclusions

In conclusion, aiming to discuss the fundamental limits of the thermal effects for high-power devices, we here establish a thermal-structural coupling model, based on which the influence of the pump power, cavity structure, and crystal size have been systematically studied. The results show that using a symmetric concentric cavity structure and increasing its radius of curvature, choosing the appropriate diamond size (increasing the length and width of the diamond crystal or decreasing the thickness of the crystal) is beneficial to alleviate the thermal effect of the device. Under ideal heat dissipation conditions, the maximum temperature rise in the crystal is ~23.4 K for DRLs, and the constant thermal times of the warming and cooling processes are ~1.5 ms and 2.5 ms, respectively, for an output power of ~2.8 kW. The results of this work provide a basis for the design of the resonant cavity structure of DRLs, which might promote the development of DRLs with high power and high beam quality.

Author Contributions: Conceptualization, Y.H.; methodology, Q.G., F.M. and L.D.; validation, C.L., X.Y. and X.F.; writing—original draft preparation, Q.G. and M.Z.; writing—review and editing, Y.H.; supervision, C.S.; funding acquisition, C.S. All authors have read and agreed to the published version of the manuscript.

Funding: This research was funded by the National Natural Science Foundation of China (NSFC) (Grant No. 11904323) and Research Funds of Zhengzhou University (Grant No. 32340305 and 32410543).

Data Availability Statement: Not applicable.

Conflicts of Interest: The authors declare no conflict of interest.

References

1. Lubin, P.; Hughes, G.B.; Bible, J.; Bublitz, J.; Arriola, J.; Motta, C.; Suen, J.; Johansson, I.; Riley, J.; Sarvian, N.; et al. Toward directed energy planetary defense. *Opt. Eng.* **2014**, *53*, 025103. [CrossRef]
2. Extance, A. Military technology: Laser weapons get real. *Nature* **2015**, *521*, 408–410. [CrossRef] [PubMed]
3. Meng, D.; Zhang, H.; Li, M.; Lin, W.; Shen, Z.; Zhang, J.; Fan, Z. Laser technology for direct IR countermeasure system. *Infrared Laser Eng.* **2018**, *47*, 1105009. [CrossRef]
4. Williams, R.J.; Spence, D.J.; Lux, O.; Mildren, R.P. High-power continuous-wave Raman frequency conversion from 1.06 μm to 1.49 μm in diamond. *Opt. Express* **2017**, *25*, 749–757. [CrossRef] [PubMed]
5. Bai, Z.; Williams, R.J.; Jasbeer, H.; Sarang, S.; Kitzler, O.; Mckay, A.; Mildren, R.P. Large brightness enhancement for quasi-continuous beams by diamond Raman laser conversion. *Opt. Lett.* **2018**, *43*, 563–566. [CrossRef] [PubMed]
6. Antipov, S.; Williams, R.J.; Sabella, A.; Kitzler, O.; Berhane, A.; Spence, D.J.; Mildren, R.P. Analysis of a thermal lens in a diamond Raman laser operating at 1.1 kW output power. *Opt. Express* **2020**, *28*, 15232–15239. [CrossRef] [PubMed]
7. McKay, A.; Kitzler, O.; Liu, H.; Fell, D.; Mildren, R.P. High average power (11 W) eye-safe diamond Raman laser. *SPIE* **2012**, *8551*, 115–122.

8. Jasbeer, H.; Williams, R.J.; Kitzler, O.; McKay, A.; Mildren, R.P. Wavelength diversification of high-power external cavity diamond Raman lasers using intracavity harmonic generation. *Opt. Express* **2018**, *26*, 1930–1941. [CrossRef]
9. Li, Y.; Bai, Z.; Chen, H.; Jin, D.; Yang, X.; Qi, Y.; Ding, J.; Wang, Y.; Lu, Z. Eye-safe diamond Raman laser. *Results Phys.* **2020**, *16*, 102853. [CrossRef]
10. Yang, X.; Kitzler, O.; Spence, D.J.; Bai, Z.; Feng, Y.; Mildren, R.P. Diamond sodium guide star laser. *Opt. Lett.* **2020**, *45*, 1898–1901. [CrossRef]
11. Williams, R.J.; Kitzler, O.; Bai, Z.; Sarang, S.; Jasbeer, H.; McKay, A.; Antipov, S.; Sabella, A.; Lux, O.; Spence, D.J.; et al. High power diamond Raman lasers. *IEEE J. Sel. Top. Quant.* **2018**, *24*, 1–14. [CrossRef]
12. Li, Y.; Ding, J.; Bai, Z.; Yang, X.; Li, Y.; Tang, J.; Zhang, Y.; Qi, Y.; Wang, Y.; Lu, Z. Diamond Raman laser: A promising high-beam-quality and low-thermal-effect laser. *High Power Laser Sci. Eng.* **2021**, *9*, 1–13. [CrossRef]
13. Mildren, R.P.; Sabella, A. Highly efficient diamond Raman laser. *Opt. Lett.* **2009**, *34*, 2811–2813. [CrossRef]
14. Spence, D.J.; Granados, E.; Mildren, R.P. Mode-locked picosecond diamond Raman laser. *Opt. Lett.* **2010**, *35*, 556–558. [CrossRef] [PubMed]
15. Lubeigt, W.; Bonner, G.M.; Hastie, J.E.; Dawson, M.D.; Burans, D.; Kemp, A.J. Continuous-wave diamond Raman laser. *Opt. Lett.* **2010**, *35*, 2994–2996. [CrossRef]
16. Sabella, A.; Piper, J.A.; Mildren, R.P. 1240 nm diamond Raman laser operating near the quantum limit. *Opt. Lett.* **2010**, *35*, 3874–3876. [CrossRef]
17. Parrotta, D.C.; Kemp, A.J.; Dawson, M.D.; Hastie, J.E. Tunable continuous-wave diamond Raman laser. *Opt. Express* **2011**, *19*, 24165–24170. [CrossRef]
18. Feve, J.P.M.; Shortoff, K.E.; Bohn, M.J.; Brasseur, J.K. High average power diamond Raman laser. *Opt. Express* **2011**, *19*, 913–922. [CrossRef]
19. Lubeigt, W.; Savitski, V.G.; Bonner, G.M.; Geoghegan, S.L.; Friel, I.; Hastie, J.E.; Dawson, M.D.; Burns, D.; Kemp, A.J. 1.6 W continuous-wave Raman laser using low-loss synthetic diamond. *Opt. Express* **2011**, *19*, 6938–6944. [CrossRef]
20. Savitski, V.G.; Reilly, S.; Kemp, A.J. Steady-State Raman Gain in Diamond as a Function of Pump Wavelength. *IEEE J. Quantum Electron.* **2013**, *49*, 218–223. [CrossRef]
21. McKay, A.; Liu, H.; Kitzler, O.; Mildren, R.P. An efficient 14.5 W diamond Raman laser at high pulse repetition rate with first (1240 nm) and second (1485 nm) Stokes output. *Laser Phys. Lett.* **2013**, *10*, 105801. [CrossRef]
22. Warrier, A.M.; Lin, J.; Pask, H.M.; Mildren, R.P.; Coutts, D.W.; Spence, D.J. Highly efficient picosecond diamond Raman laser at 1240 and 1485 nm. *Opt. Express* **2014**, *22*, 3325–3333. [CrossRef] [PubMed]
23. Wattiez, A.S.; Libert, F.; Privat, A.M.; Loiodice, S.; Fialip, J.; Eschalier, A.; Courteix, C. Evidence for a differential opioidergic involvement in the analgesic effect of antidepressants: Prediction for efficacy in animal models of neuropathic pain? *Br. J. Pharmacol.* **2011**, *163*, 792–803. [CrossRef] [PubMed]
24. Williams, R.J.; Nold, J.; Strecker, M.; Kitzler, O.; McKay, A.; Schreiber, T.; Mildren, R.P. Efficient Raman frequency conversion of high-power fiber lasers in diamond. *Laser Photonics Rev.* **2015**, *9*, 405–411. [CrossRef]
25. Williams, R.J.; Kitzler, O.; McKay, A.; Mildren, R.P. Investigating diamond Raman lasers at the 100 W level using quasi-continuous-wave pumping. *Opt. Lett.* **2014**, *39*, 4152–4155. [CrossRef]
26. Bai, Z.; Williams, R.J.; Kitzler, O.; Sarang, S.; Spence, D.J.; Mildren, R.P. 302 W quasi-continuous cascaded diamond Raman laser at 1.5 microns with large brightness enhancement. *Opt. Express* **2018**, *26*, 19797–19803. [CrossRef]
27. Antipov, S.; Sabella, A.; Williams, R.J.; Kitzler, O.; Spence, D.J.; Mildren, R.P. 1.2 kW quasi-steady-state diamond Raman laser pumped by an $M^2 = 15$ beam. *Opt. Lett.* **2019**, *44*, 2506–2509. [CrossRef]
28. Sun, R.; Wu, C.T.; Yu, M.; Yu, K.; Wang, C.; Jin, G.Y. Study on the laser crystal thermal compensation of LD end-pumped Nd: YAG 1319 nm/1338 nm dual-wavelength laser. *Laser Phys.* **2015**, *25*, 125004. [CrossRef]
29. Bai, F.; Huang, Z.; Liu, J.; Chen, X.; Wu, C.; Jin, G. Thermal analysis of double-end-pumped Tm: YLF laser. *Laser Phys.* **2015**, *25*, 075003.
30. Cui, J.; Liu, X.; He, L.; Zhang, S.; Yang, J. Investigation of the transient thermal profile of an anisotropic crystal in a pulsed and end-pumped laser. *J. Mod. Opt.* **2018**, *65*, 1847–1854. [CrossRef]
31. Zhang, H.; Wen, Y.; Zhang, L.; Zhen, F.; Liu, J.; Wu, C. Influences of Pump Spot Radius and Depth of Focus on the Thermal Effect of Tm: YAP Crystal. *Curr. Opt. Photonics* **2019**, *3*, 458–465.
32. Che, X.; Liu, J.; Wu, C.; Wang, R.; Jin, G. Thermal effects analysis of high-power slab Tm: YLF laser with dual-end-pumped based on COMSOL multiphysics. *Integr. Ferroelectr.* **2020**, *210*, 197–205.
33. Wang, K.; Fu, S.; Zhang, K.; Gao, M.; Gao, C. Thermal effects of the zig-zag Yb: YAG slab laser with composite crystals. *Appl. Phys. B* **2021**, *127*, 121. [CrossRef]
34. Nadimi, M.; Waritanant, T.; Major, A. Thermal lensing in Nd: GdVO$_4$ laser with direct in-band pumping at 912 nm. *Appl. Phys. B* **2018**, *124*, 170. [CrossRef]
35. Liu, J.L.; Chen, X.Y.; Yu, Y.J.; Wu, C.; Bai, F.; Jin, G. Analytical solution of the thermal effects in a high-power slab Tm: YLF laser with dual-end pumping. *Phys. Rev. A* **2016**, *93*, 013854. [CrossRef]
36. Wen, Y.; Wu, C.; Niu, C.; Zhang, H.; Wang, C.; Jin, G. Study on thermal effect of mid-infrared single-ended bonded Tm LuAG laser crystals. *Infrared Phys. Technol.* **2020**, *108*, 103356. [CrossRef]

37. Ren, T.; Zhao, L.; Fan, Z.; Dong, J.; Wu, C.; Chen, F.; Pan, Q.; Yu, Y. Transient thermal effect analysis and laser characteristics of novel Tm: LuYAG crystal. *Infrared Phys. Technol.* **2022**, *125*, 104238. [CrossRef]
38. Ramesh, K.N.; Sharma, T.K.; Rao, G. Latest advancements in heat transfer enhancement in the micro-channel heat sinks: A review. *Arch. Comput. Methods Eng.* **2021**, *28*, 3135–3165. [CrossRef]
39. Liu, S.; Xie, W.; Wang, Q.; Liu, Y.; Hu, N. Thermal performance of a central-jetting microchannel heat sink designed for a high-power laser crystal. *Int. J. Heat Mass Transf.* **2022**, *185*, 122409. [CrossRef]
40. Bai, Z.; Zhang, Z.; Wang, K.; Gao, J.; Zhang, Z.; Yang, X.; Wang, Y.; Lu, Z.; Mildren, R.P. Comprehensive thermal analysis of diamond in a high-power Raman cavity based on FVM-FEM coupled method. *Nanomaterials* **2021**, *11*, 1572. [CrossRef]
41. Ding, J.; Li, Y.; Chen, H.; Cai, Y.; Bai, Z.; Qi, Y.; Yan, B.; Wang, Y.; Lu, Z. Thermal modeling of an external cavity diamond Raman laser. *Opt. Laser Thchnol.* **2022**, *156*, 108578. [CrossRef]
42. Pashinin, V.P.; Ralchenko, V.G.; Bolshakov, A.P.; Ashkinazi, E.E.; Gorbashova, M.A.; Yorov, V.Y.; Konov, V.I. External-cavity diamond Raman laser performance at 1240 nm and 1485 nm wavelengths with high pulse energy. *Laser Phys. Lett.* **2016**, *13*, 065001. [CrossRef]
43. Chen, H.; Bai, Z.; Zhao, C.; Yang, X.; Ding, J.; Qi, Y.; Wang, Y.; Lu, Z. Numerical simulation of long-wave infrared generation using an external cavity diamond Raman laser. *Front. Phys.* **2021**, *9*, 671559. [CrossRef]
44. Kitzler, O.; McKay, A.; Mildren, R.P. High power CW diamond Raman laser: Analysis of efficiency and parasitic loss. In Proceedings of the 2012 Conference on Lasers and Electro-Optics (CLEO), San Jose, CA, USA, 6–11 May 2012.
45. Tu, H.; Ma, S.; Hu, Z.; Jiang, N.; Shen, Y.; Zong, N.; Yi, J.; Yuan, Q.; Wang, X.; Wang, J. Efficient monolithic diamond Raman yellow laser at 572.5 nm. *Opt. Mater.* **2021**, *114*, 110912. [CrossRef]
46. Huang, C.-Y.; Guo, B.-C.; Zheng, Z.-X.; Tsou, C.-H.; Liang, H.-C.; Chen, Y.-F. Continuous-Wave Crystalline Laser at 714 nm via Stimulated Raman Scattering and Sum Frequency Generation. *Crystals* **2022**, *12*, 1046. [CrossRef]
47. Qi, Y.; Zhang, Y.; Huo, X.; Wang, J.; Bai, Z.; Ding, J.; Li, S.; Wang, Y.; Lu, Z. Analysis on the thermal effect of Pr: YLF crystal for power scaling. *Opt. Eng.* **2022**, *61*, 046108. [CrossRef]
48. Chvykov, V. Ti: Sa Crystal Geometry Variation vs. Final Amplifiers of CPA Laser Systems Parameters. *Crystals* **2022**, *12*, 1127. [CrossRef]

Article

Tunable, High-Power, Narrow-Linewidth Diode Laser for Potassium Alkali Metal Vapor Laser Pumping

Jinliang Han [1,2], Jun Zhang [1,*], Xiaonan Shan [1], Yawei Zhang [1], Hangyu Peng [1], Li Qin [1] and Lijun Wang [1]

1. Changchun Institute of Optics, Fine Mechanics and Physics, Chinese Academy of Sciences, Changchun 130033, China
2. University of Chinese Academy of Sciences, Beijing 100049, China
* Correspondence: zhangjciomp@163.com

Abstract: This work proposes a method of compressing spectral linewidth and tuning the central wavelength of multiple high-power diode laser arrays in an external cavity feedback structure based on one volume Bragg grating (VBG). Through the combination of beam collimation and spatial beam technologies, a diode laser source producing 102.1 W at an operating current of 40 A is achieved. This laser source has a central wavelength of 766 nm and a narrow spectral linewidth of 0.164 nm. Moreover, a tuning central wavelength ranging from 776–766.231 nm is realized by precisely controlling the temperature of the VBG, and the locked central wavelength as a function of temperature shifts at the rate of approximately 0.0076 nm/°C. The results further prove that the smile under 1 µm can restrain the self-excitation effect of the emitting laser, which can influence the efficiency of the potassium alkali metal vapor laser pumping.

Keywords: diode laser array; external cavity feedback; volume Bragg grating; tuning central wavelength; narrow linewidth

Citation: Han, J.; Zhang, J.; Shan, X.; Zhang, Y.; Peng, H.; Qin, L.; Wang, L. Tunable, High-Power, Narrow-Linewidth Diode Laser for Potassium Alkali Metal Vapor Laser Pumping. *Crystals* 2022, 12, 1675. https://doi.org/10.3390/cryst12111675

Academic Editors: Ludmila Isaenko and Anna Paola Caricato

Received: 19 October 2022
Accepted: 18 November 2022
Published: 20 November 2022

Publisher's Note: MDPI stays neutral with regard to jurisdictional claims in published maps and institutional affiliations.

Copyright: © 2022 by the authors. Licensee MDPI, Basel, Switzerland. This article is an open access article distributed under the terms and conditions of the Creative Commons Attribution (CC BY) license (https://creativecommons.org/licenses/by/4.0/).

1. Introduction

Diode pumped alkali metal vapor lasers (DPALs) have attracted extensive attention in recent years due to their advantages of low quantum defect, stable high-power output, absence of stress birefringence, efficient near-infrared (IR) atomic spectrum atmospheric transmissivity, and excellent beam quality [1–4]. DPALs have been shown to have potential applications in the fields of industrial processing, medical treatment, aerospace, and military [5–8]. Several forms of gain media are used in experiments, and each material requires a specific absorption wavelength, for example, near 852 nm for cesium (Cs), 780 nm for rubidium (Rb), and 766 nm for potassium (K), with emission wavelengths of 894.3, 794.8, and 770.1 nm, respectively [9]. Based on the laser principle, we know that the quantum defect can be expressed as $(E2 - E1)/E2$, where $E1$ and $E2$ represent the absorption and emission wavelengths, respectively. Compared to traditional solid state or fiber lasers, quantum efficiencies are higher, such as 95.27% for Cs, 98.14% for Rb, and 99.47% for K compared to the rate of 75.94% for Nd:YAG. Low quantum defect is a significant factor in improving the overall efficiency of lasers and reducing the thermal effect in very high-power laser systems.

At present, high-power DPAL is still under development. One technological factor is that a typical high-power diode laser array achieves spectral linewidth of approximately 3 nm (FWHM) in the near-IR spectrum [10]. However, the absorption spectrum of DPAL is extremely narrow, thus, leading to a mismatch between the pump and absorption spectra. Another technological factor affecting the absorption efficiency is the "smile" effect of a diode laser array [11,12]. With the increase in the smile effect, the spectral linewidth expands accordingly. There are two ways to solve these problems. First, volume Bragg grating (VBG) is used to narrow the spectral linewidth of the diode laser array through external cavity feedback technology [13,14]. Second, diode laser arrays with less smile effect are selected. Over the last decade, several companies and research groups have explored the high-power narrow-linewidth diode laser for alkali metal vapor laser pumping, as shown in Table 1. For example, Zhdanov et al. demonstrated a laser diode array line narrowing using an external cavity with a holographic grating. A linewidth of 11 GHz was obtained at an operating wavelength of 852 nm, with an output power of approximately 10 W [15]. Podvyaznyy et al. presented a diode laser system that provided up to 250 W output power and an emission spectral width of 20 pm (FWHM) at a wavelength of 780 nm [16]. Yang et al. eliminated the smile effect in spectral linewidth narrowing on high-power laser diode arrays by introducing a plane reflective mirror into a Littrow configuration external cavity to enhance the correlation among emitters. Thus, they obtained a laser diode array with 35 GHz linewidth, 780 nm central wavelength, and 41 W output laser power [17]. Hao Tang et al. described a wavelength-locked and spectral-narrowed high-power diode laser with a Faraday anomalous dispersion optical filter. The central wavelength was precisely locked at 780.24 nm, and the linewidth was narrowed to 0.002 nm with 18 W output power [18]. The mainstream research of DPALs is based on the typical wavelengths of the diode laser at 852 and 780 nm. With the development of material growth and device process technology, diode lasers of 766 nm have been made commercially available. In view of the low quantum defect of K-DPAL, a 766-nm high-power narrow-linewidth laser is examined in the current study for potassium alkali metal vapor laser pumping.

Table 1. Development of DPAL pumping source.

Time	Output Power	Spectral Linewidth	Central Wavelength
2007	10 W	11 GHz	852 nm
2010	250 W	20 pm	780 nm
2011	41 W	35 GHz	780 nm
2021	18 W	0.002 nm	780 nm

Compared with the traditional external cavity feedback method, one diode laser array should be controlled by one VBG. As the VBG has angle selectivity, each VBG must be tuned precisely. This limits the external cavity feedback effect and restricts any further increase in output power. In this paper, a method of compressing spectral linewidth and tuning central wavelength of multiple high-power diode laser array is proposed by employing an external cavity feedback structure based on one VBG. A diode laser source producing 102.1 W, with a central wavelength of 766 nm and a spectral linewidth of 0.164 nm is realized at an operating current of 40 A. By precisely controlling the temperature of VBG, a tuning central wavelength ranging from 776–766.231 nm is obtained. Such a diode laser source can be applied to efficiently pump the potassium alkali metal vapor laser.

2. Experimental Design and Simulation

The external cavity feedback structure of diode laser arrays is mainly composed of laser chips, shaping lens groups, and laser feedback elements. In this paper, the pumping laser source consists of a single conduction-cooled diode laser array (CS laser) with AuSn packaging process for laser radiating, a beam transformation system (BTS) and a slow axis collimator (SAC) for beam shaping, and the VBG for selecting the mode of the incident laser and realizing optical feedback, as shown in Figure 1a. Table 2 shows the main parameters of the diode laser chip. The reflectivity of the front facet of the laser chip is set from 2–3%. The free-running single bar can achieve 40 W output power with a central wavelength of 766 ± 5 nm and a spectral linewidth of less than 3 nm at an operating current of 40 A. The single bar is packaged into a CS structure with a size of 25 mm $\times$ 25 mm $\times$ 12 mm. By using the structures of the BTS and SAC with focal lengths of 0.41 and 15 mm, respectively, the divergence angle of the fast and slow axes of the diode laser incident on a 3 mm-thick VBG with the AR coating of 0.5%, the diffraction efficiency of 15 ± 3%, and the diffraction wavelength of 765.9 ± 0.1 nm is reduced effectively. In turn, this can reduce the lower limit for the smile effect [19] and improve the feedback efficiency of the external cavity.

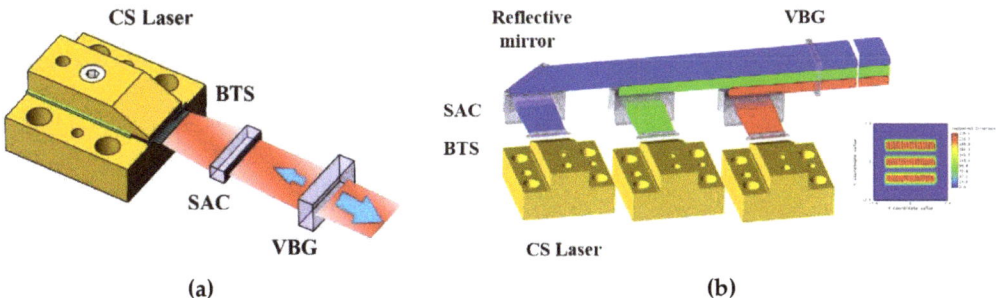

Figure 1. Schematic diagram for (**a**) the external cavity diode laser structure and (**b**) the beam shaping and spectrum control.

Table 2. Typical parameters of a laser chip.

Parameters	Unit	Specifications
Central wavelength	nm	766
Central wavelength tolerance	nm	±5
Spectral linewidth (FWHM)	nm	<3
Output power	W	≥40
Operating current	A	≤40
Operating voltage	V	<1.9
Emitter width	μm	150
Number of emitters	/	19
Filling factor	%	30
Front cavity surface coating	%	2–3

The optical stacking of multiple diode laser arrays is an efficient approach for scaling power. To obtain high-power laser output, three CS lasers are mounted onto a common staircase-like heatsink, each with a 3 mm height difference, using spatial beam combining technology. The beam size is 9 mm × 10 mm in the fast and slow axes, as shown in Figure 1b. The divergence angle of the fast axis is collimated by BTS, while the divergence angle of the slow axis is followed by SAC. Laser beams in the fast and slow axes can be exchanged by BTS at a beam spot rotation angle of 90°. Therefore, no limit is set for the minimum focal length of a SAC, and a single flat convex cylindrical lens can be used for slow axis beam collimation. In addition, the spatial beam combination of three laser beams is realized by means of three reflective mirrors. From the simulation results of Figure 2a,b, we can infer that the corresponding divergence angles of 8.4 mrad (X coordinate value) and 9.8 mrad (Y coordinate value) in the fast and slow axes are achieved. By adjusting the angle of the reflective mirror, the combined three laser beams radiate to a single VBG, which can select the mode of the incident laser and realize the optical feedback [20,21]. Only the eligible laser returns to the laser chip and couples into the internal laser field. The output laser with narrow linewidth can be realized via mode competition inside the internal laser field. However, the accuracy of the central wavelength is also critical for pumping an alkali metal vapor laser. In this study, a metal ceramic heater (MCH) is used to precisely control the VBG temperature, and the tuning of the central wavelength is simultaneously realized with a stable narrow linewidth.

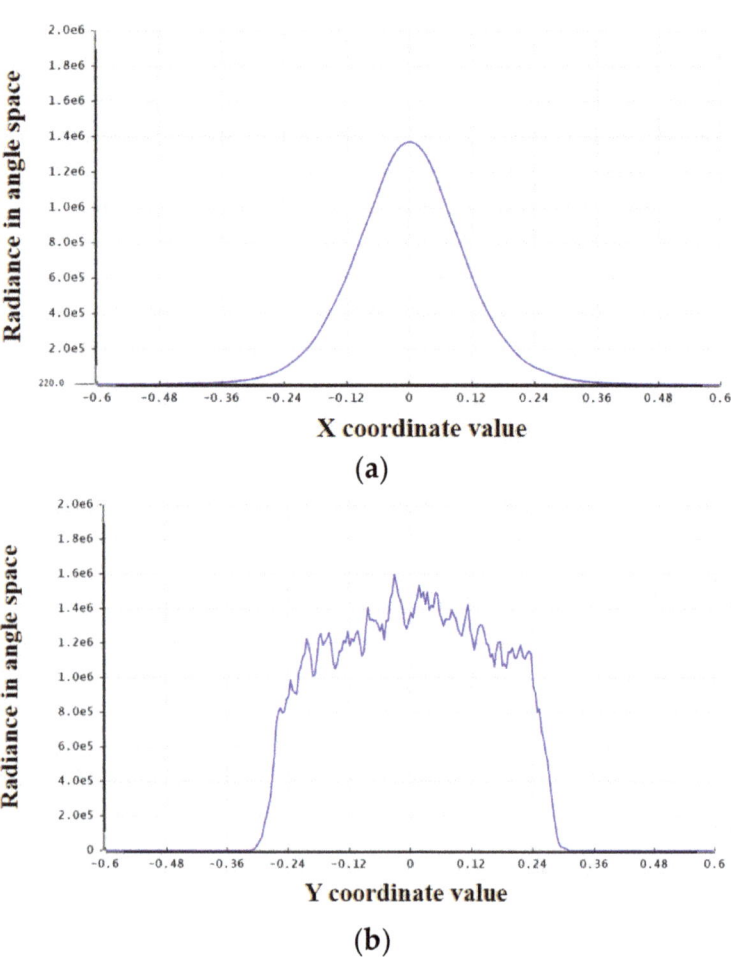

Figure 2. Simulation results of (**a**) the fast axis divergence angle and (**b**) the slow axis divergence angle.

3. Results and Discussions

The smile effect influence on the linewidth of the external cavity feedback single CS laser is studied at the cooling water temperature of 25 °C, as shown in Figure 3. The spectral characteristics of free-running CS lasers are tested at 40 A, and the normalized emission spectra at different smile values are obtained. When the smile value is increased from 1 to 3 μm, the increased smile does not affect the spectral characteristics with a linewidth of approximately 1.6 nm. However, after external cavity feedback by VBG, the corresponding narrowing linewidths become 0.152, 0.154, and 0.158 nm. Meanwhile, a smile greater than 1 μm causes the self-excitation effect of the emitting laser, which can influence the pumping efficiency of the DPAL. Furthermore, the smile effect of a CS laser affects the narrowed spectrum in the following way: the increased smile causes each emitter to radiate to the VBG at a different angle because of the angle selectivity of VBG. Hence, part of the feedback laser is unable to return to the internal cavity of the diode laser for mode competition. In the end, the narrowed spectrum of each emitter has a different central wavelength, and the total emitters are likely to broaden the spectral linewidth.

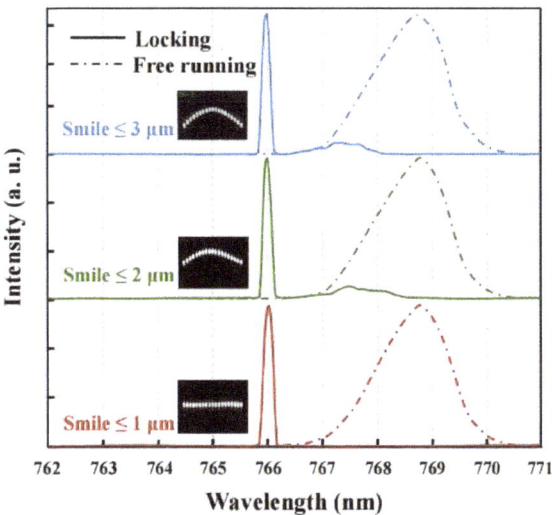

Figure 3. The influence of the smile effect on the linewidth of the external cavity feedback diode laser array. The dashed curves represent the spectral characteristics of the free-running diode laser arrays, while the solid curves represent the spectral characteristics of the external cavity diode laser arrays.

The smile effect can be effectively reduced by optical compensation method. However, a large number of optical lenses must be added to the structure, which makes the structure more complex and difficult to adjust. Therefore, in this paper, the smile effect of CS structure is mainly controlled by the packaging process. By optimizing the packaging structure, selecting the proper heatsink that can match the thermal expansion coefficient of the laser chip, compensating the stress of the laser chip in the welding process, and selecting the preset AuSn solder welding, the smile effect value can be reduced effectively. To obtain superior spectral characteristics, the CS lasers with a smile value under 1 μm are selected for the subsequent experiment.

The spectral characteristics of three CS lasers based on spatial beam combining technology are investigated under the conditions of free-running and external cavity feedback, as shown in Figure 4. The dashed curves show the typical free-running spectra at different operating currents. With the increase in operating current, the red-shift phenomenon of the central wavelength becomes prominent. The central wavelengths of 764.7, 766.08, 767.45, and 768.85 nm and the spectral linewidths of 1.215, 1.448, 1.555, and 1.635 nm are measured at the cooling water temperature of 25 °C and operating currents of 10, 20, 30, and 40 A, respectively. The solid curves show the locking spectra at different operating currents with external cavity feedback. At the same cooling condition, the central wavelengths of 765.924, 765.935, 765.975, and 766.000 nm and the spectral linewidths of 0.125, 0.139, 0.152, and 0.164 nm are measured at the operating currents of 10, 20, 30, and 40 A, respectively. From the experimental results, we can conclude that all combined CS lasers have achieved spectral locking with a narrow linewidth of less than 0.2 nm. Nevertheless, the central wavelength is shifted from 765.924 nm at 10 A to 766.000 nm at 40 A, and the locked central wavelength shift as a function of operating current has a rate of approximately 0.00253 nm/A. As the operating current increases, the laser power irradiating to the VBG generates more heat, and the diffraction central wavelength of the VBG changes to a long wavelength direction. This experimental result is in accordance with the temperature drift characteristics of VBG [22].

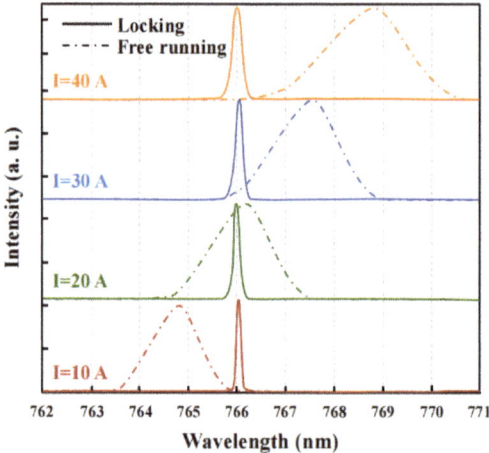

Figure 4. Spectral characteristics of the free-running and locking three CS lasers based on spatial beam combination at four different operating currents.

To control the central wavelength of the CS laser, MCH is used to change the temperature of VBG. The MCH is placed on the underside of the VBG for temperature control. The spectrum shift after controlling the temperature of VBG at different operating currents is shown in Figure 5. As can be seen, with increasing temperature, the central wavelength shows a consistent red shift phenomenon under different current conditions. Moreover, compared with increasing operating current, the controlling temperature of VBG only has a slight effect on the linewidth at the same condition. When the operating current is set to 30 A, the central wavelengths of 765.978, 766.054, 766.130, and 766.205 nm and the spectral linewidth of 0.153 nm are measured at the controlling temperatures of 25 °C, 35 °C, and 45 °C, 55 °C, respectively. When the operating current increases to 40 A, the central wavelength ranging from 766.002–766.231 nm and spectral linewidth of 0.165 nm are obtained at the controlling temperatures ranging from 25–55 °C. Furthermore, the locked central wavelength as a function of controlling temperature shifts at a rate of approximately 0.0076 nm/°C. Thus, the tunable narrow-linewidth diode laser is developed after benefiting from the temperature control technology of VBG.

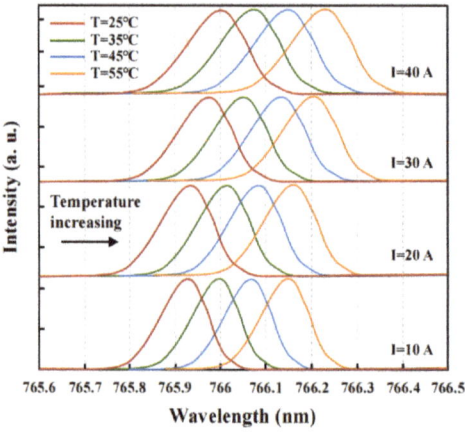

Figure 5. Tunable spectrum of locked CS lasers with increasing heating temperature at different operating currents.

Finally, the external cavity feedback structure of the multiple CS lasers based on spatial beam combining technology is constructed for high-power laser output. In the experiment, the free-running, beam-combined, and external cavity feedback powers are tested, along with electric-optical conversion and external cavity feedback efficiencies, as shown in Figure 6a. On the conditions that the water-cooling temperature is 25 °C and the operating current is 40 A, the output power of the free-running CS laser is 116.86 W with the voltage of 5.50 V. Through the beam collimation and spatial beam combining technologies, the output power is reduced to 110.86 W with 5.13% power loss. By employing a VBG for external cavity feedback and the VBG control temperature is 55 °C, the output power decreased to 102.1 W with the external cavity feedback efficiency of 92.09% and a final electric-optical conversion efficiency of 46.4%. The output power should be kept constant among the tuning range, especially when working for a long time. The output power is measured under different controlling temperatures at an operating current of 40 A. The results are shown in Figure 6b. The experimental results indicate that the laser has good output power stabilization among the tuning range. Furthermore, the tunable high-power narrow-linewidth diode laser pumping source can be obtained for potassium alkali metal vapor laser pumping.

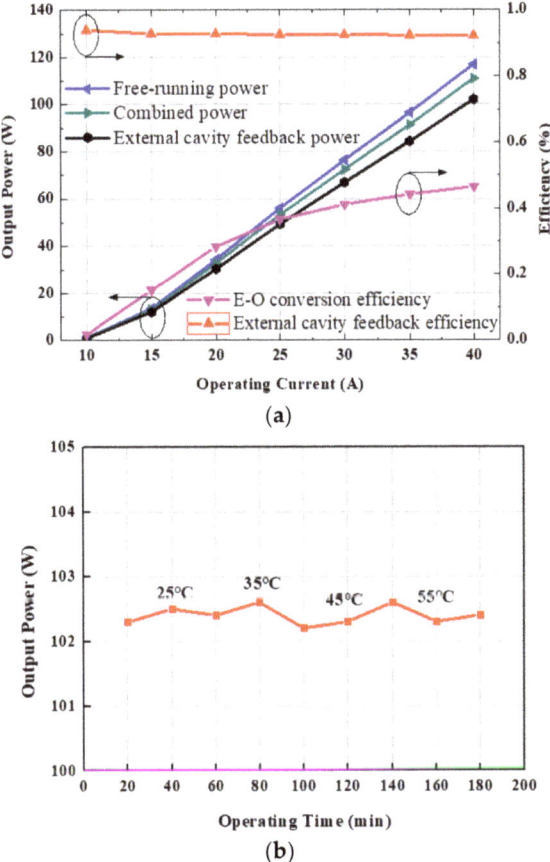

Figure 6. (a) Output power and external cavity feedback efficiency as a function of operating current; (b) output power with operating time at different temperature tuning range.

4. Conclusions

In summary, we have presented a high-power, narrow-linewidth diode laser pumping source based on external cavity feedback technology. Benefiting from the external cavity feedback structure and beam combination technology, the laser achieves a narrow linewidth of 0.164 nm and a central wavelength of 766 nm at the output power of 102.1 W. Furthermore, the external cavity feedback efficiency and electro-optical conversion efficiency exceed 92.09% and 46.4%, respectively. Moreover, tuning central wavelengths ranging from 776–766.231 nm are realized at the corresponding operating currents of 10, 20, 30, and 40 A by precisely controlling the temperature of VBG. The locked central wavelength, as a function of temperature, shifts at a rate of approximately 0.0076 nm/°C. The research results can be applied to the efficient pumping of a potassium alkali metal vapor laser.

Author Contributions: J.H. and J.Z. performed the theoretical analysis and experiment; X.S. and Y.Z. investigated the relevant literature and reports; H.P., L.Q., and L.W. were responsible for supervision. All authors contributed to the results, discussion, and manuscript writing. All authors have read and agreed to the published version of the manuscript.

Funding: Natural National Science Foundation of China (NSFC) (61991433, 62121005); Pilot Project of CAS (XDB43030302); Equipment Pre Research (2006ZYGG0304, 2020-JCJQ-ZD-245-11); Key Research and Development Project of Guangdong Province (2020B090922003).

Institutional Review Board Statement: Not applicable.

Informed Consent Statement: Not applicable.

Data Availability Statement: Not applicable.

Acknowledgments: We thank our project partners for the assistance and fruitful discussions.

Conflicts of Interest: The authors declare no conflict of interest.

References

1. Yang, J.; An, G.F.; Han, J.H.; Wang, L.; Liu, X.X.; Cai, H.; Wang, Y. Theoretical Study on Amplified Spontaneous Emission (ASE) in a V-Pumped Thin-Disk Alkali Laser. *Opt. Laser Technol.* **2021**, *142*, 107130. [CrossRef]
2. Endo, M. Peak-Power Enhancement of a Cavity-Dumped Cesium-Vapor Laser by Using Dual Longitudinal-Mode Oscillations. *Opt. Express* **2020**, *28*, 33994–34007. [CrossRef] [PubMed]
3. Hwang, J.M.; Jeong, T.; Moon, H.S. Hyperfine-State Dependence of Highly Efficient Amplification from Diode-Pumped Cesium Vapor. *Opt. Express* **2019**, *27*, 36231–36240. [CrossRef] [PubMed]
4. Auslender, I.; Yacoby, E.; Barmashenko, B.D.; Rosenwaks, S. General Model of DPAL Output Power and Beam Quality Dependence on Pump Beam Parameters: Experimental and Theoretical Studies. *JOSA B* **2018**, *35*, 3134–3142. [CrossRef]
5. Zhu, Q.; Pen, B.L.; Chen, L.; Wang, Y.J.; Zhang, X.Y. Analysis of Temperature Distributions in Diode-Pumped Alkali Vapor Lasers. *Opt. Commun.* **2010**, *283*, 2406–2410. [CrossRef]
6. Wang, Y.L.; Pan, B.L.; Zhu, Q.; Yang, J. A Kinetic Model for Diode Pumped Rubidium Vapor Laser. *Opt. Commun.* **2011**, *284*, 4045–4048.
7. Zhdanov, B.V.; Knize, R.J. Diode Pumped Alkali Lasers. In *Technologies for Optical Countermeasures VIII*; SPIE Press: Bellingham, WA, USA, 2011; pp. 31–43.
8. Gao, F.; Chen, F.; Xie, J.J.; Li, D.J.; Zhang, L.M.; Yang, G.L.; Guo, J.; Guo, L.H. Review on Diode-Pumped Alkali Vapor Laser. *Optik* **2013**, *124*, 4353–4358. [CrossRef]
9. Kissel, H.; Köhler, B.; Biesenbach, J. High-Power Diode Laser Pumps for Alkali Lasers (DPALs). In *High-Power Diode Laser Technology and Applications X*; Zediker, M.S., Ed.; SPIE Press: Bellingham, WA, USA, 2012.
10. Koenning, T.; Irwin, D.; Stapleton, D.; Pandey, R.; Guiney, T.; Patterson, S. Narrow Line Diode Laser Stacks for DPAL Pumping. In *High Energy/Average Power Lasers and Intense Beam Applications VII*; SPIE Press: Bellingham, WA, USA, 2014; pp. 67–73.
11. Talbot, C.L.; Friese, M.E.J.; Wang, D.; Brereton, I.; Heckenberg, N.R.; Rubinsztein-Dunlop, H. Linewidth Reduction in a Large-Smile Laser Diode Array. *Appl. Opt.* **2005**, *44*, 6264–9268. [CrossRef] [PubMed]
12. Liu, B.; Liu, Y.; Braiman, Y. Linewidth Reduction of a Broad-Area Laser Diode Array in a Compound External Cavity. *Appl. Opt.* **2009**, *48*, 365–370. [CrossRef] [PubMed]
13. Liu, B.; Liu, H.; Zhu, P.F.; Liu, X.S. High-Side Mode Suppression Ratio with a High-Stability External-Cavity Diode Laser Array at 976 nm in a Wide Temperature and Current Range. *Opt. Commun.* **2021**, *486*, 126792. [CrossRef]
14. Köhler, B.; Brand, T.; Haag, M.; Biesenbach, J. Wavelength Stabilized High-Power Diode Laser Modules. In *High-Power Diode Laser Technology and Applications VII*; SPIE Press: Bellingham, WA, USA, 2009; pp. 290–301.

15. Zhdanov, B.V.; Ehrenreich, T.; Knize, R.J. Narrowband External Cavity Laser Diode Array. *Electron. Lett.* **2007**, *43*, 1–2. [CrossRef]
16. Podvyaznyy, A.; Venus, G.; Smirnov, V.; Mokhun, O.; Koulechov, V.; Hostutler, D.; Glebov, L. 250W Diode Laser for Low Pressure Rb Vapor Pumping. In *High-Power Diode Laser Technology and Applications VIII*; SPIE Press: Bellingham, WA, USA, 2010; pp. 352–357.
17. Yang, Z.N.; Wang, H.Y.; Li, Y.D.; Lu, Q.S.; Hua, W.H.; Xu, X.J.; Chen, J.B. A Smile Insensitive Method for Spectral Linewidth Narrowing on High Power Laser Diode Arrays. *Opt. Commun.* **2011**, *284*, 5189–5191. [CrossRef]
18. Tang, H.; Zhao, H.; Wang, R.; Li, L.; Yang, Z.; Wang, H.; Yang, W.; Han, K.; Xu, X. 18W Ultra-Narrow Diode Laser Absolutely Locked to Rb D_2 Line. *Opt. Express* **2021**, *29*, 38728–38736. [CrossRef] [PubMed]
19. Haas, M.; Rauch, S.; Nagel, S.; Beißwanger, R.; Dekorsy, T.; Zimer, H. Beam Quality Deterioration in Dense Wavelength Beam-Combined Broad-Area Diode Lasers. *IEEE J. Quantum Eletronics* **2017**, *53*, 2600111. [CrossRef]
20. Ciapurin, I.V.; Drachenberg, D.R.; Smirnov, V.I.; Venus, G.B.; Glebov, L.B. Modeling of Phase Volume Diffractive Gratings, Part 2: Reflecting Sinusoidal Uniform Gratings, Bragg Mirrors. *Opt. Eng.* **2012**, *51*, 058001. [CrossRef]
21. Levy, J.; Feeler, R.; Junghans, J. VBG Controlled Narrow Bandwidth Diode Laser Arrays. In *High-Power Diode Laser Technology and Applications X*; SPIE Press: Bellingham, WA, USA, 2012; pp. 165–171.
22. Wang, H.Y.; Yang, Z.N.; Hua, W.H.; Liu, W.G.; Xu, X.J. Volume Bragg Grating Temperature Gradient Effect on Laser Diode Array and Stack Spectra Narrowing. In Proceedings of the XIX International Symposium on High-Power Laser Systems and Applications 2012, Istanbul, Turkey, 10–14 September 2012; SPIE Press: Bellingham, WA, USA, 2013; pp. 367–372.

Article

Long-Distance High-Power Wireless Optical Energy Transmission Based on VECSELs

Zhuo Zhang [1,2], Jianwei Zhang [1,*], Yuxiang Gong [1,2], Yinli Zhou [1], Xing Zhang [1], Chao Chen [1], Hao Wu [1], Yongyi Chen [1], Li Qin [1], Yongqiang Ning [1] and Lijun Wang [1]

[1] State Key Laboratory of Luminescence and Applications, Changchun Institute of Optics, Fine Mechanics and Physics, Chinese Academy of Sciences, Changchun 130033, China
[2] University of Chinese Academy of Sciences, Beijing 100049, China
* Correspondence: zjw1985@ciomp.ac.cn; Tel.: +86-0431-8617-6020

Abstract: Wireless charging systems are critical for safely and efficiently recharging mobile electronic devices. Current wireless charging technologies involving inductive coupling, magnetic resonance coupling, and microwave transmission are bulky, require complicated systems, expose users to harmful radiation, and have very short energy transmission distances. Herein, we report on a long-distance optical power transmission system by optimizing the external cavity structure of semiconductor lasers for laser charging applications. An ultra-long stable oscillating laser cavity with a transmission distance of 10 m is designed. The optimal laser cavity design is determined by simulating the structural parameters for stable operation, and an improved laser cavity that produces an output of 2.589 W at a transmission distance of 150 cm is fabricated. The peak power attenuation when the transmission distance increases from 50 to 150 cm is only approximately 6.4%, which proves that this wireless power transfer scheme based on a vertical external cavity surface-emitting laser can be used to realize ultra-long-distance power transmission. The proposed wireless energy transmission scheme based on a VECSEL laser is the first of its kind to report a 1.5 m transmission distance output power that exceeds 2.5 W. Compared with other wireless energy transmission technologies, this simple, compact, and safe long-distance wireless laser energy transmission system is more suitable for indoor charging applications.

Keywords: vertical external cavity surface-emitting laser; laser resonator; wireless power charging

1. Introduction

The rapid development of the 5G network and Internet of Things technology has promoted the development of automated trains, intelligent medical equipment, mobile intelligent devices, sustainable railway transportation, and other technologies, which bring a lot of convenience to people's lives [1–5]. In addition, the use of billions of mobile devices has increasingly diversified communication and entertainment systems, enhancing the lives of users [6]. However, mobile devices have to be routinely charged and carrying a charger that needs to be physically connected to a power outlet can be inconvenient. In contrast, wireless power transfer technology can transmit electrical energy from a power supply to electronic equipment without any physical connection or contact [7–9]. Therefore, research on wireless charging technology for mobile devices has increased and accelerated in recent years [10–12]. There have been many reports on wireless charging systems, among which inductive coupling, magnetic resonance coupling, and microwave radiation are the three major types [13]. Inductive coupling is safe and involves simple equipment, but the charging distance is extremely short, typically within several centimeters [14,15]. Magnetic resonance coupling can achieve efficient energy transmission, but issues such as a large coil volume and short charging distance are encountered [16,17]. In contrast, microwave radiation systems use microwaves as the medium for transmitting energy, rather than a variable magnetic field [18]. Microwave radiation systems can transmit energy across distances of thousands of meters, but their energy transmission efficiency is low [19]. Moreover, during the process of microwave radiation transmission, dense high radio frequency leakage that is harmful to the human body occurs; therefore, it is only suitable for industrial applications [20,21]. Laser beams have good directivity and can be used to realize long-distance optical power transmission through collimation. The rapid development of laser technologies has provided strong support for energy transfer technologies using light as the energy carrier [22,23]. Therefore, wireless optical power transmission technologies have been actively studied [24–26].

The unique properties of lasers make them suitable for wireless optical power transmission. The directivity of lasers allows them to have a beam diameter of only tens of centimeters at a transmission distance of several kilometers, thereby enabling long-distance energy transmission [27]. The energy of monochromatic and high power density laser beams can be efficiently absorbed and converted into electrical energy by solar cells of the corresponding wavelengths at the receiving end [28]. The resonant cavity structure of lasers also minimizes the risk of laser leakage. Once any form of barrier disrupts the laser resonator, the laser oscillation and output stop. Therefore, using a resonant cavity as an energy transmission pathway can improve the safety of laser energy transmission systems. Optically pumped vertical external cavity surface-emitting lasers (VECSELs) exhibit high power, low cost, and flexible external cavity structures, making them an ideal light source for safe and efficient wireless optical power transmission technologies [29,30].

VECSELs combine the advantages of solid-state and gas lasers to provide high output power and excellent beam quality in a simple and compact setup [31]. The unique external cavity structure of these systems allows optical elements to be inserted into the cavity for frequency conversion, mode locking, and other operations [32–35]. By adding an electro-optic modulator into the cavity, light can be used as an information and energy carrier to transmit information and power simultaneously [36–39]. By combining semiconductor band engineering designs with intracavity frequency conversion systems, VECSELs can achieve wavelength emissions from ultraviolet to near infrared wavelengths [30]. The external cavities of VECSELs can be designed to meet the needs of various applications. Therefore, wireless energy transmission technologies based on VECSELs have many potential applications, but few studies on their development have been conducted and no reports on high-power optical energy transmission systems have yet been published.

This study develops a wireless optical energy transmission system based on the external cavity characteristics of VECSELs. In order to realize safe, simple, and compact wireless power transfer, stable conditions for the laser resonant cavity are determined by simulations to allow the design of a resonant cavity that can transmit up to 10 m. Then, the size of the pump spot is matched by adjusting the cavity beam radius on the VECSEL gain chip surface, thereby increasing the output power and simplifying system debugging. Finally, we test the performance of the wireless optical transmission system.

2. System Overview

Figure 1 shows the wireless optical power transmission scheme based on a VECSEL external cavity structure. The long and straight cavity of the overall system consists of two parts. The transmitter end comprises a gain chip and a convex lens M_1 with a curvature radius of 15 cm. The output end comprises a plane mirror M_{out} and a concave lens M_2 with a curvature radius of 15 cm. The distance L_2 between the two ends is the energy transmission distance. M_1 adjusts the divergence angle of the output light in the cavity such that the beam in the cavity does not exceed the size of M_2 when transmitted to the output end. M_2 and M_{out} form the structure of the retroreflective mirror, which can reflect the incident light back to the original path. The plane mirror M_{out} has a reflectance of 97.5% in the 980-nm band. M_1 and M_2 have the same specifications, and a convex lens with a transmittance of >99.9% at 980 nm is selected to reduce the loss caused by lens reflection in the cavity. By adjusting the parameters of the laser cavity, stable laser oscillation can be maintained even if the distance L_2 between the emitter and output is increased to several meters.

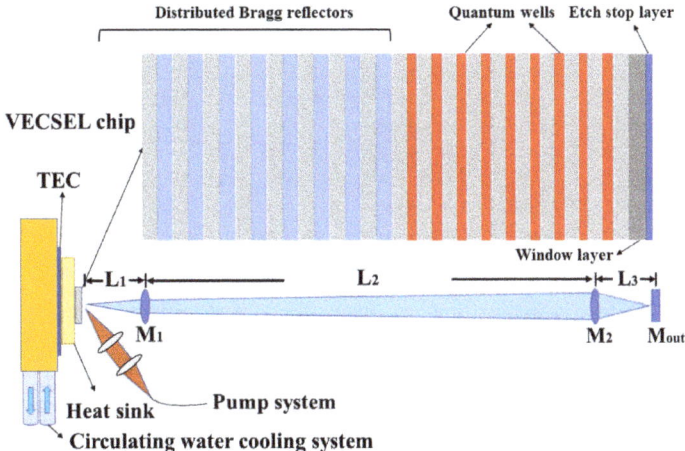

Figure 1. Schematic of the working principle in the long-cavity VECSEL.

Figure 1 shows the pump and heat dissipation systems. The pump system consists of a pump source and a focusing mirror group. The pump source provides an 808-nm pump-light output with a maximum pump power of 100 W. The output of the pump source is focused on the chip surface at an angle of 45° using a focusing mirror group that comprises two convex lenses. By adjusting the angle between the reflector group and chip, the size of the pump spot on the chip surface can be controlled. The heat dissipation system consists of a thermoelectric cooler (TEC) and circulating water cooling system. As shown in Figure 1, TEC is inserted between the copper radiator and base. TEC controls the temperature of the copper radiator, circulating water through the copper base to remove the heat generated via TEC refrigeration.

The gain chip is grown on GaAs (100) substrates using an Aixtron 200/4 MOCVD system. The etch-stop layer, window layer, active region, and distributed Bragg reflector (DBR) are successively grown on the GaAs substrate. After the structure growth is completed, the wafer is cleaved into a 3 mm × 3 mm chip. At this time, the bottom of the chip is the substrate, and the outermost layer is DBR, which is referred to as a bottom-emitting structure [40]. DBR is metallized and then soldered onto the copper heat sink using indium. The waste heat generated by the chip is rapidly dissipated through the copper radiator. A portion of the GaAs substrate is then removed by mechanical thinning, and all remaining substrates are subsequently removed by chemical etching. The GaAsP etch-stop layer is used to protect the chip structure from chemical etching. After removing the substrate, the copper heat sink is installed on the heat dissipation system.

As shown in the structural illustration of Figure 1, the Bragg reflector consists of 30 pairs of AlAs/GaAs pairs with a quarter-wavelength thickness that are designed to provide 99.9% reflectivity centered at 980 nm. The adjacent active region comprises nine 7-nm-thick InGaAs quantum wells, each of which is separated by a GaAs pump-light absorber layer. Thin GaAsP layers on both sides of QWs are used to compensate for the material strain produced by InGaAs QWs [41]. Finally, a 30-nm-thick AlGaAs window layer and a thin GaAsP etch-stop layer are grown. The role of the AlGaAs window layer is to prevent excited state carriers from escaping to the surface and performing non-radiative recombination [42].

The laser cavity scheme in Figure 1 is used to achieve long-distance stable laser oscillations and requires accurate dimensions. The distance between M_1 and the chip is L1, and the distance between M_2 and M_1 is L_2, which is the energy transmission distance. M_1 is used to adjust the beam size in the cavity to reduce the beam divergence angle, and the beam size does not exceed the lens size when reaching the M_2 surface. M_2 focuses the intracavity beam on M_{out}, and the light reflected by M_{out} converges on the chip surface through M_2 and M_1. Owing to the long cavity length, small changes in the lens position in the laser cavity will have a strong impact on the stability of the laser cavity. Therefore, we establish a theoretical model to simulate the stability of the laser cavity using the generalized ABCD matrix algorithm to obtain a more accurate laser cavity design scheme [43]. Owing to the simplicity and efficiency of the ABCD matrix when considering beam propagation, this method has been widely used to design laser resonators and analyze beam propagation [44].

Each lens in the laser cavity will affect the beam transmission inside of the cavity, and it is therefore necessary to calculate the ABCD matrix transformation after the beam matrix in the cavity passes through each lens. When a laser beam can oscillate multiple times without leakage, a stable laser cavity is achieved. Therefore, according to the stability conditions of the coaxial spherical cavity, the absolute value of the range of stability parameters calculated using the ABCD matrix is between 0 and 1 [45]. Figure 2 shows the stable working area of the laser cavity, where the unstable working area of the laser cavity is indicated in dark blue.

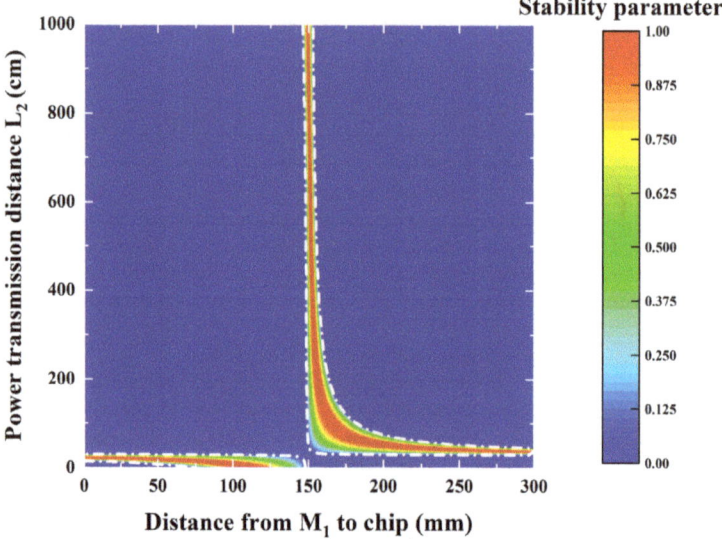

Figure 2. Influence of the cavity lengths L_1 and L_2 in VECSEL on the cavity stability. The area enclosed by the white dotted line is the working area for achieving a stable cavity.

In the graph shown in Figure 2, the abscissa is the distance L_1 from M_1 to the chip, and the ordinate is the energy transmission distance L_2. It can be seen that the laser cavity can stably function within 10 m of the transmission distance L_2 when L_1 is 155 mm. Therefore, this cavity type can indeed achieve long-distance energy transmission. Although the simulation results show that the cavity is stable, it does not necessarily achieve a high power output. The beam size on the chip surface has a large influence on the output performance of VECSEL. The beam radius of the intracavity oscillating beam on the chip surface is therefore investigated via simulation.

Figure 3 shows the variation of the intracavity beam radius on the chip surface with the cavity length L_2. As the energy transmission distance L_2 increases, the beam radius on the chip surface becomes smaller. When the transmission distance L_2 is equal to 100 cm, the beam radius on the chip surface is approximately 50 μm. As L_2 continues to increase, the beam radius on the chip surface gradually decreases and finally stabilizes at approximately 35 µm. The beam size on the chip surface matches the pump spot, and the optically pumped laser can achieve the best output under these conditions [46]. A large pump spot represents an increased output, and the pump spot size has a maximum critical value. Once the critical value is exceeded, the thermal resistance of the radiator will be greater than the thermal resistance of the chip, and the radiator will no longer function properly. According to the critical value formula, the pump spot size that the copper heat sink can support is approximately 200 µm [46]. The 35-µm intracavity beam radius on the chip surface therefore cannot support such a large pump spot size, and the laser cavity must be adjusted.

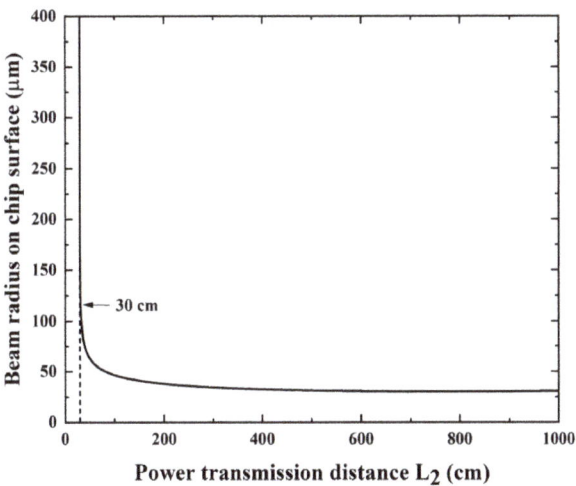

Figure 3. Beam radius of the cavity on the chip surface varying with the cavity length L_2.

Figure 4 shows the simulation results obtained after adjusting the position of the optical device in the laser cavity. As shown in Figure 4a, the stable operating region of the laser cavity after the parameter adjustment has changed significantly. Compared with the original stable cavity region, the laser cavity can also function stably at a transmission distance L_2 of 5 m. When L_2 is in the range of 0.3 to 2 m, the stable working range of the laser cavity is widened and the distance L_1 between the chip and M_1 ranges from 15 to 16 cm. This relatively wide stability range indicates that the difficulty associated with laser cavity debugging is reduced. Next, the variation of the beam radius of the chip surface with the transmission distance L_2 is next simulated in this stable working range. As shown in Figure 4b, when the transmission distance is within 0.3 to 2 m, the beam radius of the intracavity beam on the chip surface remains above 100 µm. Beyond the stable working area, the beam radius on the chip surface becomes extremely large, which indicates the leakage of the laser in the cavity. A cavity base film with this spot size is sufficient to support a large pump spot and achieve a high power output.

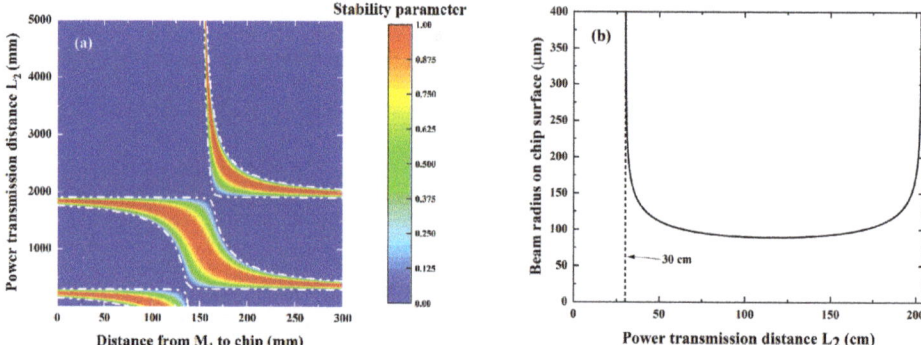

Figure 4. Simulation results after redesigning the cavity parameters. (**a**) Influence of the cavity lengths L_1 and L_2 in the VECSEL on the cavity stability. The area enclosed by the white dotted line is the working area for cavity stabilization. (**b**) Beam radius of the cavity on the chip surface varying with the cavity length L_2.

Figure 5 shows the radius variations of the beam propagation over the entire cavity when L_2 is 50, 100, and 150 cm. The position of M_1 is indicated in this figure, and the output and transmitter are framed by the black dashed lines. As the propagation distance L_2 increases, the beam radius in the range of the transmitting end does not significantly change, indicating that a compact transmitting end can be achieved. The increase in the propagation distance L_2 leads to a slight increase in the beam radius on the surface of M_2. The output end composed of M_2 and the plane mirror can completely receive and reflect all incident light, return the light to the transmitting end, and form a stable laser oscillation. Therefore, the large beam size incident on the M_2 surface can make the output end slightly deviate from the main optical axis such that the output end alignment is easier to achieve. The beam size in the output end is extremely stable and maintains the same trend. The beam radius on the output mirror is approximately 50 µm. When the transmission distance L_2 increases from 50 to 150 cm, the surface beam radius of the chip surface and output mirror remain stable. The wireless charging system can therefore maintain a stable working state over a constantly changing transmission distance.

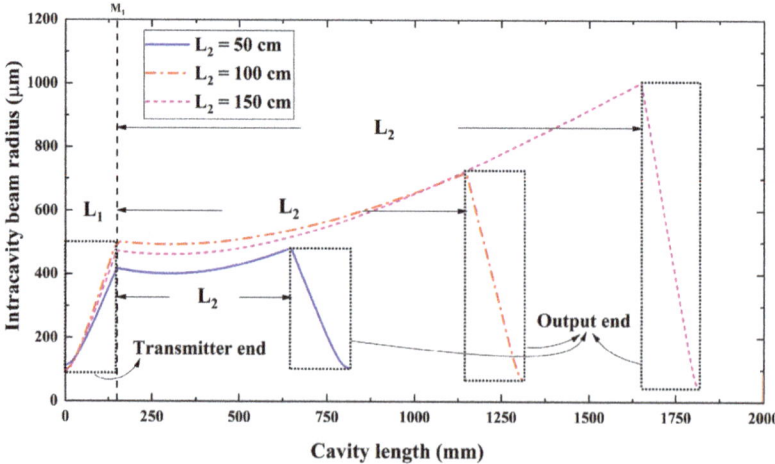

Figure 5. VECSEL internal oscillating laser beam distribution for L_4 values of 50, 100, and 150 cm.

3. Experimental Results

We determined the optimal parameters of the experimental system via simulation and designed a straight cavity that can operate stably over a long cavity length, as shown in Figure 1. Before building the straight cavity, the reflection spectrum and photoluminescence (PL) spectrum of the chip were tested. Figure 6 shows the PL and reflection spectra of the gain chip after removing the GaAs substrate. The reflection spectrum has a wide reflection band of 80 nm, extending from 940 to 1020 nm. The reflectivity decreases at 969 nm, which represents the resonance wavelength position of the Fabry–Perot (F-P) cavity [47]. The peak of the PL spectrum as modified by the microcavity is 971 nm. No side peak in the PL spectrum is present, which indicates that the chip material after strain compensation grows uniformly without serious growth defects.

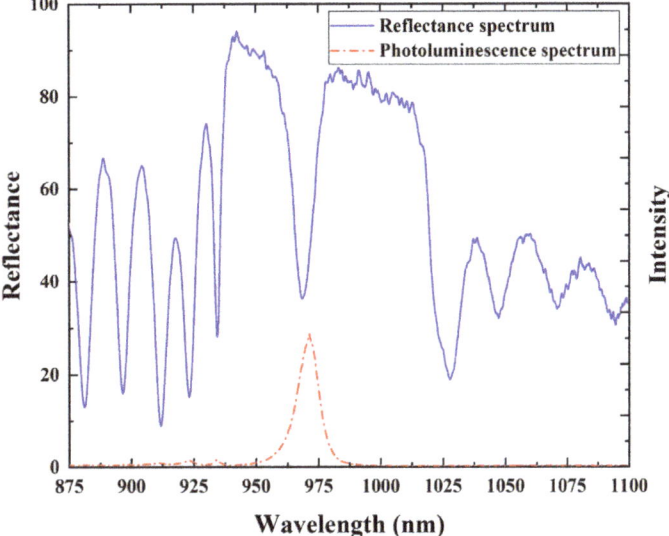

Figure 6. Gain chip reflection spectrum (solid line) and PL spectrum C (dashed red line) of the InGaAs chip obtained at 0°.

Figure 7 shows the functional relationship between the output and pump powers at different transmission distances L_2 (50, 100, and 150 cm) at a TEC control temperature of 0 °C. The output power increases linearly as the pump power is increased until thermal inversion occurs. The process of thermal inversion occurs because the pump power is too high such that the radiator cannot remove the waste heat generated by the active region at an adequate rate, and the temperature of the active region is therefore too high. The temperature drift coefficients of the cavity mode and gain peak differ [40]. Excessive temperatures lead to a large mismatch between the gain peak and cavity mode, resulting in a decrease in the output power. The slope efficiencies of the power curves do not significantly vary between different transmission distances, which indicates that the loss caused by the increase in the cavity length is small, as indicated by the variation of the peak power with the cavity length. The peak power is 1.781, 1.734, and 1.666 W at transmission distances of 50, 100, and 150 cm, respectively. When the transmission distance L_2 increases from 50 to 100 cm, the peak power decreases by 2.6%. As the transmission distance L_2 increases from 100 to 150 cm, the peak power decreases by only 3.9%. Such a small power attenuation of 6.4%/m is sufficient to prove that this cavity can support long-distance power transmission through parameter optimization.

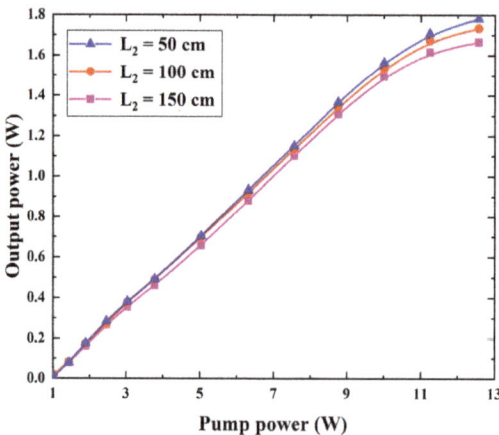

Figure 7. VECSEL output power curves for different external cavity lengths at 0 °C.

Figure 8 shows the variation of the output wavelength and full width at half maximum (FWHM) values of VECSEL with the temperature at different transmission distances. The pump power, angle, and spot size of VECSELs with different transmission distances remain unchanged. At the same temperature, there is little change in the output wavelength as the transmission distance increases. As the temperature controlled by TEC increases, the output wavelengths of different transmission distances maintain the same growth trend. As the temperature controlled by TEC is increased from −15 to 15 °C, the output wavelength shifts from 970.18 (970.57 nm at 150 cm) to 973.9 nm. The variation of the wavelength with temperature is consistent, and the temperature drift coefficient is approximately 0.12 nm/°C, which indicates that the variation of the cavity length has little effect on the output wavelength. The FWHM values of different transmission distances are <1 nm at all temperatures. A longer transmission distance L_2 is shown to result in a smaller FWHM. An increase in the cavity length leads to an increase in the cavity loss, suppression of the weaker cavity mode, and decrease in FWHM of the output wavelength. As a long cavity has an improved filtering effect on the mode with a lower intensity, a long cavity can be used to achieve a lower FWHM value.

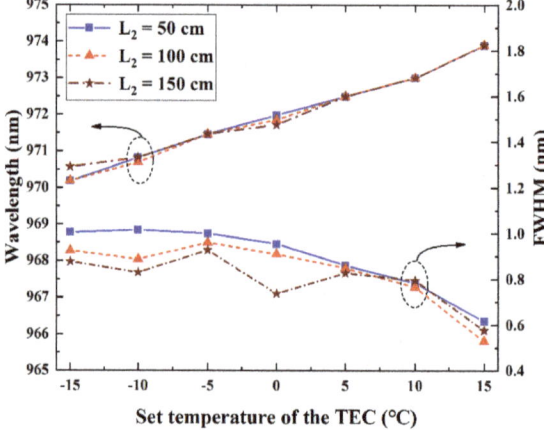

Figure 8. Output wavelength and FWHM of VECSEL varying with the temperature at different transmission distances.

Figure 9 shows the far-field modes of VECSEL at 0 °C at transmission distances L_2 of 50, 100, and 150 cm. The far-field modes at different positions show Gaussian cross-sections in both dimensions. The insets show the 2D beam profiles captured by a charge coupled device (CCD). With an increase in the transmission distance L_2, the distribution of the light beam profile remains uniformly circular. The divergence angles are 3.033°, 4.866°, and 4.095° at transmission distances L_2 of 50, 100, and 150 cm, respectively. The divergence angles of the different transmission distances are less than 5°. This shows that the output performance of VECSELs can remain stable even if the transmission distance becomes larger.

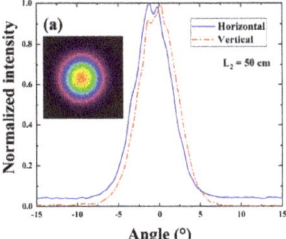
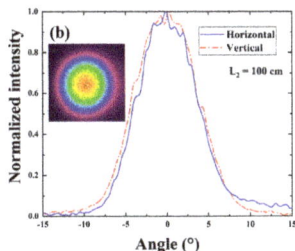
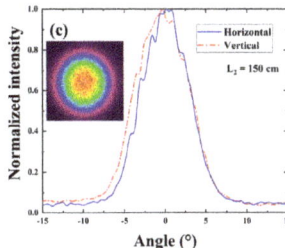

Figure 9. One-dimensional far-field modes of VECSEL measured at different transmission distances L_2 of (**a**) 50, (**b**) 100, and (**c**) 150 cm. The beam profiles of VECSEL are shown in the inset images.

Figure 10 shows the influence of the radiator temperature on the VECSEL power curve when the transmission distance L_2 is 150 cm. The power curves obtained at different temperatures exhibit the same trend, with an obvious linear growth region and thermal inversion. As the radiator temperature increases, the slope efficiency of the power curve decreases. This occurs because the loss caused by the absorption of free carriers in the semiconductor laser increases as the temperature increases. Consequently, the number of carriers overflowing from the active region increases, resulting in a decrease in the external differential quantum efficiency. At a transmission distance of 150 cm, we achieved a maximum output power of 2.589 W at a radiator temperature of −15 °C.

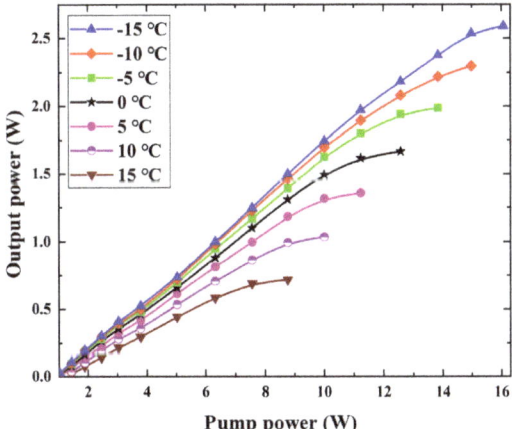

Figure 10. VECSEL output power curve for a transmission distance L_2 of 150 cm at different operating temperatures.

4. Conclusions

We designed a safe and efficient wireless laser energy transmission scheme based on the unique external cavity structure of VECSEL. The stable oscillation of the laser cavity was determined using the ABCD transfer matrix, and a stable laser cavity with a theoretical distance of 10 m was designed. To achieve a high power output and simplify the debugging process, the laser cavity parameters were adjusted to achieve a wide stable region in the laser cavity with a transmission distance of 0.3 to 2 m. The size of the fundamental mode spot on the surface of the adjusted laser cavity chip was increased to support a large pump spot and achieve a high power output. This wireless power transfer scheme yielded an output of 2.589 W at a transmission distance of 150 cm. The influence of the variation of the transmission distance L_2 on the output power was investigated, and a power reduction of approximately 6.4%/m was achieved. The beam profile of three transmission distances showed a Gaussian distribution, and the divergence angle was less than 5°.

In addition to being low cost, the optically pumped external cavity surface-emitting semiconductor laser has a small volume, high beam quality, and high output power. When an obstacle enters the laser cavity, the laser oscillation will immediately stop without causing damage. The proposed wireless energy transmission scheme based on a VECSEL laser cavity is safe and efficient and is ideal for indoor wireless charging applications. However, the current wireless energy transmission system is not modular and can only transmit energy along a straight line. In the future, we will focus on designing a small and compact modular laser system that can safely and efficiently transmit energy to multiple devices while deviating from the optical axis, providing a new strategy for enhancing the current wireless charging scheme.

Author Contributions: Project supervision, Y.N., L.W. and J.Z.; chip design, J.Z., X.Z. and C.C.; external cavity design, Z.Z. and Y.Z.; reflection spectrum measurement, Z.Z. and Y.G.; external cavity debugging, Z.Z., Y.G. and H.W.; comprehensive system performance test, Z.Z., Y.G. and Y.C.; analysis and discussion of experimental results, Z.Z., J.Z., C.C. and Y.Z.; writing—first draft preparation, Z.Z.; writing review and editing, J.Z.; fund Acquisition, J.Z., Y.N., L.Q. and L.W. All authors have read and agreed to the published version of the manuscript.

Funding: This work was funded by the National Key Research and Development Program of China (Grant no. 2018YFB2201103); the Major Program of National Natural Science Foundation of China (Grant no. 62090060); the Key Scientific and Technological Research Projects in Jilin Province (Grant no. 20220201066GX); and the National Natural Science Foundation of China (Grant nos. 61874119, 62274165, and 52172165).

Institutional Review Board Statement: Not applicable.

Informed Consent Statement: Not applicable.

Data Availability Statement: Experimental data are available upon reasonable request to the authors.

Conflicts of Interest: The authors declare no conflict of interest.

References

1. Sodhro, A.H.; Awad, A.I.; van de Beek, J.; Nikolakopoulos, G. Intelligent authentication of 5G healthcare devices: A survey. *Internet Things* **2022**, *20*, 100610. [CrossRef]
2. Singh, P.; Elmi, Z.; Meriga, V.K.; Pasha, J.; Dulebenets, M.A. Internet of Things for sustainable railway transportation: Past, present, and future. *Clean. Logist. Supply Chain.* **2022**, *4*, 100065. [CrossRef]
3. Singh, P.; Elmi, Z.; Lau, Y.; Borowska-Stefańska, M.; Wiśniewski, S.; Dulebenets, M.A. Blockchain and AI Technology Convergence: Applications in Transportation Systems. *Veh. Commun.* **2022**, *38*, 100521. [CrossRef]
4. Goswami, H.; Choudhury, H. Remote Registration and Group Authentication of IoT Devices in 5G Cellular Network. *Comput. Secur.* **2022**, *120*, 102806. [CrossRef]
5. Singh, P.; Dulebenets, M.A.; Pasha, J.; Gonzalez, E.D.S.; Lau, Y.Y.; Kampmann, R. Deployment of autonomous trains in rail transportation: Current trends and existing challenges. *IEEE Access* **2021**, *9*, 91427–91461. [CrossRef]
6. Delgado-Santos, P.; Stragapede, G.; Tolosana, R.; Guest, R.; Deravi, F.; Vera-Rodriguez, R. A survey of privacy vulnerabilities of mobile device sensors. *ACM Comput. Surv. (CSUR)* **2022**, *54*, 1–30. [CrossRef]

7. Lu, F.; Zhang, H.; Mi, C. A review on the recent development of capacitive wireless power transfer technology. *Energies* **2017**, *10*, 1752. [CrossRef]
8. Rim, C.T.; Mi, C. *Wireless Power Transfer for Electric Vehicles and Mobile Devices*; John Wiley & Sons: Hoboken, NJ, USA, 2017.
9. Sun, L.; Ma, D.; Tang, H. A review of recent trends in wireless power transfer technology and its applications in electric vehicle wireless charging. *Renew. Sustain. Energy Rev.* **2018**, *91*, 490–503. [CrossRef]
10. Zhang, S.; Qian, Z.; Wu, J.; Kong, F.; Lu, S. Wireless charger placement and power allocation for maximizing charging quality. *IEEE Trans. Mob. Comput.* **2017**, *17*, 1483–1496. [CrossRef]
11. Hui, S.Y. Planar wireless charging technology for portable electronic products and Qi. *Proc. IEEE* **2013**, *101*, 1290–1301. [CrossRef]
12. Xie, L.; Shi, Y.; Hou, Y.T.; Lou, A. Wireless power transfer and applications to sensor networks. *IEEE Wirel. Commun.* **2013**, *20*, 140–145.
13. Costanzo, A.; Dionigi, M.; Masotti, D.; Mongiardo, M.; Monti, G.; Tarricone, L.; Sorrentino, R. Electromagnetic energy harvesting and wireless power transmission: A unified approach. *Proc. IEEE* **2014**, *102*, 1692–1711. [CrossRef]
14. Wei, X.; Wang, Z.; Dai, H. A critical review of wireless power transfer via strongly coupled magnetic resonances. *Energies* **2014**, *7*, 4316–4341. [CrossRef]
15. Pantic, Z.; Lukic, S.M. Framework and topology for active tuning of parallel compensated receivers in power transfer systems. *IEEE Trans. Power Electron.* **2012**, *27*, 4503–4513. [CrossRef]
16. Kurs, A.; Moffatt, R.; Soljačić, M. Simultaneous midrange power transfer to multiple devices. *Appl. Phys. Lett.* **2010**, *96*, 044102. [CrossRef]
17. Karalis, A.; Joannopoulos, J.D.; Soljačić, M. Efficient wireless non-radiative midrange energy transfer. *Ann. Phys.* **2008**, *323*, 34–48. [CrossRef]
18. Fu, W.; Zhang, B.; Qiu, D. Study on frequency-tracking wireless power transfer system by resonant coupling. In Proceedings of the 2009 IEEE 6th International Power Electronics and Motion Control Conference, Wuhan, China, 17–20 May 2009; IEEE Publications: Piscataway, NJ, USA, 2009; Volume 2009, pp. 2658–2663. [CrossRef]
19. Garnica, J.; Chinga, R.A.; Lin, J. Wireless power transmission: From far field to near field. *Proc. IEEE* **2013**, *101*, 1321–1331. [CrossRef]
20. Huang, K.; Lau, V.K.N. Enabling wireless power transfer in cellular networks: Architecture, modeling and deployment. *IEEE Trans. Wirel. Commun.* **2014**, *13*, 902–912. [CrossRef]
21. Ladan, S.; Ghassemi, N.; Ghiotto, A.; Wu, K. Highly efficient compact rectenna for wireless energy harvesting application. *IEEE Microw. Mag.* **2013**, *14*, 117–122. [CrossRef]
22. Summerer, L.; Purcell, O. *Concepts for Wireless Energy Transmission via Laser*; Europeans Space Agency (ESA)-Advanced Concepts Team: Noordwijk, The Netherlands, 2009.
23. Man, Z.; Bao, J.; Xu, Z.; Lv, Z.; Liao, Q.; Yao, J.; Fu, H. Boosting the Efficiency of Organic Solid-State Lasers by Solvato-Tailored Assemblies. *Adv. Funct. Mater.* **2022**, 2207282. [CrossRef]
24. Kim, S.M.; Kim, S.M. Wireless optical energy transmission using optical beamforming. *Opt. Eng.* **2013**, *52*, 043205. [CrossRef]
25. Putra, A.W.S.; Tanizawa, M.; Maruyama, T. Optical wireless power transmission using Si photovoltaic through air, water, and skin. *IEEE Photonics Technol. Lett.* **2018**, *31*, 157–160. [CrossRef]
26. Ding, J.; Liu, W.; I, C.L.; Zhang, H.; Mei, H. Advanced progress of optical wireless technologies for power industry: An overview. *Appl. Sci.* **2020**, *10*, 6463. [CrossRef]
27. Jin, K.; Zhou, W. Wireless laser power transmission: A review of recent progress. *IEEE Trans. Power Electron.* **2018**, *34*, 3842–3859. [CrossRef]
28. Mukherjee, J.; Jarvis, S.; Perren, M.; Sweeney, S.J. Efficiency limits of laser power converters for optical power transfer applications. *J. Phys. D Appl. Phys.* **2013**, *46*, 264006. [CrossRef]
29. Guina, M.; Rantamäki, A.; Härkönen, A. Optically pumped VECSELs: Review of technology and progress. *J. Phys. D Appl. Phys.* **2017**, *50*, 383001. [CrossRef]
30. Kuznetsov, M. *VECSEL Semiconductor Lasers: A Path to High-Power, Quality Beam and UV to IR Wavelength by Design*; Wiley Online Library: Hoboken, NJ, USA, 2010.
31. Tropper, A.C.; Foreman, H.D.; Garnache, A.; Wilcox, K.G.; Hoogland, S.H. Vertical-external-cavity semiconductor lasers. *J. Phys. D Appl. Phys.* **2004**, *37*, R75–R85. [CrossRef]
32. Kantola, E.; Leinonen, T.; Ranta, S.; Tavast, M.; Guina, M. High-efficiency 20 W yellow VECSEL. *Opt. Express* **2014**, *22*, 6372–6380. [CrossRef]
33. Tilma, B.W.; Mangold, M.; Zaugg, C.A.; Link, S.M.; Waldburger, D.; Klenner, A.; Mayer, A.S.; Gini, E.; Golling, M.; Keller, U. Recent advances in ultrafast semiconductor disk lasers. *Light Sci. Appl.* **2015**, *4*, e310. [CrossRef]
34. Lorenser, D.; Maas, D.J.H.C.; Unold, H.J.; Bellancourt, A.-R.; Rudin, B.; Gini, E.; Ebling, D.; Keller, U. 50-GHz passively mode-locked surface-emitting semiconductor laser with 100-mW average output power. *IEEE J. Quantum Electron.* **2006**, *42*, 838–847. [CrossRef]
35. Kemp, A.J.; Maclean, A.J.; Hastie, J.E.; Smith, S.A.; Hopkins, J.-M.; Calvez, S.; Valentine, G.J.; Dawson, M.D.; Burns, D. Thermal lensing, thermal management and transverse mode control in microchip VECSELs. *Appl. Phys. B* **2006**, *83*, 189–194. [CrossRef]
36. Mansour, A.; Mesleh, R.; Abaza, M. New challenges in wireless and free space optical communications. *Opt. Lasers Eng.* **2017**, *89*, 95–108. [CrossRef]

37. Mujeeb-U-Rahman, M.; Adalian, D.; Chang, C.F.; Scherer, A. Optical power transfer and communication methods for wireless implantable sensing platforms. *J. Biomed. Opt.* **2015**, *20*, 095012. [CrossRef] [PubMed]
38. Minotto, A.; Haigh, P.A.; Łukasiewicz, Ł.G.; Lunedei, E.; Gryko, D.T.; Darwazeh, I.; Cacialli, F. Visible light communication with efficient farred/near-infrared polymer light-emitting diodes. *Light Sci. Appl.* **2020**, *9*, 70. [CrossRef] [PubMed]
39. Tavakkolnia, I.; Jagadamma, L.K.; Bian, R.; Manousiadis, P.P.; Videv, S.; Turnbull, G.A.; Samuel, I.D.W.; Haas, H. Organic photovoltaics for simultaneous energy harvesting and high-speed MIMO optical wireless communications. *Light Sci. Appl.* **2021**, *10*, 41. [CrossRef] [PubMed]
40. Tropper, A.C.; Hoogland, S. Extended cavity surface-emitting semiconductor lasers. *Prog. Quantum Electron.* **2006**, *30*, 1–43. [CrossRef]
41. Jasik, A.; Sokół, A.K.; Broda, A.; Sankowska, I.; Wójcik-Jedlińska, A.; Wasiak, M.; Trajnerowicz, A.; Kubacka-Traczyk, J.; Muszalski, J. Impact of strain on periodic gain structures in vertical external cavity surface-emitting lasers. *Appl. Phys. B* **2016**, *122*, 258. [CrossRef]
42. Holm, M.A.; Burns, D.; Ferguson, A.I.; Dawson, M.D. Actively stabilized single-frequency vertical-external-cavity AlGaAs laser. *IEEE Photonics Technol. Lett.* **1999**, *11*, 1551–1553. [CrossRef]
43. Siegman, A.E. New developments in laser resonators. In *Optical Resonators*; SPIE: Bellingham, WA, USA, 1990; Volume 1224, pp. 2–14.
44. Alda, J. Laser and Gaussian beam propagation and transformation. In *Encyclopedia of Optical Engineering*; CRC Press: Boca Raton, FL, USA, 2003; Volume 999.
45. Tromborg, B.; Osmundsen, J.; Olesen, H. Stability analysis for a semiconductor laser in an external cavity. *IEEE J. Quantum Electron.* **1984**, *20*, 1023–1032. [CrossRef]
46. Haring, R.; Paschotta, M.; Aschwanden, A.; Gini, E.; Morier-Genoud, F.; Keller, U. High-power passively mode-locked semiconductor lasers. *IEEE J. Quantum Electron.* **2002**, *38*, 1268–1275. [CrossRef]
47. Yang, H.D.; Lu, C.; Hsiao, R.; Chiou, C.; Lee, C.; Huang, C.; Yu, H.; Wang, C.; Lin, K.; Maleev, N.A.; et al. Characteristics of MOCVD- and MBE-grown InGa(N)As VCSELs. *Semicond. Sci. Technol.* **2005**, *20*, 834–839. [CrossRef]

Review

Research on Mid-Infrared External Cavity Quantum Cascade Lasers and Applications

Yuhang Ma, Keke Ding, Long Wei, Xuan Li, Junce Shi, Zaijin Li *, Yi Qu, Lin Li, Zhongliang Qiao, Guojun Liu, Lina Zeng and Dongxin Xu

Key Laboratory of Laser Technology and Optoelectronic Functional Materials of Hainan Province, Academician Team Innovation Center of Hainan Province, College of Physics and Electronic Engineering, Hainan Normal University, Haikou 571158, China
* Correspondence: lizaijin@126.com

Abstract: In this paper, we review the progress of the development and application of external cavity quantum cascade lasers (ECQCLs). We concentrated on ECQCLs based on the wide tunable range for multi-component detection and applications. ECQCLs in the mid-infrared band have a series of unique spectral properties, which can be widely used in spectroscopy, gas detection, protein detection, medical diagnosis, free space optical communication, and so on, especially wide tuning range, the tuning range up to hundreds of wavenumbers; therefore, ECQCLs show great applications potential in many fields. In this paper, the main external cavity structures of ECQCLs are reviewed and compared, such as the Littrow structure, the Littman structure, and some new structures. Some new structures include the intra-cavity out-coupling structure, multimode interference (MMI) structure, and acousto-optic modulator (AOM) control structure. At the same time, the application research of ECQCLs in gas detection, protein detection, and industry detection are introduced in detail. The results show that the use of diffraction gratings as optical feedback elements can not only achieve wide tuning, but it also has low cost, which is beneficial to reduce the complexity of the laser structure. Therefore, the use of diffraction gratings as optical feedback elements is still the mainstream direction of ECQCLs, and ECQCLs offer a further new option for multi-component detection.

Keywords: detection; ECQCL; QCL; tunable

1. Introduction

The operating wavelength of III-V PN junction semiconductor lasers based on quantum wells usually does not exceed 4 μm, which cannot meet the wavelength requirements of mid-far-infrared. Due to the existence of manufacturing defects, coupled with excessive Auger recombination loss and free carrier absorption loss, the high-quality operation of the laser cannot be guaranteed. The mid-infrared laser with a wide tunable wavelength can be produced by using nonlinear optical phenomena, which originates from the nonlinear polarizability of material. The optical parametric oscillator (OPO) is a typical example of using difference frequency generation (DFG), and it can generate a mid-infrared laser with the frequency by combining a near-infrared laser with a visible laser. The major advantage of OPO systems is their good accessibility to various pulse forms. By using an ultra-short pulse laser in femtosecond scale, it is possible to generate an extremely strong peak power of up to several MW in pulse mode. Such features are quite advantageous for nonlinear microscopy applications. However, the drawback of OPO systems is their less robustness due to its complex optical system, and the tuning range is still limited because of the wavelength accessibility of nonlinear crystals. In addition, the OPO needs a pump laser system for the optical pumping to generate the nonlinear optical process. Thus, the final form of the whole system is inevitably large and not suitable as a portable system. Distributed feedback (DFB) lasers can also generate mid-infrared wavelength and are very useful for spectroscopic sources, but their main disadvantages are their limited tuning capability and lower output power.

In order to meet the needs of mid-far-infrared wavelengths, a new type of infrared laser is required, and QCLs are a good choice, which utilize electronic transitions between quantum well subbands instead of interband optical transitions.

The emergence of QCL has created a precedent for the development of mid-far-infrared semiconductor lasers using wide-bandgap materials. Due to its narrow linewidth and high-power operation in the mid-infrared band (3–24 μm) at room temperature continuous wave (CW) conditions, it is very suitable for tracing gas sensing in mid-infrared spectroscopy. At present, quantum well lasers in the mid-infrared band lack continuous wave tunability. Gas lasers, such as CO lasers and CO_2 lasers, have a large volume and weight. Lead salt semiconductor lasers have high cooling requirements and low output laser power. QCL overcomes these shortcomings; thus, it can be used in directional infrared countermeasure, gas pollution detection, medical diagnosis, etc. ECQCLs can broaden the working wavelength, improve the beam quality and output laser power, and promote the applications of QCLs.

2. ECQCLs Basic Structure and Characteristics

2.1. Basic Structure of ECQCLs

In general, mid-infrared semiconductor lasers without any other lasers are compact and easy to use; however, their performance is limited because the emission wavelength is not a single laser mode, but it often has multiple laser modes in the spectrum. In addition, the linewidth of each mode is not narrow enough. Therefore, a form of Fabry-Perot (FP) laser is not suitable for spectroscopy applications. An ECQCL is a wavelength tunable laser, which includes an optical gain medium, an optical device for coupling the output of the gain dielectric waveguide to the free space mode of the external cavity, and a wavelength selective component, such as an interference filter or diffraction grating. Other additional optics such as polarizers, beam splitters, and prisms can also be integrated. Compared with FP lasers, ECQCL does not only benefit the tunability, but it also greatly improves the linewidth of the laser.

The basic structure of ECQCLs is usually composed of a laser external cavity and optical feedback elements, such as collimating lenses and wavelength selectors. The wavelength selectors can be a diffraction grating, a photoelectric filter, an acousto-optic filter, etc., and the laser is enhanced through the feedback of the selectors. The spectrum can be reflected by rotating the angle of the grating; thereby, the laser emission wavelength can be adjusted. The basic structure of an ECQCL is shown in Figure 1.

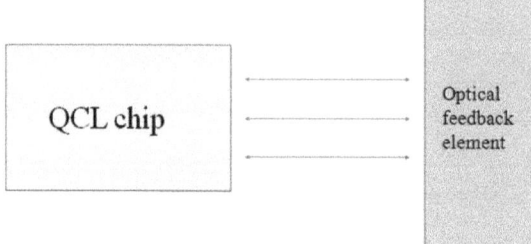

Figure 1. Basic structure of ECQCLs.

2.2. Operating Principle of QCLs

QCL is a unipolar semiconductor device that utilizes electron transitions between the subbands confined in the heterostructure conduction band for light emission. The electron distribution of a QCL needs to be designed to have a population inversion by adjusting the thicknesses of quantum wells and barriers. Injected electrons flow from the upper levels to lower levels by following the sequential wavefunctions, which represents the electron probability. During this process, the electrons optically transit at the "active region". By piling the active region of the same design in sequence, light emission can be amplified. As a consequence, the light emitted from one active region stimulates the following radiation of photons, achieving laser oscillation. During this process, the electrons flow through from one active region to the next active region while generating the stimulated emission of photons and electrons. This is why the laser device is called a quantum "cascade" laser. The more cascades are built, the more electrons can contribute to the stream for light emission. This optical amplification mechanism gives higher quantum efficiency in the laser operation.

QCLs avoid the operating principle of conventional semiconductor lasers by relying on a radically different process for laser emission, which is independent of the band gap [1]. Instead of using opposite charge carriers in semiconductors (electrons and holes) at the bottom of their respective conduction bands and valence bands, which recombine to produce light of frequency $v \approx E_g/h$ (where E_g is the energy band gap and h is Planck's constant), QCLs use only one type of charge carriers (electrons), which undergo a quantum jump between energy levels E_n and E_{n-1} to create a laser photon of frequency $(E_n-E_{n-1})/h$. The energy diagram of QCLs is shown in Figure 2. These energy levels do not naturally exist in the constituent materials of the active region, but they are artificially created by constructing the active region into nanometer-thick quantum wells. The electron motion perpendicular to the layer interface is quantified and characterized in terms of energy levels, the difference of which is determined by the thickness of the wells and the height of the energy barrier separating the wells. The implications of this new approach are profound. Based on the decoupling of lasing emissions from the bandgap by exploiting optical transitions between quantized electronic states, QCLs are equivalent to a laser with operating characteristics that are quite different from semiconductor lasers and features far superior to semiconductor lasers.

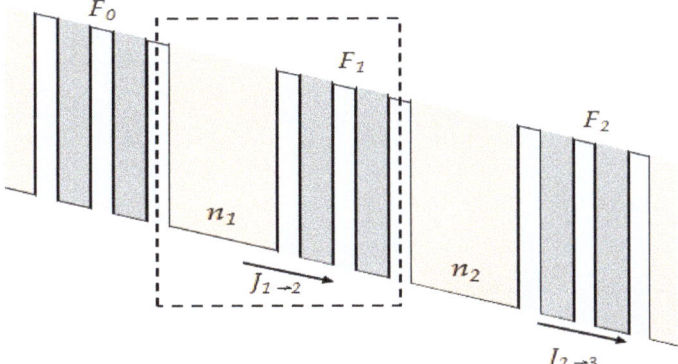

Figure 2. Energy diagram of three well structures of QCL(current densities J_m, electron densities n_m and fields F_m, module number m).

3. Research Progress of ECQCLs

The structures of ECQCLs can be mainly divided into the Littrow structure, the Littman structure, and some new structures. Some new structures include the intra-cavity out-coupling structure, the MMI structure, and the AOM control structure. This paper mainly discusses the research progress of the Littrow structure, the Littman structure, and some new structures of ECQCLs.

3.1. Littrow Structure of ECQCLs

The Littrow structure of ECQCLs is composed of a QCL gain chip, a collimating lens, and a diffraction grating, as shown in Figure 3. The QCL emits from the front end of the gain chip, then enters the diffraction grating for diffraction after collimating the beam through the collimating lens. The first-order diffraction laser returns to the gain chip along the original optical path, and the laser is output from the zero-order diffraction direction of the grating or the rear end of the gain chip.

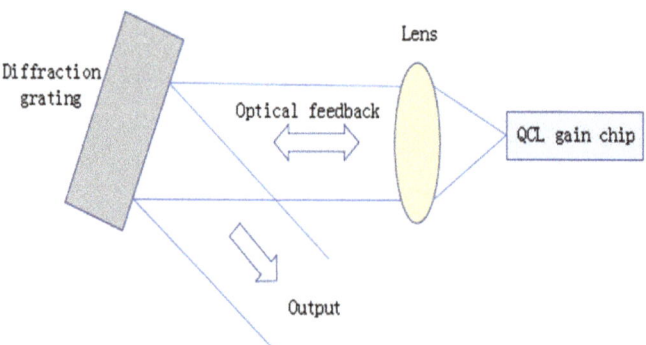

Figure 3. Littrow structure of ECQCL.

In 2006, R Maulini et al. [2] reported an ECQCL that was tuned from 8.2 µm to 10.4 µm. The ECQCL was operated in pulse mode at room temperature. For achieving tunability, the laser beam was collimated using an aspheric germanium lens and a blazed grating mounted in the Littrow structure to provide optical feedback. The zero-order reflection laser was output from the grating. The ECQCL was operated in pulse mode by using 100 ns pulse with a repetition rate of 200 kHz.

In 2009, Wysocki Gerard et al. [3] reported broadband mid-infrared laser heterodyne radiation measurements using an ECQCL. The ECQCL was operated at 8.4 µm and was able to provide tunability up to 180 cm^{-1} at $-30\,^\circ$C.

In 2014, M. Carras et al. [4] reported a 7.5 µm ECQCL spectrometer. The laser used in the spectrometer was an ECQCL. The ECQCL system used the Littrow structure, which consisted of only two optical elements, a lens for laser collimation and a diffraction grating. Rotating the diffraction grating by 1.75°, the laser achieved a tuning range of 57 cm^{-1} in a single mode emission wavelength from 7.4 µm to 7.73 µm. The wide tuning range was to be achieved over 60.4 cm^{-1}.

In 2015, Feng Xie et al. [5] reported an ultra broad tunable QCL array in the Littrow structure; the wavelength tunable range was from 6.5 to 10.4 µm, the SMSR showed 20–25 dB, and the threshold current showed 1.7–3.9 kA/cm^2.

In 2016, Zhibin Zhao et al. [6] reported a tunable ECQCL with a 7.2 µm central wavelength that was operated at room temperature. The ECQCL was implemented in a Littrow structure. The backside of the gain chip had a highly reflective coating. A two-layer anti-reflection (AR) coating composed of Al_2O_3 and ZnSe was deposited on the front side of the chip to suppress the FP mode. By using this AR coating, the single-mode tuning range of the ECQCL was reached at 128 cm^{-1}, and the wavelength was from 6.78 µm to 7.43 µm. A high SMSR was over 30 dB, and an ultra-low threshold current density was 0.89 kA/cm^2. The ECQCL was operated in CW mode at 20 °C, and an output power of 50 mW was obtained.

In 2022, Ismail Bayrakli [7] reported an ECQCL that used an FP QCL without an antireflection facet coating on the gain chip in the Littrow structure. In addition to electrical pumping, a DFB QCL was also used to optically pump the FP QCL. The spectral range was from 4.45 µm to 4.8 µm, and a wide-range dual-mode tuning range of 164 cm^{-1} was achieved.

In 2022, Ismail Bayrakli [8] reported two tunable ECQCLs that used intra-cavity and extra-cavity acousto-optic frequency shifters (AOFS), respectively. In the extra-cavity AOFS structure, the wide coarse tunable range was 256 cm^{-1}, the wavelength range was between 4.33 µm and 4.87 µm, and the fine tunable range was 0.3 cm^{-1}; it obtained within 6 ms by adjusting the injected current. In the intra-cavity AOFS structure, the wavelength was tuned rapidly at 50 µs over a broad tunable bandwidth of 33 cm^{-1} by changing the frequency of the AOF from 65 MHz to 89 MHz. The structures are shown in Figure 4.

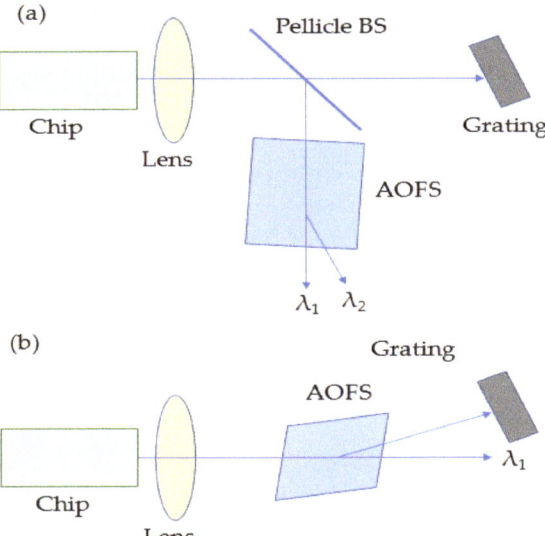

Figure 4. Two tunable structures of ECQCL: (**a**) ECQCL with extra-cavity AOFS, (**b**) ECQCL with intra-cavity AOFS. BS: Beam splitter.

3.2. Littman Structure of ECQCLs

The Littman structure added a mirror to the Littrow structure, and the Littman structure is shown in Figure 5. The first-order diffracted laser is reflected by the mirror and diffracted for the second time, and then it is fed back to the QCL gain chip and forms a resonance. Through mode competition, the first-order diffraction mode is amplified, other oscillation modes are suppressed, and the laser achieves single-mode output.

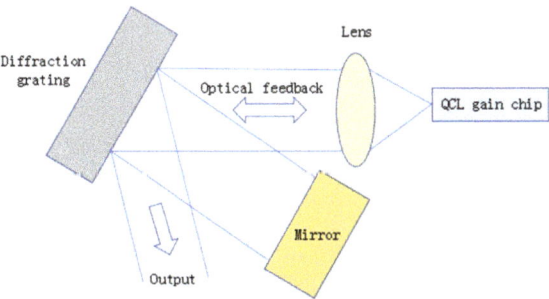

Figure 5. Littman structure of ECQCL.

In 2002, Guipeng Luo et al. [9] reported a Littman–Metcalf structure ECQCL; the operating wavelength of the ECQCL was 5.1 µm, and the wavelength tunable range was 245 nm for temperatures from 80 to 243 K.

In 2016, Wei Luo et al. [10] reported an ECQCL that used a Littman-Metcalf structure; the operating wavelength of the ECQCL was 6.9 µm, and the tunable range was from 1340 to 1640 cm^{-1}.

In 2018, Xuefeng Jia et al. [11] reported a low threshold current and fast wavelength tunable ECQCL that used a scanning galvanometer in the Littman-Metcalf cavity structure.

The ECQCL was scanned repeatedly at 100 Hz over a full tunable range of about 290 nm, from 4.46 μm to 4.75 μm, by providing a scan speed of 59.3 μm/s. The CW mode threshold current of ECQCL was as low as 250 mA for a 3 mm long QCL gain chip, and the maximum output power was 20.8 mW at 400 mA.

In 2018, Tatsuo Dougakiuchi et al. [12] reported an ECQCL based on the Littman structure. The tunable range was from 895 cm^{-1} to 990 cm^{-1}, and the output power of the ECQCL for a tunable wavelength was about 8 mW.

A Littrow structure ECQCL is usually composed of a QCL gain chip, an optical lens, and a diffraction grating. By changing the grating angle, the light wave of a certain wavelength is fed back to the QCL gain chip, which greatly increases the diffraction loss of the light wave of other wavelengths. At the same time, the overall length of the resonator is changed, and a narrow linewidth laser output with a stable wavelength is realized. A Littman structure ECQCL is usually composed of a QCL gain chip, an optical lens, a diffraction grating, and a mirror. The mirror acts as a tuner, the grating is fixed, and the incident light is returned along the incident light path by changing the angle of the mirror. After the light wave is second diffracted by the grating, the SMSR is greatly improved and the laser linewidth is further narrowed. However, its structure is more complex than a Littrow structure ECQCL, resulting in a large power loss. It is not easy to achieve high power output.

3.3. New Structures of ECQCLs

An intra-cavity out-coupling structure is a new optical structure for further improvement performances of ECQCL. By placing a BS inside the laser external cavity, much more power was directly extracted from the structure. The structure is shown in Figure 6.

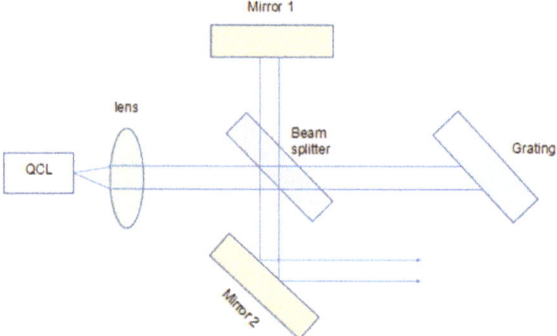

Figure 6. Intra-cavity out-coupling structure of ECQCL.

In 2018, Yohei Matsuoka et al. [13] reported an ECQCL with an intra-cavity outcoupling structure that can be tuned from 8.4 μm to 10.8 μm. Compared to the conventional Littrow structure of ECQCLs, this structure achieved higher output power and maintained a broad wavelength tunability. The maximum output power was 1 W in pulsed mode, which was more than double the output power of the Littrow structure using the same QCL gain chip.

For a single-mode laser, an excellent choice for beam splitting is the MMI [14,15], which has long been used for near-infrared beam splitting, with high splitting efficiency. The MMI structure of an ECQCL is shown in Figure 7. The MMI QCL array is collimated by collimating lens. The collimated beam is diffracted on the grating. The first-order diffracted beam is fed back to the MMI QCL array for mode selection. Wavelength tuning is achieved by changing the placement angle of the grating.

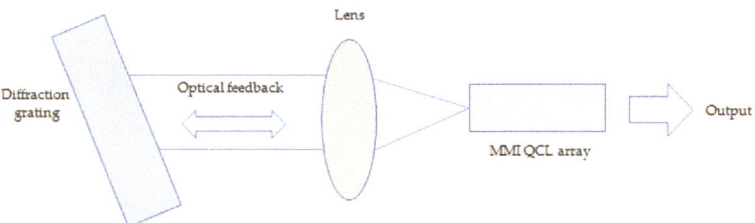

Figure 7. MMI structure of an ECQCL.

In 2021, Zeng-Hui Gu et al. [16] reported an MMI structure of an ECQCL that was designed to simplify the fabrication of QCL arrays. A wavelength tuning range of more than 60 cm^{-1} was demonstrated, and the ECQCL realized a high power and frequency tunable.

The main disadvantage of a mechanically controlled grating-based ECQCL is the relatively slow wavelength tuning, typically tens of milliseconds. Combustion and explosion diagnostics and some other infrared national defense applications require fast tuning over a broad mid-infrared range. Compared to conventional ECQCLs with mechanically controlled gratings, the use of an electrically controlled AOM enables fast wavelength tuning. In an AOM, radio-frequency acoustic waves are produced by applying an electronic signal to a piezoelectric transducer connected to an optical crystal, such as germanium, that is transparent at the wavelengths the laser needs to operate. The acoustic wave represents the phase transmission grating from which the light beam passing through the AOM crystal can be diffracted.

In 2019, Arkadiy Lyakh et al. [17] reported a new structure that combined AOM tuning with heterogeneous QCL, with the goal of developing a laser source for ultrafast tuning in a broad infrared spectral region. The tunable range of the laser was from 1990 cm^{-1} to 2250 cm^{-1}, and the schematic diagram is shown in Figure 8.

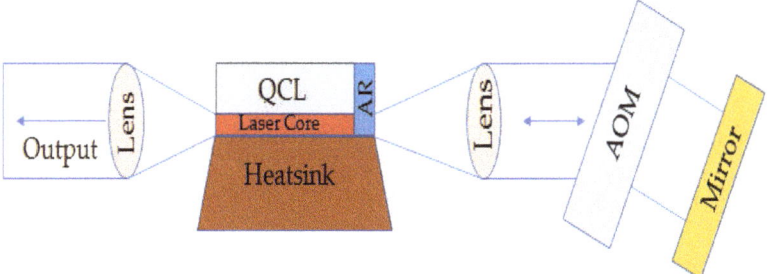

Figure 8. Structure of an AOM-controlled ECQCL.

The research development on performances of ECQCLs in recent years is listed in Table 1. Compared with the Littrow structure of ECQCLs, the Littman structure of ECQCLs and some new structures of ECQCLs have the characteristic of a wide tunable range, and they can be widely used in various fields. By changing the optical feedback elements, a wider wavelength tuning range and a higher power are realized. Therefore, some new optical feed elements or new structures can make the ECQCLs obtain a wider wavelength tunable range, a higher power, and a higher SMSR.

Table 1. The performance parameters of ECQCLs in recent years.

Type	λ (μm)	Tunable Range (μm)	Output Power (mW)	SMSR (dB)	Resolution (cm^{-1})	Year
Littman	5.1	0.245	10	40	0.062	2002 [9]
Littrow	8.2–10.4	2.2 *	147	30	-	2006 [2]
Littrow	8.4	0.352 *	>50	-	0.001	2009 [3]
Littrow	7.5	0.33 *	73	-	0.36	2014 [4]
Littrow	6.5–10.4	3.9	250	25	-	2015 [5]
Littrow	6.78–7.43	0.65 *	50	30	-	2016 [6]
Littman	6.9	1.365 *	-	-	0.14	2016 [10]
Littman	4.46–4.75 *	0.29	20.8	25	0.2	2018 [11]
Littman	10.1–11.2	1.1 *	16	-	-	2018 [12]
AOM	4.44–5.02 *	0.58 *	14	-	-	2019 [17]
Intra-cavity out-coupling	8.4–10.8	2.4 *	1000	-	-	2021 [13]
MMI	8	0.391 *	1450	-	0.25	2021 [16]
Littrow	4.45–4.80	0.35	68	>20	-	2022 [7]
Littrow	4.33–4.87	0.54	70	20–70	0.001	2022 [8]

Note: "*" denotes that the data are calculated, "-" denotes that the data are not available.

4. Applications of Mid-Infrared ECQCLs

ECQCLs have the advantages of high conversion efficiency, compact size, and high reliability [18]. Their lasing wavelength covers two important atmospheric windows of 3–5 μm and 8–14 μm; thus, it can be used in molecular detection, free space communication, and industry applications. Using tunable ECQCL spectroscopy has many advantages, including high sensitivity and selectivity. It is non-destructive, fast, and requires no sample preparation. With advances in ECQCLs in terms of tunability, output power, reliability, and operating temperature, there has been a growing interest in the use of ECQCLs for gas detection by spectroscopic groups. Molecular detection is currently the most studied and widely used field, such as environmental monitoring, emission measurement, remote sensing, medical and life science applications [19], industrial process control, safety, and basic science, all of which benefit from the technological advancements in ECQCLs. Different techniques for absorption spectroscopy were demonstrated using ECQCLs [20,21]. Various substances have unique functional groups, and specific functional groups have specific optical activities and have different absorption characteristics for different wavelength lasers. Therefore, by analyzing the spectral changes caused by the gas, the corresponding molecular content and species can be obtained. Due to the high resolution and high reliability of ECQCLs, there are important application prospects in this field. Laser-based infrared spectroscopy is an emerging key technology for analyzing solutes and monitoring reactions in liquids in real time. Compared to the traditional standard Fourier transform infrared spectroscopy (FTIR), the larger applicable path length enables the robust measurement of analytes in strongly absorbing matrices such as water. Recent advances in laser development have also provided a wide range of accessible spectral coverage, thus overcoming the inherent shortcomings of laser-based infrared spectroscopy.

4.1. Application of ECQCLs in Gas Detection

The detection of gas is important in many applications, including combustion diagnostics, industrial detection, atmospheric detection, and medical detection.

4.1.1. Nitric Oxide (NO) Detection

In 2005, G. Wysocki et al. [22] reported a mode-hopping, broadly tunable CW and thermoelectrically cooled ECQCL capable of high-resolution spectroscopic measurements. The system provides independent wavelength tracking through all three wavelength-selective elements of ECQCLs, which made it suitable for applications using gain chips. The current prototype instrument had a wavelength of 5.2 µm, a tuning range of 35 cm^{-1}, and a continuous mode-hop-free (MHF) mode tuning range of 2 cm^{-1}. The overall performance of the spectrometer system was demonstrated by direct absorption spectroscopy measurements of NO under reduced pressure. Its wavelength modulation capability, coupled with wide tunability and high spectral resolution, made the ECQCL an excellent light source for many mid-infrared spectroscopy applications, such as trace gas detection.

In 2010, V. Spagnolo et al. [23] reported a gas sensor based on quartz-enhanced photoacoustic detection and an ECQCL. It was characterized by NO absorption multiplication at 5.26 µm to monitor trace NO and studied the dependence of signal and noise on gas pressure to optimize the performance of the sensor. The tuning range of the ECQCL was 5.13–5.67 µm, and the specified MHF mode tuning range was 5.26–5.53 µm, which corresponded to 5% of its central wavelength; the output power exceeded 100 mW. By contrast, quartz-enhanced photoacoustic spectroscopy (QEPAS) technology was competitive in sensitivity, while offering a more compact sensor design and smaller size.

4.1.2. CO_2 Detection

In 2017, Ramin Ghorbani et al. [24] reported a mid-infrared tunable diode laser absorption spectroscopy sensor for the real-time detection of CO_2 in exhaled breath. The system used an ECQCL. The output wavelength tunable range was from 4.50 to 4.96 µm, and the peak output power was 160 mW.

4.1.3. Butane (C_4H_{10}) Detection

In 2013, D. Mammez et al. [25] reported the commercialization of an ECQCL in a photoacoustic spectrometer with an emission wavelength of 10.5 µm. The spectrometer could measure in a wide spectral range of 60 cm^{-1}, which means that the spectra of complex molecules as well as the entire absorption bands of small molecules could be recorded. The wide tuning range of this photoacoustic spectrometer light source demonstrates the possibility of detecting complex small molecules such as CO_2 and C_4H_{10}. Figure 9 shows the structure diagram of spectral detection.

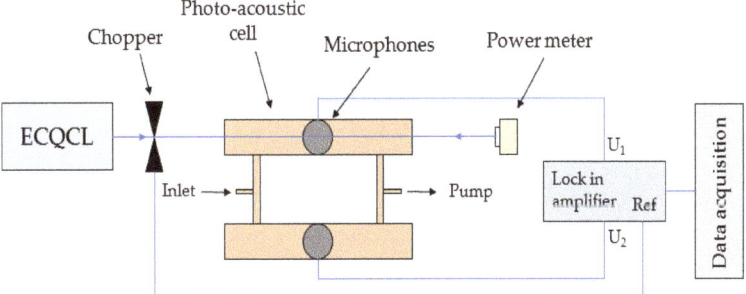

Figure 9. Structure diagram of spectral detection.

4.1.4. Acetylene (C_2H_2) Detection

In 2018, Abhijit Maity et al. [26] reported a mid-infrared detection strategy using an ECQCL, and the working wavelength was from 7.5 µm to 8 µm. The C_2H_2 detection had a noise limit of three parts per billion (ppb), and the integration time was 110 s. The current high resolution ECQCL system was further validated in the C_2H_2 concentration range of 0.1–1000 ppm, which showed good promise in practical sensing applications.

4.1.5. Nitrous Oxide(N_2O) Detection

In 2019, Faisal Nadeem et al. [27] reported a mid-infrared trace gas detection system that was used to detect N_2O. An ECQCL was used in the system, and the working wavelength was 7.7 μm, the output power was 40 mW, and the tunable range was about 320 cm^{-1}.

4.1.6. Methane (CH_4) Detection

In 2017, Abhijit Maity et al. [28] reported mid-infrared CW cavity ring down spectroscopy (CRDS) technology and MHF ECQCL technology operating at 7.5 μm. The authors validated the ECQCL-based high-resolution CW-CRDS system by measuring the $^{12}CH_4$ and $^{12}CH_4$ isotopes of CH_4 as a reference molecule. By probing the asymmetric bending (υ4 band), vibrations of the ^{12}C and ^{13}C isotopes of CH_4 in the sample from bonds centered at 7.534 μm and 7.502 μm, respectively. The current high-resolution CW-CRDS system could further utilize the spectral region covering 7.5–8 μm to trace several other molecular species and their isotopes.

In 2018, Xiaojuan Cui et al. [29] introduced a compact laser absorption sensor system associated with a 152 m long absorption cell for the simultaneous detection of N_2O, CH_4, and the second harmonic of H_2O vapor. An 8 μm ECQCL was the excitation laser source, and three adjacent absorption lines, N_2O, CH_4, and H_2O, at 7.968 μm and 7.969 μm were simultaneously aimed at 7.965 μm. At the optimum pressure of 50 Torr, the lowest detection limit was achieved with 1 s integration time, 0.9 ppb for N_2O, 4.8 ppb for CH_4, and 31 ppm for H_2O.

In 2021, Qianhe Wei et al. [30] reported a gas sensor based on a tunable 7.6 μm CW MHF ECQCL (from 7.4 μm to 7.7 μm) CRDS technique. The sensor could detect CH_4 and N_2O in ambient air.

4.1.7. Chlorodifluoromethane ($CHClF_2$) Detection

$CHClF_2$ is one of the most abundant HCFCs in the atmosphere. Due to its relatively low ozone depletion potential in chlorine-containing haloalkanes, it is often used as a replacement for the high ozone depleting CFC-11 and CFC-12. $CHClF_2$ is commonly used as a refrigerant in air conditioning systems. Although developed countries are phasing out $CHClF_2$ due to high global warming potential, the use of $CHClF_2$ continues to increase due to a high demand in developing countries. In 2019, Sheng Zhou et al. [31] reported a sensor that used QEPAS and ECQCL to detect $CHClF_2$ with unresolved rotation-vibration absorption lines. The spectral range was from 7.04 μm to 8.13 μm.

4.1.8. Hydrogen Sulfide (H_2S) Detection

In 2017, Michal Nikodem et al. [32] reported on a quantum cascade laser-based spectroscopic system for the detection of H_2S in the mid-infrared of 7.2 μm, and the wavelength tunable range was from 7 to 8.2 μm.

In 2019, Mithun Pal et al. [33] reported a sensor that used a mid-infrared ECQCL cavity ring-down spectroscopy to simultaneously monitor the ^{32}S, ^{33}S, and ^{34}S isotopes of H_2S. It verified the possibility of this system for tracking the characteristics of sulfur isotopes in compounds in practical applications. Nine independent transition lines of $H_2^{32}S$ and $H_2^{33}S$ isotopes in the current MHF ECQCL tuning range were further explored for the trace monitoring of the H_2S single isotope. At a pressure of 30 Torr and an integration time of 255 s, the lowest detection limit was 20 ppb. It provided a new method that combined the unique spectral features of 7.5 μm, the high sensitivity of CRDS technology, the high resolution of ECQCL, and a wide MHF tunability.

4.1.9. Sulfur Dioxide (SO_2)

In 2020, Xukun Yin et al. [34] reported a ppb-level SO_2 photoacoustic sensor using a 7.41 μm ECQCL to suppress the absorption-desorption effect. For the first time, a CW ECQCL combined with a customized differential photoacoustic cell was employed to detect trace SO_2 in mid-infrared. The ECQCL current was set at 700 mA and operated in MHF mode. The maximum power was about 60 mW in the wavelength between 7.35 μm and 7.45 μm, satisfying the excitation wavelength requirement. When the input current was greater than 600 mA, the ECQCL started to emit light. Subsequently, the ECQCL output power with the large inner diameter of the two resonators and the differential photoacoustic cell structure could reduce the background noise and response time, resulting in the best detection limit at the ppb level.

4.1.10. Benzene (C_6H_6) Detection

In 2020, Mohammad Khaled Shakfa et al. [35] reported a new mid-infrared laser diagnostic instrument for C_6H_6 measurement. The structure is shown in Figure 10, and this instrument consisted of an ECQCL and a CO_2 gas laser, directional patterned GaAs crystal. The laser system emitted a laser that ranged from 12.64 μm to 15 μm in the mid-infrared region. The laser was tuned to the Q-branch transition peak around 14.838 μm, and it showed that the cross section of C_6H_6 was very sensitive to pressure. Experiments were carried out after reflecting the shock wave to determine the absorption cross section of C_6H_6 at 553–1473 K and 1.17–2.48 bar. This new detection method was been confirmed in the reaction shock tube experiment of propargyl iodide to C_6H_6.

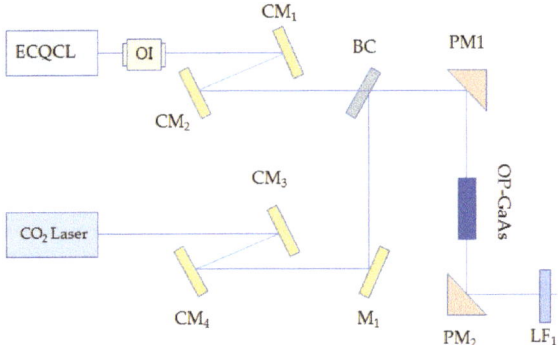

Figure 10. Schematic drawing of the mid-infrared laser setup.

4.1.11. Volatile Organic Compounds (VOCs) Detection

Ethanol and acetone are the most common VOCs, are good organic solvents in plastics, rubber, paint, and other industries, and are also commonly used detergents in laboratories [36]. However, continued exposure to these VOCs can lead to feelings of irritation, many other uncomfortable symptoms, and even the risk of cancer, and VOCs readily react with atmospheric oxidants, leading to ozone pollution and atmospheric acidification. Therefore, studying the real-time monitoring of VOCs is crucial for optimizing the living environment and protecting human health.

In 2016, Juan Sun et al. [37] reported a tunable diode laser absorption spectroscopy system based on a broad band ECQCL near 7.78 μm that was used to study VOC measurements. The tunable wavelength range of the ECQCL was from 6.96 μm to 8.85 μm.

In 2018, Ningwu Liu et al. [38] reported a broadband ECQCL-based sensor used for the open-path sensing of multiple VOCs. The ECQCL had a tuning range of 6.96–8.85 μm, and the laser could generate different pulse repetition rates (up to 3 MHz) in the pulse range from 20 ns to 350 ns while maintaining a duty cycle of about 15%. The ECQCL sensor was successfully used for the long-range detection of mixed plumes of three VOCs at a distance of 40 m, proving its suitability for leak plumes in the safety field. Preliminary alcohol, acetone, ether open circuit detection, and identification tests demonstrated the high potential of the ECQCL sensor for monitoring chemical leaks in the safety field.

4.1.12. Alkanes Detection

In 2019, Robet Heinrich et al. [39] reported a multi-component spectrum of hydrocarbons based on a CW ECQCL spectrometer, which provided a tunability of 6–11 μm to measure the first seven alkanes and their mixtures. Gas spectra were obtained in the range 6.756–6.944 μm at a reduced pressure of 50 mbar and a temperature of 323 K. Spectral accuracy up to ± 0.001 cm^{-1} was achieved by the linearization of the measurement wavelength using a custom made highly temperature-stabilized air spaced etalon. The high resolution of 0.001 cm^{-1} produced heavy alkane (C_3–C_5) spectra with unprecedented richness details and allowed to distinguish narrow spectral features of light alkanes (C_1–C_2).

4.2. Protein Detection

Due to the limited emission wavelength range of ECQCLs, the spectral coverage is limited compared to FTIR, but its significant advantage is that it is stable and convenient in the detection of amides [40]. The large optical path can be used to directly measure the infrared absorption spectrum of water-based solutions, such as body fluids (blood, serum, breast milk), foods (commercial milk), etc. Compared to earlier laser-based infrared lasers, the expanded spectral coverage, including the most prominent protein infrared band, provides advantages for qualitative and quantitative studies of proteins. In the future, ECQCLs will be used to study dynamic secondary structure changes and stoichiometry-based protein quantification in complex matrices.

In 2019, Milagros Montermurro et al. [41] reported a rapid analysis system of commercial milk proteins using ECQCL mid-infrared spectroscopy. In the system, a thermoelectrically cooled ECQCL with a repetition rate of 100 kHz and a pulse width of 5000 ns was used. All spectra were recorded in the spectral tuning range of 5.78–6.8 μm, covering the amide I and amide II regions of the protein, with a scan speed of 1200 cm^{-1}s^{-1}. The mid infrared ECQCL was focused on the detector element by a gold-coated off-axis parabolic mirror with a focal length of 43 mm. The operating temperature was -78 °C.

In 2020, Alicja Dabrowska et al. [42] reported a Mach-Zehnder interferometer-based sensor for detecting the dispersive spectroscopy of proteins. This is also the first time that the refractive index spectrum of a protein was measured with such high speed and resolution over such a broad spectral range. The thermoelectrically cooled ECQCL could be tuned in the range of 5.78–6.8 μm. Dispersive spectroscopy achieves a figure of merit similar to established high-end FTIR spectroscopy at the same acquisition time. In the same year, a mid-infrared transmission setup for the analysis of protein amide I and amide II bands in aqueous solution was studied using the ECQCL.

In 2021, Schwaighofer Andreas et al. [43] reported a commercial room temperature operating broadband ECQCL infrared spectroscopy with a spectral coverage of 5.65–7.4 μm combined with FTIR spectroscopy that was compared and demonstrated for its application in measuring a protein secondary structure in water and for monitoring the lipase-catalyzed saponification of triacetin. For the obtained limits of detection, ECQCL-based spectrometers performed better than research-grade FTIR spectrometers with liquid nitrogen cooled detectors. The device monitored the enzymatic hydrolysis of triacetin by lipase, demonstrating the advantage of broad spectral coverage for the subsequent monitoring of complex chemical reactions that cannot be readily obtained by FTIR spectroscopy without the use of liquid nitrogen cooling.

4.3. Industry Detection

In 2020, Mark C. Phillips et al. [44] reported a swept-wavelength ECQCL that was used to perform the standoff detection of combustion gases in a plume generated from an outdoor high-explosive open detonation. The swept-ECQCL system was located at a standoff distance of 830 m from a 41 kg charge of LX-14 (polymer-bonded high explosive) and was used to measure the infrared transmission or absorption through the post-detonation plume as it propagated through the beam path. The swept-ECQCL was operated continuously to record broadband absorption spectra at a 200 Hz rate over a spectral range from 2050 cm^{-1}

to 2230 cm^{-1}. The fitting of measured spectra was used to determine time-resolved column densities of CO, CO$_2$, H$_2$O, and N$_2$O.

In 2020, Anaïs Parrot et al. [45] reported an ECQCL mid-infrared reflectance spectroscopy that was used to discriminate silicate and carbonate minerals in a standoff measurement setting. The tunable ECQCL source that was used allowed measurement from the 5.2 µm to 13.4 µm wavelength, where the fundamental vibrational bands of silicates and carbonates were observed. Mid-infrared reflectance spectroscopy using compact ECQCL sources allowed rapid spectral measurements at standoff distances and high spatial resolution. It showed the potential of ECQCL mid-infrared reflectance spectroscopy for in the field mining applications.

In 2022, Francis Vanier et al. [46] designed an ECQCL-based mid-infrared spectrometer. The light source consisted of four ECQCLs with spectral coverage ranging from 5.2 µm to 13.4 µm wavelengths. The performance of a mid-infrared reflectance spectroscopy device based on a tunable ECQCL module was described. The results assessed the quality and usability of spectra of mineral mixtures obtained using ECQCL-based mid-infrared spectroscopy, completing the first step in mineral characterization using ECQCL-based mid-infrared spectroscopy.

The research development on applications of ECQCLs in recent years is listed in Table 2. The recent availability of ECQCLs provides a promising new avenue for multicomponent detection and protein detection. Based on the recent development of advanced instrumentation, including compact and robust ECQCL systems, the promotion of this high resolution mid-infrared spectroscopy for industrial applications has been rapidly realized.

Table 2. The applications of ECQCLs and the detection substance.

λ (µm)	Tuning Range (µm)	Detect Matter	Application Field	Year
5.2	0.09 *	NO	Gas detection	2005 [22]
5.26	0.54 *	NO	Gas detection	2010 [23]
10.5	0.62 *	C$_4$H$_{10}$	Gas detection	2013 [25]
7.78	1.89	VOCs	Gas detection	2016 [37]
4.50–4.96	0.46	CO$_2$	Gas detection	2017 [24]
7.2	1.2	H$_2$S	Gas detection	2017 [32]
7.5–8	0.5	CH$_4$	Gas detection	2017 [28]
7.5–8	0.5	C$_2$H$_2$	Gas detection	2018 [26]
7.91–8.17 *	0.26 *	CH$_4$	Gas detection	2018 [29]
6.96–8.85	1.89	VOCs	Gas detection	2018 [38]
7.7	1.92 ^{d}	NO$_2$	Gas detection	2019 [27]
7.04–8.13 *	1.09 *	CHClF$_2$	Gas detection	2019 [31]
7.46–7.95 *	0.49 *	H$_2$S	Gas detection	2019 [33]
6–11	5	Alkane	Gas detection	2019 [39]
5.78–6.80 *	1.02 *	Milk	Protein detection	2019 [41]
7.20–7.52	0.32	SO$_2$	Gas detection	2020 [34]
12.64–15.00 *	2.36 *	C$_6$H$_6$	Gas detection	2020 [35]
5.78–6.80 *	1.02 *	Protein	Protein detection	2020 [42]
4.48–4.88	0.4	Explosives	Industry detection	2020 [44]
5.2–13.4	8.2	Silicate and carbonate minerals	Industry detection	2020 [45]
7.40–7.75 *	0.35 *	CH$_4$	Gas detection	2021 [30]

Table 2. *Cont.*

λ (μm)	Tuning Range (μm)	Detect Matter	Application Field	Year
5.64–7.40 *	1.76 *	Protein, Enzymatic activity	Protein detection	2021 [43]
5.2–13.4	8.2	Mineral	Industry detection	2022 [46]

Note: "*" denotes that the data are calculated.

5. Summary

In the field of mid-infrared ECQCL laser research [47], several open questions invite major research investment. Such fundamental issues include the development of sub-picosecond ECQCLs, high-pulse-energy pulsed ECQCLs, ECQCLs used to detect materials, and ECQCLs used as consumer electronics. Such devices would highlight the significant potential of ECQCLs and open up new fields for research and applications.

Mid-infrared techniques are a very powerful tool for molecular spectroscopy because many molecular vibrational modes lie in this wavelength range [48,49]. An ECQCL is a mid-infrared tunable ECQCL that can cover any part of this spectral range. Therefore, ECQCLs have great application potential as industrial-scale standard light sources.

After more than 20 years of rapid development, QCLs are becoming the most important mid-far-infrared light sources [50]. The advantages of QCLs are vividly reflected in the fields of high-power devices, low-power single mode devices, high-speed tunable devices, and broadband optical frequency comb devices. In the near future, quantum cascade lasers will play an increasingly important role in infrared countermeasures, gas sensing, and free space communication. With the continuous optimization of the external cavity structure, ECQCLs will bring a wider tuning range and narrower linewidth in the future, which will shine in environmental monitoring, medical treatment, infrared countermeasures, etc.

Author Contributions: Conceptualization, Y.M., Z.L. and K.D.; methodology, X.L. and J.S.; writing—original draft preparation, X.L. and Z.L.; writing—review and editing, L.W., L.L. and L.Z.; visualization, G.L. and Y.Q.; supervision, Z.Q. and D.X.; funding acquisition, G.L. All authors have read and agreed to the published version of the manuscript.

Funding: This work was supported in part by a specific research fund fort the Innovation Platform for Academicians of Hainan Province under Grant YSPTZX202034 and Grant YSPTZX202127; in part by the Major Science and Technology Program of Hainan Province of China under Grant ZDKJ 2019005; in part by Scientific Research Projects of Higher Education Institutions in Hainan Province under Grant hnky2020-24, Grant Hnjg2021ZD-22, Grant hnky2020ZD-12; in part by the Hainan Provincial Natural Science Foundation of China under Grant 622RC671, Grant 120MS031, Grant 2019RC190, Grant 2019RC192; in part by the National Natural Science Foundation of China under Grant 61774024, Grant 61864002, Grant 11764012, Grant 62174046,Grant 62064004 and Grant 61964007; in part by the Key Research and Development Projects in Hainan Province under Grant ZDYF2020020, Grant ZDYF2020036, and Grant ZDYF2020217; in part by the Open Fund for Innovation and Entrepreneurship of college students under Grant 202111658021X, Grant 202111658022X, Grant 202111658023X, Grant 202111658013.

Data Availability Statement: Not applicable.

Acknowledgments: The authors thank Hao Chen, Yanbo Liang, and Xing Mu for helping with this article.

Conflicts of Interest: Not applicable.

References

1. Capasso, F. High-performance mid-infrared quantum cascade lasers. *Opt. Eng.* **2010**, *49*, 1102. [CrossRef]
2. Maulini, R.; Mohan, A.; Giovannini, M.; Faist, J.; Gini, E. External cavity quantum-cascade laser tunable from 8.2 to 10.4 μm using a gain element with a heterogeneous cascade. *Appl. Phys. Lett.* **2006**, *88*, 2834–2883. [CrossRef]

3. Wysocki, G.; Weidmann, D. Applications of external cavity quantum cascade lasers: Broadband mid-IR laser heterodyne radiometry. In Proceedings of the Conference on Lasers and Electro-Optics, Optical Society of America, Baltimore, MD, USA, 31 May–5 June 2009.
4. Mammez, D.; Vallon, R.; Parvitte, B.; Mammez, M.-H.; Carras, M.; Zéninari, V. Development of an external cavity quantum cascade laser spectrometer at 7.5 μm for gas detection. *Appl. Phys. A* **2014**, *116*, 951–958.
5. Xie, F.; Caneau, C.; Leblanc, H.; Ho, M.T.; Zah, C. Ultra-broad gain quantum cascade lasers tunable from 6.5 to 10.4 μm. *Opt. Lett.* **2015**, *40*, 4158–4161. [CrossRef]
6. Zhao, Z.; Wang, L.; Jia, Z.; Zhang, J.; Liu, F.; Zhuo, N.; Zhai, S. Low-threshold external-cavity quantum cascade laser around 7.2 μm. *Opt. Eng.* **2016**, *55*, 046116. [CrossRef]
7. Bayrakli, I. Optically and electrically pumped grating-coupled external cavity quantum cascade laser. *Opt. Quantum Electron.* **2022**, *54*, 1–7. [CrossRef]
8. Bayrakli, I. External cavity quantum cascade lasers without anti-reflection coating with intra-cavity and extra-cavity acoustic-optic frequency shifter for fast standoff detection. *Opt. Laser Technol.* **2022**, *148*, 107747. [CrossRef]
9. Luo, G.; Peng, C.; Le, H.; Pei, S.-S.; Lee, H.; Hwang, W.-Y.; Ishaug, B.; Zheng, J. Broadly Wavelength-Tunable External Cavity Mid-Infrared Quantum Cascade Lasers. *IEEE J. Quantum Electron.* **2002**, *38*, 486–494.
10. Wei, L.; Chuan-Xi, D. A broadband pulsed external-cavity quantum cascade laser operating near 6.9 μm. *Chin. Phys. Lett.* **2016**, *33*, 024207. [CrossRef]
11. Jia, X.; Wang, L.; Jia, Z.; Zhuo, N.; Zhang, J.; Zhai, S.; Liu, J.; Liu, S.; Liu, F.; Wang, Z. Fast Swept-Wavelength, Low Threshold-Current, Continuous-Wave External Cavity Quantum Cascade Laser. *Nanoscale Res. Lett.* **2018**, *13*, 341. [CrossRef]
12. Dougakiuchi, T.; Kawada, Y.; Takebe, G. Continuous multispectral imaging of surface phonon polaritons on silicon carbide with an external cavity quantum cascade laser. *Appl. Phys. Express* **2018**, *11*, 032001. [CrossRef]
13. Matsuoka, Y.; Peters, S.; Semtsiv, M.P.; Masselink, W.T. External-cavity quantum cascade laser using intra-cavity out-coupling. *Opt. Lett.* **2018**, *43*, 3726–3729. [CrossRef] [PubMed]
14. Zhou, W.; Slivken, S.; Razeghi, M. Phase-locked, high power, mid-infrared quantum cascade laser arrays. *Appl. Phys. Lett.* **2018**, *112*, 181106. [CrossRef]
15. Zhou, W.; Wu, D.; Lu, Q.-Y.; Slivken, S.; Razeghi, M. Single-mode, high-power, mid-infrared, quantum cascade laser phased arrays. *Sci. Rep.* **2018**, *8*, 14866. [CrossRef]
16. Gu, Z.-H.; Zhang, J.-C.; Wang, H.; Yang, P.-C.; Zhuo, N.; Zhai, S.-Q.; Liu, J.-Q.; Wang, L.-J.; Liu, S.-M.; Liu, F.-Q.; et al. Tunable characteristic of phase-locked quantum cascade laser arrays. *Chin. Phys. B* **2021**, *30*, 104201. [CrossRef]
17. Lyakh, A.; Azim, A.; Loparo, Z.E.; Thurmond, K.; Vasu, S.S. Design of External Cavity Quantum Cascade Lasers for Combustion and Explosion Diagnostics. In Proceedings of the 2019 IEEE Research and Applications of Photonics in Defense Conference (RAPID), Miramar Beach, FL, USA, 19–21 August 2019.
18. Niu, S.; Liu, J.; Zhang, J.; Zhuo, N.; Zhai, S.; Wang, X.; Wei, Z. Single-Mode Fabry-Pérot Quantum Cascade Lasers at λ∼10.5 μm. *J. Mater. Sci. Chem. Eng.* **2020**, *8*, 85–91.
19. Risby, T.; Solga, S. Current status of clinical breath analysis. *Appl. Phys. A* **2006**, *85*, 421–426.
20. Wysocki, G.; Lewicki, R.; Curl, R.; Tittel, F.; Diehl, L.; Capasso, F.; Troccoli, M.; Höfler, G.; Bour, D.; Corzine, S.; et al. Widely tunable mode-hop free external cavity quantum cascade lasers for high resolution spectroscopy and chemical sensing. *Appl. Phys. B* **2008**, *92*, 305–311. [CrossRef]
21. Kosterev, A.; Wysocki, G.; Bakhirkin, Y.; So, S.; Lewicki, R.; Fraser, M.; Tittel, F.; Curl, R. Application of quantum cascade lasers to trace gas analysis. *Appl. Phys.* **2008**, *90*, 165–176. [CrossRef]
22. Wysocki, G.; Curl, R.; Tittel, F.; Maulini, R.; Bulliard, J.; Faist, J. Widely tunable mode-hop free external cavity quantum cascade laser for high resolution spectroscopic applications. *Appl. Phys. B* **2005**, *81*, 769–777. [CrossRef]
23. Spagnolo, V.; Kosterev, A.A.; Dong, L.; Lewicki, R.; Tittel, F.K. NO trace gas sensor based on quartz-enhanced photoacoustic spectroscopy and external cavity quantum cascade laser. *Appl. Phys. B* **2010**, *100*, 125–130. [CrossRef]
24. Ramin, G.; Schmidt, F.M. Real-time breath gas analysis of CO and CO_2 using an EC-QCL. *Appl. Phys. B* **2017**, *123*, 144.
25. Mammez, D.; Stoeffler, C.; Cousin, J.; Vallon, R.; Mammez, M.; Joly, L.; Parvitte, B.; Zéninari, V. Photoacoustic gas sensing with a commercial external cavity-quantum cascade laser at 10.5 μm. *Infrared Phys. Technol.* **2013**, *61*, 14–19. [CrossRef]
26. Maity, A.; Pal, M.; Maithani, S.; Banik, G.D.; Pradhan, M. Wavelength modulation spectroscopy coupled with an external-cavity quantum cascade laser operating between 7.5 and 8 μm. *Laser Phys. Lett.* **2018**, *15*, 045701. [CrossRef]
27. Nadeem, F.; Khodabakhsh, A.; Mandon, J.; Cristescu, S.S.; Harren, F.J.M. Detection of N_2O Using An External-Cavity Quantum Cascade Laser. *OSA Contin.* **2019**, *2*, 2667–2682. [CrossRef]
28. Maity, A.; Pal, M.; Banik, G.D.; Maithani, S.; Pradhan, M. Cavity ring-down spectroscopy using an EC-QCL operating at 7.5 m for direct monitoring of methane isotopes in air. *Laser Phys. Lett.* **2017**, *14*, 115701. [CrossRef]
29. Cui, X.; Dong, F.; Zhang, Z.; Sun, P.; Xia, H.; Fertein, E.; Chen, W. Simultaneous detection of ambient methane, nitrous oxide, and water vapor using an external-cavity quantum cascade laser. *Atmos. Environ.* **2018**, *189*, 125–132. [CrossRef]
30. Wei, Q.; Li, B.; Wang, J.; Zhao, B.; Yang, P. Impact of Residual Water Vapor on the Simultaneous Measurements of Trace CH_4 and N_2O in Air with Cavity Ring-Down Spectroscopy. *Atmosphere* **2021**, *12*, 221. [CrossRef]
31. Zhou, S.; Xu, L.; Zhang, L.; He, T.; Liu, N.; Liu, Y.; Yu, B.; Li, J. External cavity quantum cascade laser-based QEPAS for chlorodifluoromethane spectroscopy and sensing. *Appl. Phys. B* **2019**, *125*, 125. [CrossRef]

32. Nikodem, M.; Krzempek, K.; Stachowiak, D.; Wysocki, G. Quantum cascade laser-based analyzer for hydrogen sulfide detection at sub-parts-per-million levels. *Opt. Eng.* **2017**, *57*, 011019. [CrossRef]
33. Pal, M.; Maithani, S.; Maity, A.; Pradhan, M. Simultaneous monitoring of ^{32}S, ^{33}S and ^{34}S isotopes of H_2S using cavity ring-down spectroscopy with a mid-infrared external-cavity quantum cascade laser. *J. Anal. At. Spectrom.* **2019**, *34*, 860–866. [CrossRef]
34. Yin, X.; Wu, H.; Dong, L.; Li, B.; Ma, W.; Zhang, L.; Yin, W.; Xiao, L.; Jia, S.; Tittel, F.K. ppb-level SO_2 photoacoustic sensors with a suppressed absorption-desorption effect by using a 7.41 μm external-cavity quantum cascade laser. *ACS Sens.* **2020**, *5*, 549–556. [CrossRef] [PubMed]
35. Shakfa, M.K.; Mhanna, M.; Jin, H.; Liu, D.; Djebbi, K.; Marangoni, M.; Farooq, A. A mid-infrared diagnostic for benzene using a tunable difference-frequency-generation laser. *Proc. Combust. Inst.* **2021**, *38*, 1787–1796. [CrossRef]
36. Sun, J.; Ding, J.; Liu, N.; Yang, G.; Li, J. Detection of multiple chemicals based on external cavity quantum cascade laser spectroscopy. *Spectrochim. Acta Part A Mol. Biomol. Spectrosc.* **2018**, *191*, 532–538. [CrossRef]
37. Sun, J.; Liu, N.; Deng, H.; Ding, J.; Sun, J.; Zhang, L.; Li, J. Laser absorption spectroscopy based on a broadband external cavity quantum cascade laser. *Int. Conf. Opt. Photonics Eng. SPIE* **2017**, *10250*, 378–382.
38. Liu, N.; Zhou, S.; Zhang, L.; Yu, B.; Fischer, H.; Ren, W.; Li, J. Standoff detection of VOCs using external cavity quantum cascade laser spectroscopy. *Laser Phys. Lett.* **2018**, *15*, 085701. [CrossRef]
39. Heinrich, R.; Popescu, A.; Strzoda, R.; Hangauer, A.; Höfling, S. High resolution quantitative multi-species hydrocarbon gas sensing with a cw external cavity quantum cascade laser based spectrometer in the 6–11 μm range. *J. Appl. Phys.* **2019**, *125*, 134501. [CrossRef]
40. Schwaighofer, A.; Montemurro, M.; Freitag, S.; Kristament, C.; Culzoni, M.J.; Lendl, B. Beyond FT-IR Spectroscopy EC-QCL based mid-IR Transmission Spectroscopy of Proteins in the Amide I and Amide II Region. *Anal. Chem.* **2018**, *90*, 7072–7079. [CrossRef]
41. Montemurro, M.; Schwaighofer, A.; Schmidt, A.; Culzoni, M.J.; Mayer, H.K.; Lendl, B. High-throughput quantitation of bovine milk proteins and discrimination of commercial milk types by external cavity-quantum cascade laser spectroscopy and chemometrics. *Analyst* **2019**, *144*, 5571–5579. [CrossRef]
42. Dabrowska, A.; Schwaighofer, A.; Lindner, S.; Lendl, B. Mid-IR refractive index sensor for detecting proteins employing an external cavity quantum cascade laser-based Mach-Zehnder interferometer. *Opt. Express* **2020**, *28*, 36632–36642. [CrossRef]
43. Schwaighofer, A.; Akhgar, C.K.; Lendl, B. Broadband laser-based mid-IR spectroscopy for analysis of proteins and monitoring of enzyme activity. *Spectrochim. Acta Part A Mol. Biomol. Spectrosc.* **2021**, *253*, 119563. [CrossRef] [PubMed]
44. Phillips, M.C.; Harilal, S.S.; Yeak, J.; Jason Jones, R.; Wharton, S.; Bernacki, B.E. Standoff detection of chemical plumes from high explosive open detonations using a swept-wavelength external cavity quantum cascade laser. *J. Appl. Phys.* **2020**, *128*, 163103. [CrossRef]
45. Parrot, A.; Vanier, F.; Blouin, A. Standoff mid-infrared reflectance spectroscopy using quantum cascade laser for mineral identification. *SPIE Future Sens. Technol.* **2020**, *11525*, 217–225.
46. Vanier, F.; Parrot, A.; Padioleau, C.; Blouin, A. Mid-Infrared Reflectance Spectroscopy Based on External Cavity Quantum Cascade Lasers for Mineral Characterization. *Appl. Spectrosc.* **2022**, *76*, 361–368. [CrossRef] [PubMed]
47. Yao, Y.; Hoffman, A.J.; Gmachl, C.F. Mid-infrared quantum cascade lasers. *Nat. Photonics* **2012**, *6*, 432–439. [CrossRef]
48. Miczuga, M.; Kopczyński, K. Application of cascade lasers to detection of trace gaseous atmospheric pollutants. *Laser Technol. Prog. Appl. Lasers. SPIE* **2016**, *10159*, 295–300.
49. Tittel, F.K.; Dong, L.; Lewicki, R.; Liu, K.; Spagnolo, V. Mid-infrared quantum cascade laser based trace gas technologies: Recent progress and applications in health and environmental monitoring. In Proceedings of the 2011 International Quantum Electronics Conference (IQEC) and Conference on Lasers and Electro-Optics (CLEO) Pacific Rim Incorporating the Australasian Conference on Optics, Lasers and Spectroscopy and the Australian Conference on Optical Fibre Technology, Sydney, Australia, 28 August–1 September 2011.
50. Almond, N.W.; Qi, X.; Degl'Innocenti, R.; Kindness, S.J.; Michailow, W.; Wei, B.; Braeuninger-Weimer, P.; Hofmann, S.; Dean, P.; Indjin, D.; et al. External cavity terahertz quantum cascade laser with a metamaterial/graphene optoelectronic mirror. *Appl. Phys. Lett.* **2020**, *117*, 041105. [CrossRef]

Review

Development of Solution-Processed Perovskite Semiconductors Lasers

Nan Zhang *, Quanxin Na, Qijie Xie and Siqi Jia

Department of Mathematics and Theories, Peng Cheng Laboratory, No. 2, Xingke 1st Street, Nanshan, Shenzhen 518055, China
* Correspondence: zhangn06@pcl.ac.cn

Abstract: Lead halide perovskite is a new photovoltaic material with excellent material characteristics, such as high optical absorption coefficient, long carrier transmission length, long carrier lifetime and low defect state density. At present, the steady-state photoelectric conversion efficiency of all-perovskite laminated cells is as high as 28.0%, which has surpassed the highest efficiency of monocrystalline silicon cells (26.7%). In addition to its excellent photovoltaic properties, perovskite is also a type of direct bandgap semiconductor with low cost, solubilization, high fluorescence quantum efficiency and tunable radiation wavelength, which brings hope for the realization of electrically pumped low-cost semiconductor lasers. In recent years, a variety of perovskite lasers have emerged, ranging from the type of resonator, the wavelength and pulse width of the pump source, and the preparation process. However, the current research on perovskite lasers is only about the type of resonator, the type of perovskite and the pump wavelength, but the performance of the laser itself and the practical application of perovskite lasers are still in the initial stages. In this review, we summarize the recent developments and progress of solution-processed perovskite semiconductors lasers. We discuss the merit of solution-processed perovskite semiconductors as lasing gain materials and summarized the characteristics of a variety of perovskite lasers. In addition, in view of the issues of poor stability and high current density required to achieve electrically pumped lasers in perovskite lasers, the development trend of perovskite lasers in the future is prospected.

Keywords: perovskite lasers; perovskite semiconductors; solution process

Citation: Zhang, N.; Na, Q.; Xie, Q.; Jia, S. Development of Solution-Processed Perovskite Semiconductors Lasers. *Crystals* **2022**, *12*, 1274. https://doi.org/10.3390/cryst12091274

Academic Editor: Anna Paola Caricato

Received: 28 July 2022
Accepted: 2 September 2022
Published: 8 September 2022

Publisher's Note: MDPI stays neutral with regard to jurisdictional claims in published maps and institutional affiliations.

Copyright: © 2022 by the authors. Licensee MDPI, Basel, Switzerland. This article is an open access article distributed under the terms and conditions of the Creative Commons Attribution (CC BY) license (https:// creativecommons.org/licenses/by/ 4.0/).

1. Introduction

Since semiconductor lasers came out in the 1960s [1,2], after nearly 60 years of development, their materials, processes and properties have made rapid changes, and their application scope and market scale have been expanding. Semiconductor lasers have gradually become an irreplaceable part of people's daily life and work and have also ranked among the world's high-end scientific and technological research. The three elements of laser generation are gain medium, pump source and optical resonant cavity. Among them, the gain medium provides the energy level structure for forming the laser, which is the internal cause of the laser generation; the pump source provides the excitation energy required to form the laser emission, which is the external cause of the laser generation; the optical resonant cavity provides feedback amplification for the laser to make stimulated emission. The intensity, directionality and monochromaticity of the lasers have been further improved.

Throughout the development of semiconductor lasers, it can be found that the development of optical gain materials has played a very important role. Lots of different materials have been developed and tested as the laser gain material. Laser gain medium materials can be divided into the following categories: solid gain media, gas gain media, liquid gain media, inorganic semiconductors, etc. [3]. Compared with other types of lasers, semiconductor lasers have attracted much attention because of their advantages of small size, fast response, low power consumption and high-efficiency [3,4]. Historically, this has been enabled using semiconductor lasers made from crystalline inorganic semiconductors such as II–VI or III–V compounds-materials heavily used in modern electronics and optoelectronics. However, crystalline inorganic semiconductors materials also have some inherent disadvantages. For example, the spectral coverage of the laser is limited due to the limited variety of inorganic semiconductors and the difficulty of the lattice doping [5]. In addition, the emission of inorganic semiconductor materials is derived from band edge radiation, the emission peak is usually narrow, and the wavelength adjustment ability is poor. In addition, most crystalline inorganic semiconductors require a complex and high-cost high-temperature fabrication process, which restricts the further development of crystalline inorganic semiconductors lasers [6].

Low-temperature solution-processed semiconductors are an emerging class of optoelectronic materials that can be processed in ink form through the wet chemistry [7]. They are technologically attractive due to their unique merits, such as facile solution processibility, lightweight, low cost and high mechanical flexibility. In addition to the above benefits offered by solution-processed semiconductors, they have the merit that their optoelectronic properties can be tailored. There are several ways to control the optical band gaps and energy levels of semiconductor materials. Solution-processed semiconductors cover organic materials, metal-halide perovskites (MHPs), and inorganic nanocrystals and quantum dots; each class of materials takes on (to lesser or greater extents) an optoelectronic tunability [7]. Solution-processed semiconductors materials as optical gain media have many incomparable advantages over traditional crystalline inorganic semiconductor materials: (i) large absorption and radiation cross-sections are conducive to high optical gain [8], (ii) abundant excited state process is conducive to the construction of a four-level system to achieve population inversion, thereby reducing the laser threshold [9] and it is also convenient to realize dynamic control of laser wavelength [10], (iii) there are many kinds of solution-processed semiconductors, which can achieve the light emission of the full spectrum from ultraviolet to near-infrared [11], (iv) solution method is easy to process and suitable for the preparation of large-area devices [5,12]. Therefore, solution-processed semiconductors gain medium is very promising to become an ideal choice for the next generation of semiconductor lasers.

In the decades following the creation of the world's first lasers in the 1960s, solution-processed semiconductors materials including organic semiconductors (polymers), perovskites, inorganic nanocrystals and quantum dots have accounted for a large proportion of the development in this field. Progress in the controlling and understanding of these materials sciences have led to the most advanced performance of selected applications, and so far as to commercial deployments. Dominated by various weak interactions between molecules, organic molecules can self-assemble or be processed into a variety of regular micro-nano structures under mild conditions [13]. These regular micro-nano structures can be used as high-quality optical microcavity to provide structural support for the realization of low threshold laser [5,8,14]. Recent advances indicate a remarkable potential of colloidal quantum dots as an optical gain medium capable of operating under both optical and electrical pumping [15]. Latest studies include the demonstration of optically pumped continuous-wave colloidal quantum dot laser [16], the realization of optical gain by electrically pumped quantum dots [17], and the demonstration of dual-functional devices working as an optically pumped laser and an electrically excited light-emitting diodes (LEDs) [18,19].

Hybrid organic-inorganic halide perovskites have recently emerged as a potential new class of optoelectronic materials. The high brightness and tunable bandgap of perovskites have made it an attractive candidate for a new series of optical gain medium for low-cost semiconductor lasers [20]. So far, a variety of high-performance micro-/nanolasers have been demonstrated including 2D Ruddlesden-Popper perovskites [21], perovskite single crystals [22], and thin films [23]. Perovskite lasers have been exhibited in some architectures: resonators based on a Fabry-Perot (FP) cavity formed with parallel edge facets [22]; ring resonators in microspheres or nanoplatelets [24–27]; and random lasing in scattering films [20,28]. All the above configurations have achieved the multimode lasing over the full amplified spontaneous emission (ASE) bandwidth. However, the high-cost and high-energy-consuming synthesis approaches such as chemical vapor deposition and molecular beam epitaxy may undermine their practical applications. By comparison, the facile solution-processable lasers can not only reduce costs but also extend laser-related applications to flexible generations. In brief, despite the remarkable progress made in solution-processed perovskite semiconductors lasers, challenges and opportunities remain both basic science and a device engineering perspective. This review first introduces the structure and characteristics of perovskite materials and then states the development of perovskite laser based on different gain mediums. Finally, it summarizes the development of solution-processed perovskite semiconductors lasers, expounds its existing problems, and gives its own views on the trends of perovskite lasers in the future.

2. Perovskite Semiconductor

2.1. Crystal Structure

Any material with the same crystal structure as $CaTiO_3$ is collectively referred to as perovskite structure, which widely exists in nature. The general chemical formula of perovskite material is ABX_3, where a and B are two different cations, and X is the anion combined with them. Figure 1 shows a typical perovskite structure. The B cation is located in the center of the octahedron composed of X ions and embedded in the tetragonal body with the A-site ion as the apex. In common perovskite materials, A ion can be either organic or inorganic, such as CH_3NH^{3+}, $CH(NH_2)^{3+}$, Cs^+ and Rb^+. B is a transition metal ion, such as Fe^{2+}, Mn^{2+}, Sn^{2+}, and Pb^{2+}. X is an oxygen or halogen ion.

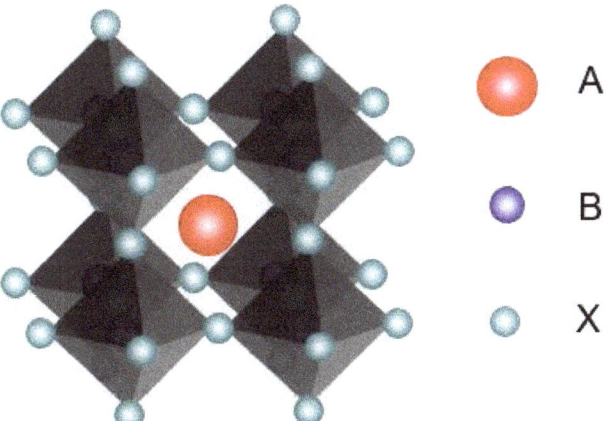

Figure 1. Perovskite crystal structure.

2.2. Luminescence Properties

Perovskite is a direct bandgap semiconductor, which can control spectral tuning by substitution or mixing of halide components and cations. The luminous wavelength ranges from 390 nm to 790 nm and can be extended to 820 nm by mixing methyl and formamidine. $MASnX_3$ perovskite semiconductor has tunable emission wavelengths in excess of 900 nm but is more sensitive to air and illumination. $CsPbX_3$ quantum dots have also been extensively studied in recent years [29]. These perovskite nanoparticles provide a spectral range spanning 410–700 nm through halide substitution and quantum tuning, as shown in the Figure 2a,b. Their narrow photoluminescence (PL) spectrum and continuous spectral tunability enable a solid color distribution on CIE chromaticity maps that exceeds the national Television Systems Council (NTSC) standard. Some studies have shown that

CsPbBr$_3$ perovskite quantum dots exhibit less blinking than other quantum dot systems, and excitons are insensitive to the size of quantum dots. Perovskite semiconductor has excellent optical absorption, with the absorption coefficient exceeding 10^4 cm^{-1} near the band edge, which can efficiently convert light into electric current and correspondingly be used as a gain material in lasers. In laser applications, the low Stokes shift reduces heat loss during the down-conversion process. Band gaps with minimal charge-trapping defects improve the efficiency of interband radiative recombination, which is critical for light-emitting devices.

Tunable emission wavelength is a beneficial characteristic of perovskite materials. The substitution of perovskite cation or halogen ion can change the bandwidth of perovskite material, and then realize the tuning of emission wavelength from visible to infrared. Since the Pb-X bond of perovskite crystals is related to the energy band structure, the band gap decreases sequentially from chlorine to bromine to iodine replacement, so the tunable emission wavelength of perovskite materials can be achieved by the replacement of halide ions [30]. In addition, the continuous tuning of the emission wavelength can also be achieved by mixing halogen elements to regulate perovskite semiconductor (see Figure 2c–e) [31,32].

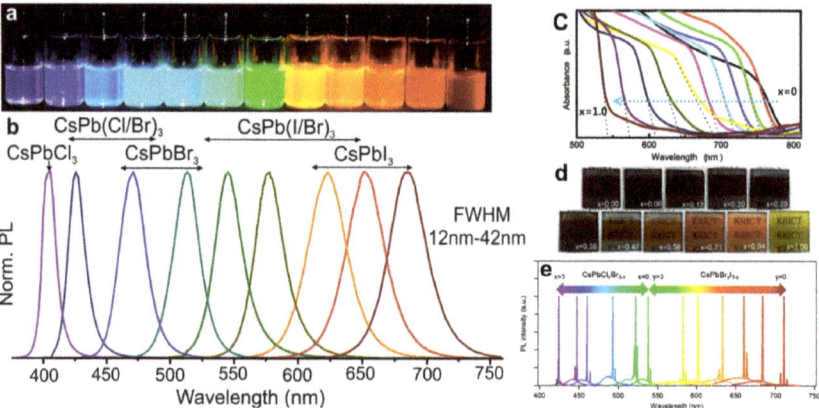

Figure 2. Colloidal perovskite CsPbX$_3$ nanocrystals (X = Cl, Br, I) exhibit size–and composition-tunable bandgap energies covering the entire visible spectral region with narrow and bright emission: (**a**) colloidal solutions in toluene under UV lamp (λ = 365 nm) [29]. Reprinted with permission from [29]. Copyright © 2015 American Chemical Society. (**b**) representative PL spectra (λ$_{exc}$ = 400 nm for all but 350 nm for CsPbCl$_3$ samples) [29]. Reprinted with permission from [29]. Copyright © 2015 American Chemical Society. (**c**) Photographs and UV–vis absorption spectra of MAPb(I$_{1-x}$Br$_x$)$_3$ [31]. Reprinted with permission from [31]. Copyright © 2013 American Chemical Society. (**d**) Photographs of 3D TiO$_2$/MAPb(I$_{1-x}$Br$_x$)$_3$ bilayer nanocomposites on FTO glass substrates [31]. Reprinted with permission from [31]. Copyright © 2013 American Chemical Society. (**e**) Broad wavelength-tunable lasing from single–crystal nanowires (NWs) of CsPbX$_3$ (X = Cl, Br, and I) [32]. Reprinted with permission from [32]. Copyright © 2016 American Chemical Society.

2.3. Gain Properties

Laser is a process in which a gain material that can provide feedback in a cavity is excited to form a population inversion to generate optical radiation. The process of semiconductor stimulated radiation to achieve optical gain: a photon incident on the semiconductor material undergoes an electronic transition and generates a stimulated radiation photon identical to itself at the same time. The photoexcited states near the energy band of perovskite semiconductors affect charge transport and light emission. There are free carriers and excitons at the band edge. The exciton binding energy reflects the Coulomb interaction of photoexcited electron-hole pairs. The strength of the action determines the balance of the two excited particles. Unlike the exciton binding energies of conventional organic semiconductors (hundreds of millielectron volts) and inorganic semiconductors (several millielectron volts), the exciton binding energies of perovskite materials lie in between. From different experimental methods and results, there is a wide distribution range (from several millielectron volts to hundreds of millielectron volts) of its exciton binding energies by changing the stacking, structure and cation of perovskite materials [33]. The mechanism that causes the stimulated emission of perovskite semiconductors remains to be resolved. When the exciton binding energy of the semiconductor is smaller than the thermal fluctuation energy, it will easily dissociate into free carriers; otherwise, the free carriers will form excitons. For the light emission model, the exciton binding energy of perovskite is generally larger, which can obtain high quantum yields at relatively low carrier $A = \pi r^2$ relatively large exciton binding energies are important for stable lasing at room temperature.

Optical gain is used to describe the process in which the intensity of the light incident on the gain medium increases exponentially with distance. Optical loss refers to photon scattering, non-radiative recombination and edge scattering when light is transmitted in a semiconductor medium. To achieve laser output, the gain must be greater than loss, that is, there is a positive net gain. In order to further describe the laser gain characteristics of perovskite, the optical net gain model is introduced. Due to the change in the pump spot length, the emission intensity of the sample also changes, and the net gain model formula is established according to the gain loss:

$$I = A \frac{\exp(gL_g - 1)}{g} \tag{1}$$

where I is the output light intensity, A is a constant, g is the gain coefficient, L_g is the pump fringe length and the gain coefficient can be obtained by data fitting. Sutherland et al. obtained the net gain coefficient of perovskite MAPbI$_3$ films on silicon spheres as 103~147 cm^{-1} and the gain bandwidth as 36~64 meV by the method of variable stripe length [26]. Through the continuous efforts of the above teams, the highest net gain coefficient of perovskite so far measured on the MAPbI$_3$ film obtained by atomic layer deposition is 2770~4030 cm^{-1} [34], which is much higher than that of colloidal quantum dots and conjugated polymer films, and close to that of traditional GaAs semiconductors. The low defect density of perovskite can reduce the non-radiative recombination rate [23], thus reducing the excitation threshold and has the advantages of large absorption coefficient and high fluorescence quantum yield, which makes perovskite semiconductor as an optical gain material bring beneficial development potential for high-performance semiconductor lasers. Table 1 summarized the physical properties of perovskite semiconductors in the review article [35].

Table 1. Physical properties of perovskite semiconductors.

Materials	Hole/Electron Mobility (cm^2/Vs)	Intrinsic Carrier Concentration (cm^{-3})	Carrier Lifetime (ns)	Intrinsic Resistivity (Ωm)	Thermal Conductivity (W/mK)	Dielectric Constant
MAPbBr$_3$	20–60/20–60	5×10^9–5×10^{10} [36]	41 [36]	~10^8	0.1–1.4	~5.7@530 nm
MAPbI$_3$	136/197 [37]	10^{16}–10^{18} [38]	22 [36]	~10^{10} [36] ~10^5 [39] ~10^9 [40]	1–3 [38]	~4.7–9@vislble [41]
CsPbX$_3$	100–240/80–290 [42]	10^{15}–10^{17} [42]	1.3 (X = Br) [43]	2.1×10^{10} X = Br [43]	0.3 X = I 0.5 X = Cl	4.1–4.5 [44] X = Cl 3.2–5 [44] X = Br 5–12.7 [44] X = I

2.4. Carrier Dynamics

Emerging perovskite semiconductors have similar properties to traditional inorganic direct bandgap semiconductors, so many of the theories of traditional semiconductors are also applicable to perovskite semiconductors. The photophysical processes that determine the photoelectric properties of semiconductors mainly include carrier excitation, relaxation, recombination and transport. These processes usually occur in a very short time, ranging from tens of femtoseconds to a few nanoseconds. After the carriers relax to the bottom of the conduction band or the top of the valence band, they are in a non-equilibrium state, which needs to be restored by radiative and non-radiative recombination processes. These two recombination processes release energy in the form of photons and thermal energy, respectively. The whole mechanism of carrier recombination can be divided into three types: single-molecule recombination, bimolecular recombination and Auger recombination, whose dynamics follow the following differential equation [45]:

$$\frac{dn}{dt} = -k_1 n - k_2 n^2 - k_3 n^3 \qquad (2)$$

where n is the carrier density and k_1, k_2 and k_3 are the unimolecular, bimolecular and Auger recombination rate constants, respectively.

Monomolecular recombination refers to a recombination process involving only one particle. In semiconductors, excitons consisting of a bound electron-hole pair all constitute a particle. Therefore, both cases of exciton recombination (radiative recombination) and trap state recombination (a single electron or hole trapped by a trap state) belong to single-molecule recombination. Conventional semiconductors prepared by solution methods have the disadvantages of high electronic disorder and a large number of bulk defects and surface traps [23,46,47], while the prepared perovskites have only limited density of trap states, making it easier to realize single-molecule recombination based on the radiative recombination process. The density of trap states strongly depends on the preparation conditions and surface treatment of the sample. At low pump intensities, the variation of trap state density can lead to different recombination lifetimes of single molecule recombination assisted by trap state. For perovskite lasers, it is of great significance to improve the crystallinity and purity of the samples to enhance the radiative recombination process of single molecules. Bimolecular recombination is the recombination of two particles, which is a recombination process involving free electrons and holes. This process is intrinsic photon-radiation recombination, and its dependence on material processing is much lower than that of single molecule recombination assisted by trapping states. For the perovskite laser, enhancing the bimolecular recombination process can also improve the luminescence efficiency.

Auger recombination is a many-body recombination process in which the recombination of an electron with a hole is accompanied by the transfer of energy and momentum to a third particle, either an electron or a hole. Therefore, Auger recombination is strongly dependent on carrier density. As shown in the third term of Equation (2), the Auger recombination (non-radiative recombination) effect can only be detected if the pump intensity is sufficiently large. For applications with high charge densities such as lasers, Auger recombination processes can cause large energy losses. As shown in Equation (2), the mechanism of carrier recombination depends on carrier density and time. Combined with ultrafast spectroscopy technology for global fitting, the values of k_1, k_2 and k_3 of any material can be obtained, but a certain recombination mechanism can dominate under different pump intensities. At low pump intensities, the photoexcited minority carrier density is much smaller than the total majority carrier density, and multiparticle recombination is suppressed, so the first term in Equation (2) dominates. Under these conditions, the dynamics of carrier recombination are almost unimolecular and exhibit near uniexponential decay. At high pump intensities, the density of photogenerated carriers is large and the free electron-hole bimolecular recombination dominates. The dynamics of carrier recombination and pump intensity decay in a power-law pattern with a long tail. With time delay, the bimolecular decay dynamics will continue until the carrier density drops to the single-molecule recombination density, at which point the single-molecule type recombination will reappear and appear as a long exponential tail on the decay curve. At higher pump intensities, Auger recombination involving multiple particles will dominate, and there are few relevant studies. For the perovskite laser, the stimulated radiation process requires a high pump intensity, and the Auger recombination effect is also significant, resulting in a large energy loss [48].

2.5. Stability

The stability of materials is an important factor for practical application in devices. Although metal halide perovskites have excellent lasing properties, they are less stable [49]. The metal halide perovskite can be degraded in water, oxygen, light and heat. In the case of MAPbI$_3$, water reacts with MAPbI$_3$ and decomposes to produce MAI and PbI$_2$, in which the MAI produces volatile methylamine and hydrogen iodide. Oxygen penetrates the perovskite through iodide vacancies, trapping electrons and forming a highly reactive superoxide anion. Superoxide anion decomposes MAPbI$_3$ into PbI$_2$, I and H$_2$O. The simultaneous presence of oxygen and light greatly accelerates the degradation of MAPbI$_3$. The thermal decomposition of MAPbI$_3$ is carried out by chemical decomposition followed by the sublimation of MAI and HI. In order to improve the stability of metal halide perovskite lasers, researchers have developed a variety of methods. Exposure of perovskites to oxygen and moisture can be avoided by device encapsulation [50,51], preventing irreversible loss of volatile species from light and heat. Perovskite lasers show better stability under the protection of polymers, boron nitride films, and DBR cavities. The stability of the perovskite laser can be improved by reducing the thermal degradation of perovskite by increasing heat dissipation, wherein the sapphire substrate with high thermal conductivity can be used to assist the heat dissipation [52,53]. A perovskite laser encapsulated with boron nitride film with high thermal conductivity can accelerate heat dissipation. This method effectively combined the above two schemes [54]. In addition, the synthesis of perovskite single crystals with low trap density can inhibit the degradation of perovskite structure induced by oxygen and light, which is also one of the effective methods to achieve laser stability [55].

3. Perovskite Semiconductor Lasers

With the continuous improvement of the performance of perovskite semiconductor materials, research reports on perovskite lasers finally appeared in 2014 [23]. Since then, studies of hybrid and all-inorganic perovskite semiconductors have shown that optically pumped lasing emission can be achieved throughout the visible spectrum at room temperature [56,57]. However, there are still many challenges in the realization of electrically pumped perovskite semiconductor lasers until now. According to the different crystal types, perovskite lasers are generally divided into the polycrystalline thin-film types and low-dimensional single crystal type.

3.1. Polycrystalline Thin-Film Perovskite Lasers

The organic-inorganic hybrid perovskite material has the advantages of high intensity, high stability and easy control because of its large exciton binding energy, long carrier diffusion length and considerable quantum yield [58]. In 2014, Guichuan Xing and co-workers reported the PL properties of polycrystalline perovskite thin films under different pump light intensities [23]. With the increase of the pump light intensity, the full width half maximum (FWHM) of the luminescence peak gradually narrowed, and when it exceeded the threshold intensity 10~14 µJ/cm^2, the luminescence intensity (IPL) increased rapidly, the luminescence peak became sharp, and a lasing phenomenon occurred [23]. Furthermore, when the pump light irradiated the perovskite thin film, the transition from spontaneous emission to amplified spontaneous emission was realized with the increase of pump light intensity. This indicated that organic-inorganic hybrid perovskite semiconductor materials can also be used as the gain medium to achieve lasering. In addition, Stranks et al. found that the cholesteric liquid crystal (CLC) could be used as a mirror to construct a multilayer film structure as shown in Figure 3a [59]. The lasing threshold was as low as about 7 µJ/cm^2 (see Figure 3b), and the FWHM of the luminescence peak also reached 1.24 nm [59]. The inset of Figure 3b shows this flexible device, which lays the foundation for the development of new semiconductor lasers.

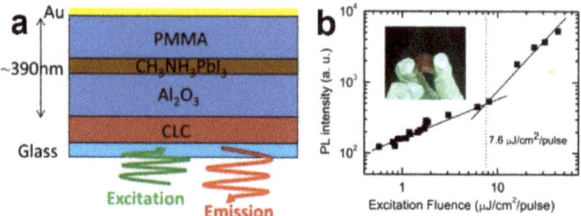

Figure 3. (**a**) Schematic diagram of laser with CLC reflector [59]. Reprinted with permission from [59]. Copyright © 2015 American Chemical Society. (**b**) Extracted emission intensity from a device stack fabricated on a flexible 80% CLC reflector following photoexcitation at a range of fluences (532 nm, 5 ns pulses, 100-Hz repetition rate) [59]. The ASE transition fluence is determined to be 7.6 µJ/cm^2/pulse. Inset: photograph of the flexible device. Reprinted with permission from [59]. Copyright © 2015 American Chemical Society.

In addition to the multi-mode lasers based on mirror microcavities reported above, many studies have also been carried out on single-mode perovskite lasers based on polycrystalline thin films in recent years. Guy L. Whitworth et al. used solution processed $CH_3NH_3PbI_3$ perovskite as gain semiconductor and UV nanoimprinted polymer as the gratings to fabricate distributed feedback (DFB) lasers, their schematic diagram of the device shown in Figure 4a [20]. Figure 4b shows that these perovskite lasers based on the solution process achieved a laser threshold of 4 µJ/cm^2 and 0.4 nm of FWHM. Cha's group reported an optically pumped single-mode laser with a two-dimensional square lattice photonic crystal (PhC) backbone structure that obtained the laser thresholds of ~200 µJ/cm^2 in pulse energy density at room temperature [60]. In 2019, optically pumped lasing was achieved from the perovskites $(PEA)_2Cs_{n-1}Pb_nBr_{3n+1}$ microcrystal film by the spin-coating technique [21]. This study reported optical pumping distributed Bragg reflectors (DBR) lasers based on a sandwich structure consisting of perovskite/PMMA/perovskite, which was shown in Figure 4c. As increasing the pump fluence above 500 µJ/cm^2, lasering occurred at the peak of 532 nm with a narrow FWHM of ~0.8 nm in Figure 4d. When pumped by a nanosecond pulsed laser (355 nm, pulse width 8 ns, 1 kHz), the lasing phenomenon can only occur within the separation pattern in their experiment. This study suggested that the lasers pumped by nanosecond pulses were the key basis for realizing continuous-wave pumped optical and electrical lasers [21].

3.2. Single Crystals Perovskite Thin Film Lasers

Compared with polycrystalline thin-film perovskite materials, low-dimensional single crystal perovskite nanomaterials can take advantage of the regular shape and smooth interface formed by themselves to form a good optical resonator, thus achieving efficient management of incident light through excitation resonance effect [61]. Moreover, the laser based on low dimensional single crystal perovskite nanostructure also has the advantages of high-quality factor, small volume and low threshold value [62]. In 2020, Y. Zhong et.al acquired large-scale $CsPbBr_3$ single-crystals films (SCFs) on sapphire substrates and achieved ASE from the $CsPbBr_3$ SCFs with a low threshold of 8 µJ/cm^2 at room temperature [63], their highest values of optical gain up to 1255 ± 160 cm^{-1} in Figure 5a. However, these large-scale $CsPbBr_3$ SCFs were fabricated by the chemical vapor-phase epitaxy deposition method rather than the solution process. Tian's group prepared a high-quality $CH_3NH_3PbCl_3$ single-crystalline film as gain material in solution confined between a pair of DBR, which naturally formed an optical microcavity [64]. Figure 5b showed a threshold energy density of about 211 µJ/cm^2 in these deep-blue DBR perovskite lasers, at which the FWHM decreased from around 20 nm to 0.38 nm.

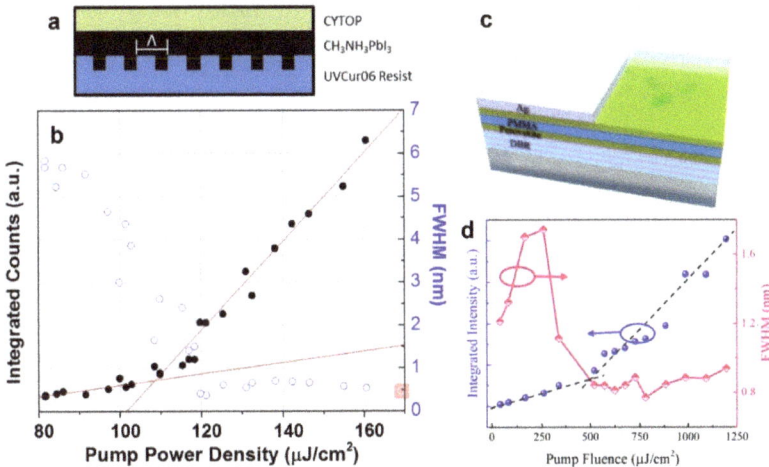

Figure 4. (a) Schematic of perovskite DFB laser with encapsulating CYTOP layer [20]. (b) Device power characteristic for low and high energy data points [20]. (c) Schematic diagram of cavity structure [21]. Reprinted with permission from [21]. Copyright © 2019 AIP Publishing. (d) The plots of integrated intensity (blue ball) and FWHM (pink diamond) of the cavity mode and the resulting lasing peak as a function of the pump fluence [21]. Reprinted with permission from [21]. Copyright © 2019 AIP Publishing.

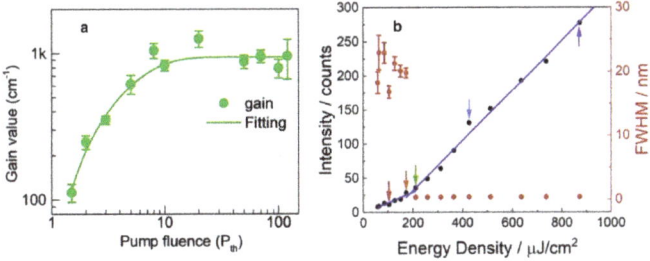

Figure 5. (a) Power-dependent optical gain of $CsPbBr_3$ SCFs [63]. Reprinted with permission from [63]. Copyright © 2020 American Chemical Society. (b) The intensities and FWHMs of the 422.4 nm peak as a function of the pump fluences [64]. Reprinted with permission from [64]. Copyright © 2020 Elsevier.

3.3. Single Crystals Perovskite Nanowires Lasers

A variety of preparation methods for perovskite nanowires have been proposed, and the most commonly used method is the solution process. One-dimensional semiconductor nanowires are nanoscale in the radial direction, which can be used as gain material and can naturally form a Fabry-Perot cavity. The light wave can be reflected back and forth between the two ends of the nanowires to continuously gain and form a stable standing wave with the same frequency and phase, thus realizing the laser emission [62]. Zhu's group showed room-temperature and wavelength-tunable lasing from a single-crystal perovskite nanowires [22]. Polycrystalline films of lead acetate were grown on a glass substrate and immersed in a high concentration of MAX (X = Cl, Br, I) isopropanol solution. The lead acetate layer reacted with the MAX solution to form single-crystal $MAPbX_3$ nanowires with approximately rectangular cross sections. With the increase of pumping energy, the luminescence behavior of nanowires gradually changes from spontaneous radiation to stimulated amplification radiation, and the line width of the luminescence peak narrowed. In this study, the mixed $MAPbBr_xI_{3-x}$ and $MAPbCl_xBr_{3-x}$ nanowires were obtained by adjusting the ratio of halogen materials in the precursor solution. The emission wavelength of the mixed perovskite nanowires can cover the near-infrared to visible wavelengths and the minimum laser threshold was 220 nJ/cm^2, and the laser quantum yield can reach 100%. To improve the stability of perovskite nanowires, Y. P. Fu's group used FA instead of MA to successfully obtain FAPbX$_3$ NWs, this lasing from single-crystal lead perovskite NWs was shown in Figure 6a [65]. Under the excitation of a femtosecond laser with a wavelength of 402 nm, the lasing threshold of FAPbX$_3$ NWs is 6.2 µJ/cm^2, the emission peak is 824 nm, the quality factor is 1554, and the FWHM is 0.53 nm (see Figure 6b).

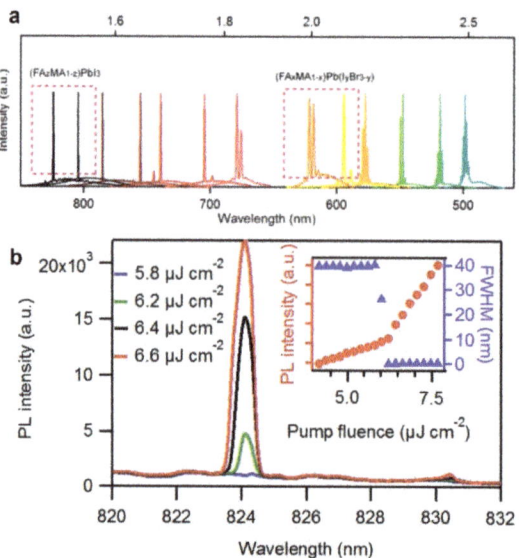

Figure 6. (a) Broad wavelength-tunable lasing from single–crystal lead perovskite NWs [65]. Reprinted with permission from [65]. Copyright © 2016 American Chemical Society. (b) NW emission spectra around the lasing threshold. Inset: Integrated PL intensity and FWHM of emission peak as a function [65]. Reprinted with permission from [65]. Copyright © 2016 American Chemical Society.

H. C. Yu et al. prepared MAPbI$_3$ NWs on the surface of Ag film and separated them with MgF$_2$ to form a surface plasmon laser [66]. The device had a laser threshold of 13.5 µJ/cm^2 and an FWHM of 5 nm under femtosecond laser irradiation with a wavelength of 400 nm, and it can maintain good performance at a high temperature of 43.6 °C. In 2018, Jiang's group used a gas-liquid transfer recrystallization method for synthesizing inorganic perovskite (CsPbX$_3$) NWs at a room temperature [67]. A femtosecond laser (1 kHz, 35 fs, 400 nm) was applied to measure the lasing behavior of NWs. This study indicated that the CsPbX$_3$ NWs perovskite lasing with a single mode, a low threshold of 12.33 µJ/cm^2 and an ultra-narrow linewidth of 0.09 nm, which is less reported in the inorganic perovskite system. Moreover, the CsPbBr$_3$ perovskite NWs are also used to achieve continuous-wave (CW) operation by polariton lasing at cryogenic temperature (77 K) with an excitation threshold of 6 kW/cm^2 [68]. Figure 7a displays fluorescence spectra from a CsPbBr$_3$ NW at CW excitation power densities. The intensity of this series of small peaks continues to grow until the threshold excitation power is about 6 kW/cm^2. In these 2–3 modes, maintaining their modal spacing becomes dominant and increases much faster than the other modes in the spectrum. This is even more evident when we curve the plotting of fluorescence intensity fitted in the energy window including the dominant modes (2.32–2.33 eV) [68]. As shown in Figure 7b, the slope above 6 kW/cm^2 is 9 times that below this threshold in accordance with polarized lasing. At three typical temperatures: 77 K, 171 K, and 295 K, lasing behavior was observed by the nonlinear growth in emission intensity of a few sharp peaks. It can be observed from Figure 7c–e that when the temperature is increased from 77 K to 295 K, the mode spacing increases by about an order of magnitude. This variation of the mode spacing with temperature is independent of the thermal expansion [68].

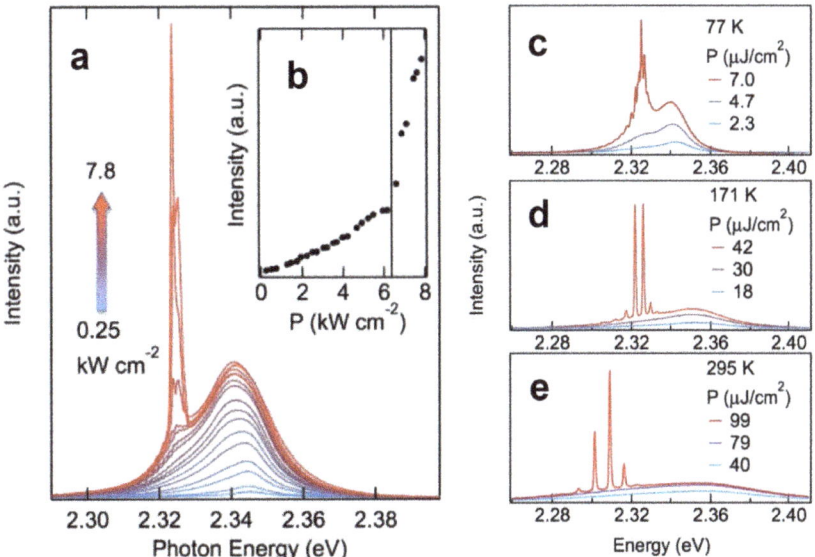

Figure 7. (**a**) PL spectra of a 20 µm long NW obtained with increasing excitation light power densities in the range of 0.25–7.8 kW/cm^2 [68]. Reprinted with permission from [68]. Copyright © 2017 WILEY–VCH Verlag GmbH & Co. KGaA, Weinheim. (**b**) Integrated power density plotted against the power density. Reprinted with permission from [68]. Copyright © 2017 WILEY–VCH Verlag GmbH & Co. KGaA, Weinheim. PL spectra under pulse laser excitation (454 nm, ≈60 fs, 0.5 MHz) of an L = 13 µm CsPbBr$_3$ NWs at: (**c**) 77 K with 2.3, 4.7 and 7.0 µJ/cm^2; (**d**) 171 K with 18, 30, and 42 µJ/cm^2; (**e**) 295 K with 40, 79, and 99 µJ/cm^2 [68]. Reprinted with permission from [68]. Copyright © 2017 WILEY–VCH Verlag GmbH & Co. KGaA, Weinheim.

3.4. Single Crystals Perovskite Microplates/Nanoplatelets Lasers

Two-dimensional (2D) single-crystals perovskite microplates/nanoplatelets have great application potential in whispering gallery mode (WGM) micro-nano lasers. Perovskite microplates/nanoplatelets are two-dimensional micro-nano materials. Usually, cubic or tetraconal halogen perovskites have highly symmetrical isotropic crystal structures and tend to grow into cube-shaped perovskites without any ligands or surfactants. In this optical resonant cavity, light wave can form continuous total internal reflection and form a stable propagation mode under certain conditions, which is usually called a WGM. Therefore, this type of resonant cavity is a good choice for making perovskite micro-nano lasers. In 2015, Tyagi et al. prepared MAPbBr$_3$ nanoplatelets with a thickness as low as 1.2 nm for the first time by solution method with the aid of surface inhibitors [69]. Figure 8a indicated the purified product consists of a colloidal solution of 2D nanoplatelets of submicron level dimensions. The absorption spectrum of the 2D nanoplatelets was dominated by a single sharp exciton absorption feature at 431 nm in Figure 8b, which occurred by 503 meV compared to the bulk exciton absorption at 525 nm blueshift. Liao et al. synthesized MAPbBr$_3$ microdisk with transverse sizes of 1–10 µm by the self-growth method in the solution [27]. Under the excitation of femtosecond laser at the wavelength of 400 nm, the laser threshold was (3.6 ± 0.5) µJ/cm^2, the FWHM was 1.1 nm, and the quality factor was 430 as shown in Figure 8c,d. By adjusting the symmetry of the microplate's shape, the output can be converted from a multimode laser to a single-mode laser. Liu's group studied the output mode of the perovskite laser [70], and these $CH_3NH_3PbBr_3$ perovskite microplates were synthesized by a simple one-step in the solution process.

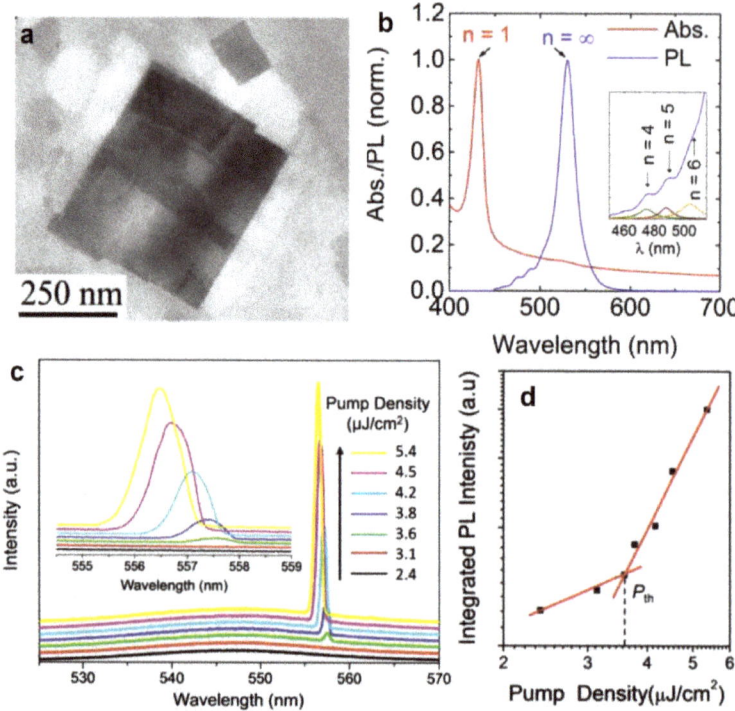

Figure 8. (**a**,**b**) TEM, absorption and PL of the purified $MAPbBr_3$ nanoplatelet solution [69]. Reprinted with permission from [69]. Copyright © 2015 American Chemical Society. (**c**) PL spectra of a square microdisk [27]. Reprinted with permission from [27]. Copyright © 2015 WILEY-VCH Verlag GmbH & Co. KGaA, Weinheim. (**d**) Integrated PL intensity as a function of excitation density [27]. Reprinted with permission from [27]. Copyright © 2015 WILEY-VCH Verlag GmbH & Co. KGaA, Weinheim.

With the continuous in-depth study of perovskite semiconductor materials, researchers have found that perovskite materials also exhibit nonlinear optical properties. Wei Zhang et al. simultaneously prepared $MAPbBr_3$ 1D microwires (MWs) and 2D square microplates (MPs) using a liquid phase synthesis method. Excited under a pulsed laser with a wavelength of 900 nm, both $MAPbBr_3$ MWs and MPs showed a two-photon absorption [71]. A maximum Q factor of about 920 was obtained by varying the MPs at the edge length. Figure 9a indicated that the FWHM at 552.3 nm narrowed rapidly from 22 to 0.6 nm with the increasing of the two-photon-pumped (TPP) fluence around the onset power, achieving a high quality (Q ≈ 920) and a low-threshold (E_{th} ≈ 62 µJ/cm^2) lasing action. In 2016, BinYang and their team used femtosecond laser excitation of $MAPbBr_3$ microdisks to study two-photon pump-amplified spontaneous emission and obtained their tunable emission spectrum from 500 nm to 570 nm [72]. They demonstrated that the photoluminescence properties of the microdisks were dominated by the reabsorption effect under two-photon excitation. In addition, it was found that the interband emission from the near-surface region and the photocarrier diffusion from the near-surface region to the inner region were important for single-photon excitation. The aforementioned two dynamic processes were illustrated in Figure 9b,c. In 2017, the linear and nonlinear light emission characteristics of $MAPbBr_3$ microplates with different sizes were investigated, and their lasing performance was characterized by two-photon excitation at 800 nm (150 fs, 1 kHz) [73]. Figure 9d clearly showed that the lasing threshold decreased linearly as the lateral dimension of the microplate decreased from 90 μm to 20 μm. Yisheng Gao et al. synthesized a high-quality MAPbBr$_3$ perovskite microstructure by solution precipitation [74]. The insets of Figure 9e display the high-resolution SEM images of the microplate and microrod. Under the intense laser pumping at 1240 nm, 100 fs, and 1 kHz, an obvious optical limit effect could be observed. Interband photoluminescence was observed at 540 nm. By increasing the pump density, three-photon excitation lasing in MAPbBr$_3$ perovskite microplate was achieved for the first time at room temperature. The measured three-photon absorption coefficient γ was 2.26×10^{-5} cm^3/GW^2, which was obtained by fitting the data in Figure 9e. Through further observation of the three-photon excited whispering gallery mode laser, it was found that the hybrid lead halide perovskite also had a very large fifth-order nonlinearity, which was of great significance for practical applications such as optical switches.

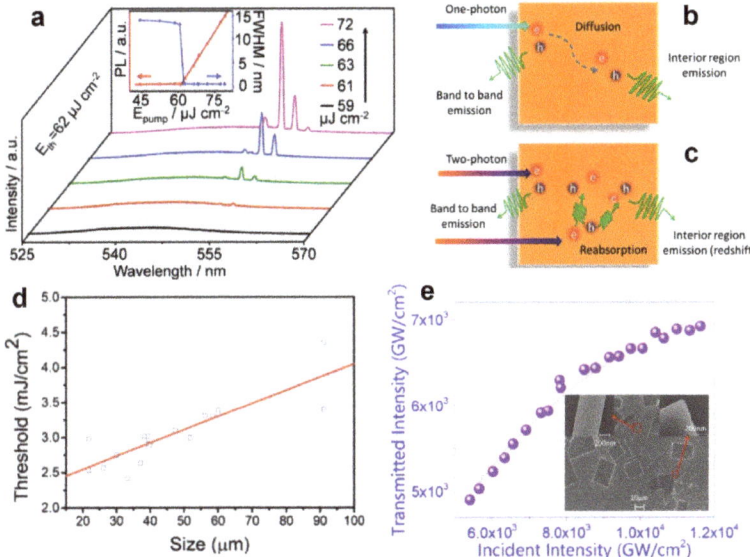

Figure 9. (**a**) Bright-field (up) and PL (middle) images of a single wire above the thresholds excited uniformly with a pulsed laser [71]. Reprinted with permission from [71]. Copyright © 2016 WILEY−VCH Verlag GmbH & Co. KGaA, Weinheim. Dynamic PL model: (**b**) Band− to−band emission and diffusion effect under one-photon excitation; (**c**) Band−to−band emission effect of near-surface regions and reabsorption effect of interior regions under two−photon excitation [72]. Reprinted with permission from [72]. Copyright © 2016 American Chemical Society. (**d**) Microplate size (20–90 μm) dependent on two-photon pumped coherent light emission threshold [73]. Reprinted with permission from [73]. Copyright © 2017 WILEY−VCH Verlag GmbH & Co. KGaA, Weinheim. (**e**) The transmission of an ultrashort pulse at 1240 nm as a function of incident power. Inset: top−view SEM image of the synthesized microstructures [74].

3.5. Perovskite Quantum Dots Lasers

Perovskite quantum dots (QDs) refer to perovskite materials that are less than 100 nm in three dimensions. They have garnered recent attention due to their unique versatility as laser gain materials, such as low cost, easy synthesis process, tunable emission wavelength and high photoluminescence quantum yield. It is one of the materials with potential development and is expected to replace the traditional II–VI, III–V and IV–VI colloidal QDs. When perovskite QDs do not have an external cavity, multiple scattering between QDs produces amplification, causing the lasing modes to fluctuate randomly. In 2015, Yakunin et al. demonstrated a low-threshold ASE from caesium lead halide perovskites

CsPbX$_3$ (X = Cl, Br or I, or mixed Cl/Br and Br/I systems) nanocrystals with an optical gain coefficient of ~450 cm^{-1} and threshold of ~5.3 µJ/cm^2 [24] (see Figure 10d). As shown in Figure 10a–c, the ASE from CsPbX$_3$ NCs was tuned from 440 to 700 nm. At last, they achieved random lasing from CsPbX$_3$ NCs films without the optical cavity and WGM lasing employing a silica sphere as the resonant cavity. In addition to realizing perovskite QDs lasing-based silica spheres and micro capillaries, well-designed DBR can also serve as an optical resonant cavity [75–77]. Sun and co-workers first realized a vertical-cavity surface-emitting laser (VCSEL) based on perovskite QDs, displaying a low threshold of ~9 µJ/cm^2 and favorable stability. Their device architecture was a sandwiched structure of DBR/CsPbBr$_3$ QDs/DBR in Figure 10e. This low lasing threshold can result from the large absorption cross-section of the perovskite QDs, high PLQY, low Auger loss, and the good match between the gain profile and the stop band of the DBRs [75]. In 2017, Huang et al. fabricated a perovskite VECSL with an ultralow threshold of ~0.39 µJ/cm^2 [77]. These VCSELs consisted of a CsPbBr$_3$ QD thin film and two highly reflective DBRs. Spectacularly, the realization of all-inorganic CsPbBr$_3$ QDs contributed to high device stability and enabled stable device operation under both femtosecond and quasi-CW nanosecond pulse pumping at ambient conditions [77].

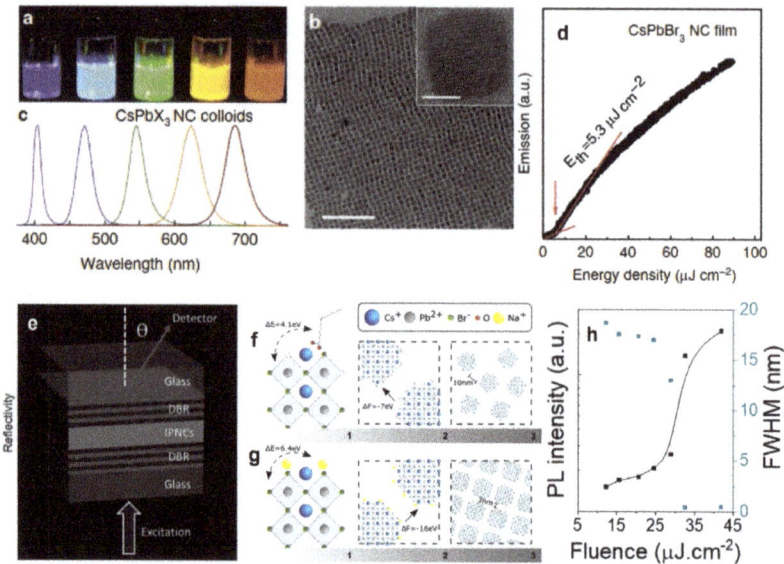

Figure 10. (**a**) Stable dispersions in toluene under excitation by an ultraviolet lamp [24]. (**b**) TEM images of CsPbBr$_3$ QDs [24]; corresponding scale bars are 100 and 5 nm. (**c**) PL spectra of the CsPbBr$_3$ QDs solutions [24]. (**d**) Threshold for the intensity of the ASE band of the CsPbBr$_3$ QDs film [24]. (**e**) Device architecture of the CsPbBr$_3$ QDs VCSEL [75]. Reprinted with permission from [75]. Copyright © 2017 WILEY–VCH Verlag GmbH & Co. KGaA, Weinheim. Self–assembly interactions of quantum dots without (**f**) and with (**g**) passivated sodium ligands [78]. Reprinted with permission from [78]. Copyright © 2021 Zhou et al. Advanced Science published by Wiley–VCH GmbH. (**h**) Power-dependent PL and FWHM as a function of excitation fluence [78]. Reprinted with permission from [78]. Copyright © 2021 Zhou et al. Advanced Science published by Wiley–VCH GmbH.

In 2021, Edward H. Sargent and co-workers reported a self-assembly passivation method that relied on sodium—an assembly director that enhanced the attractive forces between nearby CsPbBr$_3$ QDs and induced the formation of high-quality cubic superlattices [78]. Figure 10f,g showed the effect of atomic-size ligands on the self-assembly interactions of CsPbBr$_3$ QDs. These self-assembly perovskite quantum-dot superlattices structures were utilized as the resonant cavity and the gain medium, realizing nanosecond-sustained lasing with a threshold of 25 µJ/cm^2 in Figure 10h. In 2022, Zhang's group developed a new approach to realize multicolor lasing in the special structure of the perovskite QDs superlattice [79]. The alloy superlattice samples based on perovskite QDs were approximately 10 times more stable than perovskite single-crystal alloy NWs with poor band gap stability, exhibiting significant PL spectral changes within 3 days [80]. Furthermore, the carrier transport dynamics demonstrated the energy transport process in the alloy superlattice, which elucidated the core difficulty of achieving a multicolor perovskite lasers [79].

3.6. Others

In addition to the conventional perovskite lasers mentioned above, several other types of perovskite lasers based on the solution process have been reported in recent years. Wang et al. prepared the CH$_3$NH$_3$PbBr$_3$ perovskite microrod using the solution-processed one-step precipitation method [81]. This perovskite microrod formed a whispering gallery mode microcavity, which was different from the Fabry-Perot cavity. This structure was excited by a femtosecond laser at a wavelength of 400 nm, with a lasing threshold of 2.37 µJ/cm^2, an FWHM as low as 0.1 nm, and a quality factor as high as 5000. Surface-plasmon (SP) is an excited state with a large enhancement of the electromagnetic field localized at the metal–dielectric interface [82]. SP can provide a powerful platform to tailor the spontaneous emission, thus lasing the low-dimensional perovskite structures in a nanoscale regime [83]. SPs arise from a metal layer or conducting layer, and transfer along the semiconductor-metal interface. In 2017, Wang and co-workers demonstrated that the laser threshold of CsPbBr$_3$ perovskite microrod with Al nanoparticles (NPs) layer was drastically decreased by more than 20%, and the output intensity was significantly increased by more than an order of magnitude due to plasmonic resonances [84]. Wu et al. proposed a new approach to improve the ASE performance of MAPbI$_3$ perovskite film via utilization of Au nanorods-doped PMMA [85]. These MAPbI$_3$ films were prepared by the modified two-step process. Finally, the ASE threshold of the MAPbI$_3$ perovskite films was obviously decreased from 26.5 to 16.9 µJ/cm^2, which mainly resulted from the surface passivation of the PMMA layer. In the same year, the reduction of the lasing threshold of CsPbBr$_3$ perovskite nanocubes was also realized via the surface plasmonic effect of Au NPs by Leng's group [86]. Table 2 summarizes the representative works on solution-processed perovskite lasers in recent years.

Table 2. Performances comparison of solution-processed perovskite semiconductors lasers.

Materials	Structure	Laser Mode	Wavelength	Pump Laser	Threshold	FWHM	Year
MAPbX$_3$	Polycrystalline thin film	ASE	390–790 nm	600 nm, 150 fs	44 kW/cm^2	N.A.	2014 [23]
MAPbI$_3$	Polycrystalline thin film	ASE	780 nm	530 nm, 4 ns	76 µJ/cm^2	1.24	2015 [59]
MAPbI$_3$	Polycrystalline thin film	DFB	784 nm	515 nm, 200 fs	4 µJ/cm^2	0.4	2016 [20]
MAPbI$_3$	Polycrystalline thin film	PhC	780 nm	532 nm, 400 ps	200 µJ/cm^2	N.A.	2016 [60]
(PEA)$_2$Cs$_{n-1}$Pb$_n$Br$_{3n+1}$	Polycrystalline thin film	VCSEL	532 nm	355 nm, 8 ns	500 µJ/cm^2	0.8	2019 [21]
MAPbCl$_3$	Single crystals thin film	VCSEL	414–435 nm	355 nm, 8 ns	211 µJ/cm^2	0.38	2020 [64]
MAPbX$_3$	Single crystals NWs	FP	500–780 nm	402 nm, 150 fs	220 nJ/cm^2	0.22	2015 [22]
(FA$_x$MA$_{1-x}$)Pb(Br$_{3-y}$I$_y$)	Single crystals NWs	FP	490–824 nm	402 nm, 150 fs	2.6 µJ/cm^2	0.24	2016 [65]
MAPbI$_3$	Single crystals NWs	FP	776–784 nm	400 nm, 120 fs	13.5 µJ/cm^2	5	2016 [66]
CsPbX$_3$	Single crystals NWs	FP	420–650 nm	405 nm, CW	12.3 µJ/cm^2	0.09	2018 [67]
MAPbCl$_x$Br$_{3-x}$	Microdisk	WGM	525–557 nm	400 nm, 150 fs	3.6 µJ/cm^2	1.1	2015 [27]
MAPbBr$_3$	Microplates	FP&WGM	552.3 nm	900 nm, 150 fs	62 µJ/cm^2	0.6	2016 [71]

Table 2. Cont.

Materials	Structure	Laser Mode	Wavelength	Pump Laser	Threshold	FWHM	Year
MAPbBr$_3$	Microdisks	ASE	500–570 nm	1064 nm, 10 ns	2.2 mJ/cm^2	N.A.	2016 [72]
CsPbX$_3$	Quantum dots	ASE	440–700 nm	400 nm, 100 fs	6 µJ/cm^2	N.A.	2015 [24]
CsPbX$_3$	Quantum dots	VCSEL	440–700 nm	400 nm, 100 fs	9 µJ/cm^2	0.6	2017 [75]
CsPbBr$_3$	Quantum dots	VCSEL	522 nm	400 nm, 50 fs 355 nm, 5 nm	0.39 µJ/cm^2 98 µJ/cm^2	0.9	2017 [77]
CsPbBr$_3$	Quantum dots	ASE	536 nm	355 nm, 2 ns	25 µJ/cm^2	0.4	2021 [78]
CsPbBr$_3$	Quantum dots	ASE	480–508 nm	400 nm, 40 fs	30 µJ/cm^2	0.13	2022 [79]
MAPbBr$_3$	NCs	SP	554 nm	800 nm, 100 fs	10 µJ/cm^2	3	2021 [87]
CsPbBr$_3$	NCs	SP	532 nm	400 nm, 250 fs	46.8 µJ/cm^2	20.9	2022 [88]

In 2020, Atwate and co-workers demonstrated that single-mode up-conversion plasmonic lasing from MAPbBr$_3$ perovskite NCs with a low threshold (10 µJ/cm^2) and a calculated ultrasmall mode volume (~0.06 λ^3) at 6 K [87]. The MAPbBr$_3$ perovskite NCs were synthesized by doping MAPbBr$_3$ film with chlorine as shown in Figure 11a,b. Figure 11a showed a MAPbBr$_3$ perovskite NCs integrated with an Al$_2$O$_3$/TiN (5 nm/80 nm) plasmonic cavity and inset of SEM image. The device was deposited on a silicon-based substrate in which TiN served as a promising resonance-tunable plasmonic platform. In 2021, Lan's group reported polycrystalline CsPbBr$_3$ NPs composed of QDs on a thin Au film, exhibiting optically-controlled quantum size effect [89]. However, it is regrettable that their CsPbBr$_3$ NPs were synthesized by using chemical vapor deposition rather than the solution. Most recently, Lin et al. proposed an on-chip fabricated hybrid photon-plasma system consisting of a perovskite laser structure coupled to a long-range surface plasmon polariton (LRSPP) waveguide, obtaining a low threshold and propagation length in excess of 100 µm [88]. Perovskite NCs were synthesized by a solution method [90]. In this system, the CsPbBr$_3$ NCs were drop-casted on the patterned samples and then the NCs were forced into the pattern. When the pump energy density exceeded 46.8 µJ/cm^2, the emission intensity increased and the nonlinearity of FWHM decreased. These results suggested that SPs could not only improve the performance of perovskite lasers but also enable different applications in optical communications and sensor-related devices.

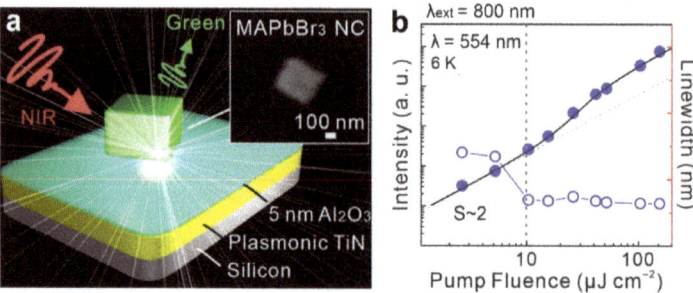

Figure 11. (a) Schematic of the up-conversion plasmonic MAPbBr$_3$ perovskite NCs lasers. Inset: SEM image of MAPbBr$_3$ perovskite NCs [87]. Reprinted with permission from [87]. Copyright © 2021 American Chemical Society. (b) PL intensity and line width vs pumped fluence at the lasing peak of 554 nm [87]. Reprinted with permission from [87]. Copyright © 2021 American Chemical Society.

4. Conclusions

Perovskite semiconductor materials have the advantages of high optical gain, large absorption coefficient, long carrier life, solution processing and so on, which is an excellent gain material to achieve low threshold laser. The combination of perovskite materials and optical microcavities can further reduce the laser threshold value. At present, it is necessary to further study the stability, optical gain, threshold and other basic properties of perovskite materials. In this review, we summarized the recent developments and progress of solution-processed perovskite semiconductors lasers. We discussed the merit of solution-processed perovskite semiconductors as lasing gain materials and summarized the characteristics of a variety of perovskite lasers. Research progress in recent years has shown that the application of perovskite and microcavity structures in lasers has a positive effect on reducing the laser threshold. Moreover, the preparation method of the solution process, as well as the flexible devices, are the unique advantages of perovskite lasers compared to other conventional semiconductor lasers.

Despite the great progresses made in perovskite semiconductor materials and perovskite semiconductor lasers, there are still many issues to be solved. Firstly, from the perspective of materials, whether it is chemical vapor deposition or solution crystallization, the prepared perovskite single crystal samples have a certain degree of randomness. To realize the large-scale and commercialization of perovskite lasers, we should also explore a method to prepare homogeneous and reproducible perovskite samples. Furthermore, since most of the perovskite samples in use today contain the heavy metal Pb, which is harmful to the human body and unfriendly to the environment, it is necessary to strengthen the search for new perovskite systems with good optoelectronic properties and less lead or lead-free. Secondly, the Q value of perovskite laser emission still needs to be improved. Further improvement of film forming quality and Q value of optical microcavity are the next important research directions. Compared with traditional III-V compounds semiconductor materials, organic and inorganic hybrid perovskite materials have relatively poor stability, and the organic molecular layer in the material structure is very sensitive to non-polar solvents. There are problems of easy decomposition in air and easy dissolution in water and organic solvents. Finally, but more importantly, a theoretical explanation of the photophysics of perovskite NCs is required to better explain the quantum size effect of perovskite crystals, which can guide research directions for regulating their electronic, optical and defective nature [82]. In response to the above problems, researchers have proposed a variety of solutions, including a small amount of doping to improve the stability of the material phase, reduce the defects of perovskite crystals, design pure inorganic perovskite, and improve the packaging process to encapsulate the polymer layer on the device surface. Therefore, the future trend of the perovskite-based laser is to optimize cavity design and improve device stability to achieve perovskite lasing under continuous optical or even electric pumping at room temperature, which is still key in this field.

Author Contributions: Conceptualization, N.Z. and S.J.; writing—original draft preparation, N.Z.; writing—review and editing, N.Z.; supervision, Q.X.; project administration, N.Z.; funding acquisition, Q.N. and Q.X. All authors have read and agreed to the published version of the manuscript.

Funding: This work is supported by the National Natural Science Foundation of China (NSFC) (62105173, 62105174).

Institutional Review Board Statement: Not applicable.

Informed Consent Statement: Not applicable.

Data Availability Statement: Not applicable.

Acknowledgments: The authors thank Shengnan Wang and Yaqi Liao for helping with this article.

Conflicts of Interest: The authors declare no conflict of interest.

References

1. Hall, R.N.; Fenner, G.E.; Kingsley, J.; Soltys, T.; Carlson, R. Coherent light emission from GaAs junctions. *Phys. Rev. Lett.* **1962**, *9*, 366. [CrossRef]
2. Nathan, M.I.; Dumke, W.P.; Burns, G.; Dill, F.H., Jr.; Lasher, G. Stimulated emission of radiation from GaAs p-n junctions. *Appl. Phys. Lett.* **1962**, *1*, 62–64. [CrossRef]
3. Xia, H.; Hu, C.; Chen, T.; Hu, D.; Zhang, M.; Xie, K. Advances in Conjugated Polymer Lasers. *Polymers* **2019**, *11*, 443. [CrossRef] [PubMed]
4. McGehee, M.D.; Heeger, A.J. Semiconducting (conjugated) polymers as materials for solid-state lasers. *Adv. Mater.* **2000**, *12*, 1655–1668. [CrossRef]
5. Zhao, J.; Yan, Y.; Zhao, Y.; Yao, J. Research progress on organic micro/nanoscale lasers. *Sci. Sin. Chim.* **2018**, *48*, 127–142. [CrossRef]
6. Clark, J.; Lanzani, G. Organic photonics for communications. *Nat. Photonics* **2010**, *4*, 438–446. [CrossRef]
7. de Arquer, F.P.G.; Armin, A.; Meredith, P.; Sargent, E.H. Solution-processed semiconductors for next-generation photodetectors. *Nat. Rev. Mater.* **2017**, *2*, 16100. [CrossRef]
8. Zhang, C.; Yan, Y.; Zhao, Y.S.; Yao, J. From molecular design and materials construction to organic nanophotonic devices. *Acc. Chem. Res.* **2014**, *47*, 3448–3458. [CrossRef]
9. Gierschner, J.; Varghese, S.; Park, S.Y. Organic single crystal lasers: A materials view. *Adv. Opt. Mater.* **2016**, *4*, 348–364. [CrossRef]
10. Yan, Y.; Zhao, Y.S. Organic nanophotonics: From controllable assembly of functional molecules to low-dimensional materials with desired photonic properties. *Chem. Soc. Rev.* **2014**, *43*, 4325–4340. [CrossRef]
11. Fang, H.H.; Yang, J.; Feng, J.; Yamao, T.; Hotta, S.; Sun, H.B. Functional organic single crystals for solid-state laser applications. *Laser Photonics Rev.* **2014**, *8*, 687–715. [CrossRef]
12. Sohler, W.; Rue, R. Organic solid-state integrated amplifiers and lasers. *Laser Photonics Rev.* **2012**, *6*, 419–462.
13. Li, Y.J.; Yan, Y.; Zhao, Y.S.; Yao, J. Construction of Nanowire Heterojunctions: Photonic Function-Oriented Nanoarchitectonics. *Adv. Mater.* **2016**, *28*, 1319–1326. [CrossRef] [PubMed]
14. Zhang, W.; Yao, J.; Zhao, Y.S. Organic micro/nanoscale lasers. *Acc. Chem. Res.* **2016**, *49*, 1691–1700. [CrossRef] [PubMed]
15. Jung, H.; Ahn, N.; Klimov, V.I. Prospects and challenges of colloidal quantum dot laser diodes. *Nat. Photonics* **2021**, *15*, 643–655. [CrossRef]
16. Fan, F.; Voznyy, O.; Sabatini, R.P.; Bicanic, K.T.; Adachi, M.M.; McBride, J.R.; Reid, K.R.; Park, Y.-S.; Li, X.; Jain, A. Continuous-wave lasing in colloidal quantum dot solids enabled by facet-selective epitaxy. *Nature* **2017**, *544*, 75–79. [CrossRef] [PubMed]
17. Lim, J.; Park, Y.-S.; Klimov, V.I. Optical gain in colloidal quantum dots achieved with direct-current electrical pumping. *Nat. Mater.* **2018**, *17*, 42–49. [CrossRef]
18. Park, Y.-S.; Roh, J.; Diroll, B.T.; Schaller, R.D.; Klimov, V.I. Colloidal quantum dot lasers. *Nat. Rev. Mater.* **2021**, *6*, 382–401. [CrossRef]
19. Roh, J.; Park, Y.-S.; Lim, J.; Klimov, V.I. Optically pumped colloidal-quantum-dot lasing in LED-like devices with an integrated optical cavity. *Nat. Commun.* **2020**, *11*, 1–10. [CrossRef]
20. Whitworth, G.L.; Harwell, J.R.; Miller, D.N.; Hedley, G.J.; Zhang, W.; Snaith, H.J.; Turnbull, G.A.; Samuel, I.D.W. Nanoimprinted distributed feedback lasers of solution processed hybrid perovskites. *Opt. Express* **2016**, *24*, 23677–23684. [CrossRef]
21. Zhai, W.; Tian, C.; Yuan, K.; Ge, C.; Zhao, S.; Yu, H.; Li, Y.; Chen, W.; Ran, G. Optically pumped lasing of segregated quasi-2D perovskite microcrystals in vertical microcavity at room temperature. *Appl. Phys. Lett.* **2019**, *114*, 131107. [CrossRef]
22. Zhu, H.; Fu, Y.; Meng, F.; Wu, X.; Gong, Z.; Ding, Q.; Gustafsson, M.V.; Trinh, M.T.; Jin, S.; Zhu, X. Lead halide perovskite nanowire lasers with low lasing thresholds and high quality factors. *Nat. Mater.* **2015**, *14*, 636–642. [CrossRef] [PubMed]
23. Xing, G.; Mathews, N.; Lim, S.S.; Yantara, N.; Liu, X.; Sabba, D.; Grätzel, M.; Mhaisalkar, S.; Sum, T.C. Low-temperature solution-processed wavelength-tunable perovskites for lasing. *Nat. Mater.* **2014**, *13*, 476–480. [CrossRef] [PubMed]
24. Yakunin, S.; Protesescu, L.; Krieg, F.; Bodnarchuk, M.I.; Nedelcu, G.; Humer, M.; De Luca, G.; Fiebig, M.; Heiss, W.; Kovalenko, M.V. Low-threshold amplified spontaneous emission and lasing from colloidal nanocrystals of caesium lead halide perovskites. *Nat. Commun.* **2015**, *6*, 1–9.
25. Zhang, Q.; Ha, S.T.; Liu, X.; Sum, T.C.; Xiong, Q. Room-temperature near-infrared high-Q perovskite whispering-gallery planar nanolasers. *Nano Lett.* **2014**, *14*, 5995–6001. [CrossRef]
26. Sutherland, B.R.; Hoogland, S.; Adachi, M.M.; Wong, C.T.; Sargent, E.H. Conformal organohalide perovskites enable lasing on spherical resonators. *ACS Nano* **2014**, *8*, 10947–10952. [CrossRef]
27. Liao, Q.; Hu, K.; Zhang, H.; Wang, X.; Yao, J.; Fu, H. Perovskite microdisk microlasers self-assembled from solution. *Adv. Mater.* **2015**, *27*, 3405–3410. [CrossRef]
28. Dhanker, R.; Brigeman, A.; Larsen, A.; Stewart, R.; Asbury, J.B.; Giebink, N.C. Random lasing in organo-lead halide perovskite microcrystal networks. *Appl. Phys. Lett.* **2014**, *105*, 151112. [CrossRef]
29. Protesescu, L.; Yakunin, S.; Bodnarchuk, M.I.; Krieg, F.; Caputo, R.; Hendon, C.H.; Yang, R.X.; Walsh, A.; Kovalenko, M.V. Nanocrystals of cesium lead halide perovskites ($CsPbX_3$, X= Cl, Br, and I): Novel optoelectronic materials showing bright emission with wide color gamut. *Nano Lett.* **2015**, *15*, 3692–3696. [CrossRef]
30. Huang, L.-y.; Lambrecht, W.R. Electronic band structure, phonons, and exciton binding energies of halide perovskites $CsSnCl_3$, $CsSnBr_3$, and $CsSnI_3$. *Phys. Rev. B* **2013**, *88*, 165203. [CrossRef]

31. Noh, J.H.; Im, S.H.; Heo, J.H.; Mandal, T.N.; Seok, S.I. Chemical management for colorful, efficient, and stable inorganic–organic hybrid nanostructured solar cells. *Nano Lett.* **2013**, *13*, 1764–1769. [CrossRef]
32. Fu, Y.; Zhu, H.; Stoumpos, C.C.; Ding, Q.; Wang, J.; Kanatzidis, M.G.; Zhu, X.; Jin, S. Broad wavelength tunable robust lasing from single-crystal nanowires of cesium lead halide perovskites (CsPbX3, X= Cl, Br, I). *ACS Nano* **2016**, *10*, 7963–7972. [CrossRef] [PubMed]
33. Jiang, Y.; Wang, X.; Pan, A. Properties of excitons and photogenerated charge carriers in metal halide perovskites. *Adv. Mater.* **2019**, *31*, 1806671. [CrossRef] [PubMed]
34. Sutherland, B.R.; Hoogland, S.; Adachi, M.M.; Kanjanaboos, P.; Wong, C.T.; McDowell, J.J.; Xu, J.; Voznyy, O.; Ning, Z.; Houtepen, A.J. Perovskite thin films via atomic layer deposition. *Adv. Mater.* **2015**, *27*, 53–58. [CrossRef] [PubMed]
35. Gao, W.; Yu, S.F. Reality or fantasy-Perovskite semiconductor laser diodes. *Ecomat* **2021**, *3*, e12077. [CrossRef]
36. Shi, D.; Adinolfi, V.; Comin, R.; Yuan, M.; Alarousu, E.; Buin, A.; Chen, Y.; Hoogland, S.; Rothenberger, A.; Katsiev, K. Low trap-state density and long carrier diffusion in organolead trihalide perovskite single crystals. *Science* **2015**, *347*, 519–522. [CrossRef]
37. Frost, J.M. Calculating polaron mobility in halide perovskites. *Phys. Rev. B* **2017**, *96*, 195202. [CrossRef]
38. He, Y.; Galli, G. Perovskites for solar thermoelectric applications: A first principle study of CH3NH3AI3 (A= Pb and Sn). *Chem. Mater.* **2014**, *26*, 5394–5400. [CrossRef]
39. Ye, T.; Wang, X.; Li, X.; Yan, A.Q.; Ramakrishna, S.; Xu, J. Ultra-high Seebeck coefficient and low thermal conductivity of a centimeter-sized perovskite single crystal acquired by a modified fast growth method. *J. Mater. Chem. C* **2017**, *5*, 1255–1260. [CrossRef]
40. Pisoni, A.; Jacimovic, J.; Barisic, O.S.; Spina, M.; Gaál, R.; Forró, L.; Horváth, E. Ultra-low thermal conductivity in organic–inorganic hybrid perovskite CH3NH3PbI3. *J. Phys. Chem. Lett.* **2014**, *5*, 2488–2492. [CrossRef]
41. Yang, A.; Bai, M.; Bao, X.; Wang, J.; Zhang, W. Investigation of Optical and Dielectric Constants of Organic-InorganicCH3NH3PbI3 Perovskite Thin Films. *J. Nanomed. Nanotechnol.* **2016**, *7*, 5–9.
42. Kang, Y.; Han, S. Intrinsic carrier mobility of cesium lead halide perovskites. *Phys. Rev. Appl.* **2018**, *10*, 044013. [CrossRef]
43. Yang, Z.; Xu, Q.; Wang, X.; Lu, J.; Wang, H.; Li, F.; Zhang, L.; Hu, G.; Pan, C. Large and ultrastable all-inorganic CsPbBr3 monocrystalline films: Low-temperature growth and application for high-performance photodetectors. *Adv. Mater.* **2018**, *30*, 1802110. [CrossRef] [PubMed]
44. Ahmad, M.; Rehman, G.; Ali, L.; Shafiq, M.; Iqbal, R.; Ahmad, R.; Khan, T.; Jalali-Asadabadi, S.; Maqbool, M.; Ahmad, I. Structural, electronic and optical properties of CsPbX3 (X = Cl, Br, I) for energy storage and hybrid solar cell applications. *J. Alloys. Compd.* **2017**, *705*, 828–839. [CrossRef]
45. Li, C.; Wang, A.; Deng, X.; Wang, S.; Yuan, Y.; Ding, L.; Hao, F. Insights into ultrafast carrier dynamics in perovskite thin films and solar cells. *ACS Photonics* **2020**, *7*, 1893–1907. [CrossRef]
46. Stranks, S.D.; Burlakov, V.M.; Leijtens, T.; Ball, J.M.; Goriely, A.; Snaith, H.J. Recombination kinetics in organic-inorganic perovskites: Excitons, free charge, and subgap states. *Phys. Rev. Appl.* **2014**, *2*, 034007. [CrossRef]
47. Yamada, Y.; Nakamura, T.; Endo, M.; Wakamiya, A.; Kanemitsu, Y. Photocarrier recombination dynamics in perovskite CH3NH3PbI3 for solar cell applications. *J. Am. Chem. Soc.* **2014**, *136*, 11610–11613. [CrossRef]
48. Tian, J. *Investigation of Pulsed Lasing Properties and Mechanisms in Lead-Based Perovskite Micro/Nano-Structures*; East China Normal University: Shanghai, China, 2021.
49. Boyd, C.C.; Cheacharoen, R.; Leijtens, T.; McGehee, M.D. Understanding Degradation Mechanisms and Improving Stability of Perovskite Photovoltaics. *Chem. Rev.* **2019**, *119*, 3418–3451. [CrossRef]
50. Wei, Y.; Cheng, Z.; Lin, J. An overview on enhancing the stability of lead halide perovskite quantum dots and their applications in phosphor-converted LEDs. *Chem. Soc. Rev.* **2019**, *48*, 310–350. [CrossRef] [PubMed]
51. Huang, S.; Xu, Y.; Xie, M.; Xu, H.; He, M.; Xia, J.; Huang, L.; Li, H. Synthesis of magnetic CoFe2O4/g-C3N4 composite and its enhancement of photocatalytic ability under visible-light. *Colloids Surf. A* **2015**, *478*, 71–80. [CrossRef]
52. Jia, Y.; Kerner, R.A.; Grede, A.J.; Rand, B.P.; Giebink, N.C. Continuous-wave lasing in an organic–inorganic lead halide perovskite semiconductor. *Nat. Photonics* **2017**, *11*, 784–788. [CrossRef]
53. Jia, Y.; Kerner, R.A.; Grede, A.J.; Brigeman, A.N.; Rand, B.P.; Giebink, N.C. Diode-pumped organo-lead halide perovskite lasing in a metal-clad distributed feedback resonator. *Nano Lett.* **2016**, *16*, 4624–4629. [CrossRef] [PubMed]
54. Li, G.; Chen, K.; Cui, Y.; Zhang, Y.; Tian, Y.; Tian, B.; Hao, Y.; Wu, Y.; Zhang, H. Stability of perovskite light sources: Status and challenges. *Adv. Opt. Mater.* **2020**, *8*, 1902012. [CrossRef]
55. Wu, X.; Hong, X.; Luo, Z.; Hui, K.S.; Chen, H.; Wu, J.; Hui, K.; Li, L.; Nan, J.; Zhang, Q. The effects of surface modification on the supercapacitive behaviors of novel mesoporous carbon derived from rod-like hydroxyapatite template. *Electrochim. Acta* **2013**, *89*, 400–406. [CrossRef]
56. Fujiwara, K.; Zhang, S.; Takahashi, S.; Ni, L.; Rao, A.; Yamashita, K. Excitation dynamics in layered lead halide perovskite crystal slabs and microcavities. *ACS Photonics* **2020**, *7*, 845–852. [CrossRef]
57. Wang, X.; Chen, H.; Zhou, H.; Wang, X.; Yuan, S.; Yang, Z.; Zhu, X.; Ma, R.; Pan, A. Room-temperature high-performance CsPbBr3 perovskite tetrahedral microlasers. *Nanoscale* **2019**, *11*, 2393–2400. [CrossRef]
58. Wang, N.; Si, J.; Jin, Y.; Wang, J.; Huang, W.; University, Z.; University, Z. Solution-Processed Organic-Inorganic Hybrid Perovskites:A Class of Dream Materials Beyond Photovoltaic Applications. *Acta Chim. Sin.* **2015**, *73*, 171–178. [CrossRef]

59. Stranks, S.D.; Wood, S.M.; Wojciechowski, K.; Deschler, F.; Saliba, M.; Khandelwal, H.; Patel, J.B.; Elston, S.J.; Herz, L.M.; Johnston, M.B.; et al. Enhanced Amplified Spontaneous Emission in Perovskites Using a Flexible Cholesteric Liquid Crystal Reflector. *Nano Lett.* **2015**, *15*, 4935–4941. [CrossRef]
60. Cha, H.; Bae, S.; Lee, M.; Jeon, H. Two-dimensional photonic crystal bandedge laser with hybrid perovskite thin film for optical gain. *Appl. Phys. Lett.* **2016**, *108*, 181104. [CrossRef]
61. Huo, C.; Wang, Z.; Li, X.; Zeng, H. Low-dimensional metal halide perovskites: A kind of microcavity laser materials. *Chin. J. Lasers* **2017**, *44*, 120–131.
62. Ji, X.; Li, G.; Cui, Y.; Kong, W.; Liu, Y.; Hao, Y. Research Progress in Organic-Inorganic Hybridized Perovskite Lasers. *Semicond. Technol.* **2018**, *43*, 401–413, 442.
63. Zhong, Y.; Liao, K.; Du, W.; Zhu, J.; Shang, Q.; Zhou, F.; Wu, X.; Sui, X.; Shi, J.; Yue, S.; et al. Large-Scale Thin $CsPbBr_3$ Single-Crystal Film Grown on Sapphire via Chemical Vapor Deposition: Toward Laser Array Application. *ACS Nano* **2020**, *14*, 15605–15615. [CrossRef]
64. Tian, C.; Zhao, S.; Guo, T.; Xu, W.; Li, Y.; Ran, G. Deep-blue DBR laser at room temperature from single-crystalline perovskite thin film. *Opt. Mater.* **2020**, *107*, 110130. [CrossRef]
65. Fu, Y.; Zhu, H.; Schrader, A.W.; Liang, D.; Ding, Q.; Joshi, P.; Hwang, L.; Zhu, X.Y.; Jin, S. Nanowire Lasers of Formamidinium Lead Halide Perovskites and Their Stabilized Alloys with Improved Stability. *Nano Lett.* **2016**, *16*, 1000–1008. [CrossRef]
66. Yu, H.; Ren, K.; Wu, Q.; Wang, J.; Lin, J.; Wang, Z.; Xu, J.; Oulton, R.F.; Qu, S.; Jin, P. Organic–inorganic perovskite plasmonic nanowire lasers with a low threshold and a good thermal stability. *Nanoscale* **2016**, *8*, 19536–19540. [CrossRef]
67. Jiang, L.; Liu, R.; Su, R.; Yu, Y.; Xu, H.; Wei, Y.; Zhou, Z.-K.; Wang, X. Continuous wave pumped single-mode nanolasers in inorganic perovskites with robust stability and high quantum yield. *Nanoscale* **2018**, *10*, 13565–13571. [CrossRef]
68. Evans, T.J.; Schlaus, A.; Fu, Y.; Zhong, X.; Atallah, T.L.; Spencer, M.S.; Brus, L.E.; Jin, S.; Zhu, X.Y. Continuous-wave lasing in cesium lead bromide perovskite nanowires. *Adv. Opt. Mater.* **2018**, *6*, 1700982. [CrossRef]
69. Tyagi, P.; Arveson, S.M.; Tisdale, W.A. Colloidal organohalide perovskite nanoplatelets exhibiting quantum confinement. *J. Phys. Chem. Lett.* **2015**, *6*, 1911–1916. [CrossRef]
70. Liu, S.; Sun, W.; Gu, Z.; Wang, K.; Zhang, N.; Xiao, S.; Song, Q. Tailoring the lasing modes in $CH_3NH_3PbBr_3$ perovskite microplates via micro-manipulation. *RSC Adv.* **2016**, *6*, 50553–50558. [CrossRef]
71. Zhang, W.; Peng, L.; Liu, J.; Tang, A.; Hu, J.S.; Yao, J.; Zhao, Y.S. Controlling the Cavity Structures of Two-Photon-Pumped Perovskite Microlasers. *Adv. Mater.* **2016**, *28*, 4040–4046. [CrossRef]
72. Yang, B.; Mao, X.; Yang, S.; Li, Y.; Wang, Y.; Wang, M. Low Threshold Two-Photon-Pumped Amplified Spontaneous Emission in $CH_3NH_3PbBr_3$ Microdisks. *Acs Appl. Mater. Interfaces* **2016**, *8*, 19587–19592. [CrossRef] [PubMed]
73. Wei, Q.; Du, B.; Wu, B.; Guo, J.; Li, M.; Fu, J.; Zhang, Z.; Yu, J.; Hou, T.; Xing, G. Two-Photon Optical Properties in Individual Organic–Inorganic Perovskite Microplates. *Adv. Opt. Mater.* **2017**, *5*, 1700809. [CrossRef]
74. Gao, Y.; Wang, S.; Huang, C.; Yi, N.; Wang, K.; Xiao, S.; Song, Q. Room temperature three-photon pumped $CH_3NH_3PbBr_3$ perovskite microlasers. *Sci. Rep.* **2017**, *7*, 1–6. [CrossRef]
75. Wang, Y.; Li, X.; Nalla, V.; Zeng, H.; Sun, H. Solution-processed low threshold vertical cavity surface emitting lasers from all-inorganic perovskite nanocrystals. *Adv. Funct. Mater.* **2017**, *27*, 1605088. [CrossRef]
76. Chen, S.; Nurmikko, A. Stable green perovskite vertical-cavity surface-emitting lasers on rigid and flexible substrates. *ACS Photonics* **2017**, *4*, 2486–2494. [CrossRef]
77. Huang, C.-Y.; Zou, C.; Mao, C.; Corp, K.L.; Yao, Y.-C.; Lee, Y.-J.; Schlenker, C.W.; Jen, A.K.; Lin, L.Y. $CsPbBr_3$ perovskite quantum dot vertical cavity lasers with low threshold and high stability. *ACS Photonics* **2017**, *4*, 2281–2289. [CrossRef]
78. Zhou, C.; Pina, J.M.; Zhu, T.; Parmar, D.H.; Chang, H.; Yu, J.; Yuan, F.; Bappi, G.; Hou, Y.; Zheng, X. Quantum Dot Self-Assembly Enables Low-Threshold Lasing. *Adv. Sci.* **2021**, *8*, 2101125. [CrossRef]
79. Chen, L.; Zhou, B.; Hu, Y.; Dong, H.; Zhang, G.; Shi, Y.; Zhang, L. Stable Multi-Wavelength Lasing in Single Perovskite Quantum Dot Superlattice. *Adv. Opt. Mater.* **2022**, *10*, 2200494. [CrossRef]
80. Tang, B.; Hu, Y.; Lu, J.; Dong, H.; Mou, N.; Gao, X.; Wang, H.; Jiang, X.; Zhang, L. Energy transfer and wavelength tunable lasing of single perovskite alloy nanowire. *Nano Energy* **2020**, *71*, 104641. [CrossRef]
81. Wang, K.; Sun, S.; Zhang, C.; Sun, W.; Gu, Z.; Xiao, S.; Song, Q. Whispering-gallery-mode based $CH_3NH_3PbBr_3$ perovskite microrod lasers with high quality factors. *Mater. Chem. Front.* **2017**, *1*, 477–481. [CrossRef]
82. Hu, Z.P.; Liu, Z.Z.; Zhan, Z.J.; Shi, T.C.; Du, J.; Tang, X.S.; Leng, Y.X. Advances in metal halide perovskite lasers: Synthetic strategies, morphology control, and lasing emission. *Adv. Photonics* **2021**, *3*, 034002. [CrossRef]
83. Li, C.; Liu, Z.; Shang, Q.; Zhang, Q. Surface-plasmon-assisted metal halide perovskite small lasers. *Adv. Opt. Mater.* **2019**, *7*, 1900279. [CrossRef]
84. Wang, S.; Fang, J.; Zhang, C.; Sun, S.; Wang, K.; Xiao, S.; Song, Q. Maskless fabrication of aluminum nanoparticles for plasmonic enhancement of lead halide perovskite lasers. *Adv. Opt. Mater.* **2017**, *5*, 1700529. [CrossRef]
85. Wu, X.; Jiang, X.-F.; Hu, X.; Zhang, D.-F.; Li, S.; Yao, X.; Liu, W.; Yip, H.-L.; Tang, Z.; Xu, Q.-H. Highly stable enhanced near-infrared amplified spontaneous emission in solution-processed perovskite films by employing polymer and gold nanorods. *Nanoscale* **2019**, *11*, 1959–1967. [CrossRef]
86. Yang, J.; Liu, Z.; Hu, Z.; Zeng, F.; Zhang, Z.; Yao, Y.; Yao, Z.; Tang, X.; Du, J.; Zang, Z. Enhanced single-mode lasers of all-inorganic perovskite nanocube by localized surface plasmonic effect from Au nanoparticles. *J. Lumin.* **2019**, *208*, 402–407. [CrossRef]

87. Lu, Y.-J.; Shen, T.L.; Peng, K.-N.; Cheng, P.-J.; Chang, S.-W.; Lu, M.-Y.; Chu, C.W.; Guo, T.-F.; Atwater, H.A. Upconversion plasmonic lasing from an organolead trihalide perovskite nanocrystal with low threshold. *ACS Photonics* **2020**, *8*, 335–342. [CrossRef]
88. Lin, H.-C.; Lee, Y.-C.; Lin, C.-C.; Ho, Y.-L.; Xing, D.; Chen, M.-H.; Lin, B.-W.; Chen, L.-Y.; Chen, C.-W.; Delaunay, J.-J. Integration of On-Chip Perovskite Nanocrystal Laser and Long-Range Surface Plasmon Polariton Waveguide with Etching-Free Process. *Nanoscale* **2022**, *14*, 10075–10081. [CrossRef]
89. Li, S.; Yuan, M.; Zhuang, W.; Zhao, X.; Tie, S.; Xiang, J.; Lan, S. Optically-Controlled Quantum Size Effect in a Hybrid Nanocavity Composed of a Perovskite Nanoparticle and a Thin Gold Film. *Laser Photonics Rev.* **2021**, *15*, 2000480. [CrossRef]
90. Xing, D.; Lin, C.C.; Ho, Y.L.; Kamal, A.S.A.; Wang, I.T.; Chen, C.C.; Wen, C.Y.; Chen, C.W.; Delaunay, J.J. Self-Healing Lithographic Patterning of Perovskite Nanocrystals for Large-Area Single-Mode Laser Array. *Adv. Funct. Mater.* **2021**, *31*, 2006283. [CrossRef]

Review

Optical Crystals for 1.3 μm All-Solid-State Passively Q-Switched Laser

Yanxin Shen [1,2], Xinpeng Fu [1], Cong Yao [1,2], Wenyuan Li [1,2], Yubin Wang [1], Xinrui Zhao [1,2], Xihong Fu [1,*] and Yongqiang Ning [1]

[1] State Key Laboratory of Luminescence and Applications, Changchun Institute of Optics, Fine Mechanics and Physics, Chinese Academy of Sciences, Changchun 130033, China; shenyanxin20@163.com (Y.S.); fuxp@ciomp.ac.cn (X.F.); yaocong@ucas.ac.cn (C.Y.); liwenyuan@ucas.ac.cn (W.L.); wangyb@ciomp.ac.cn (Y.W.); zhaoxinrui@ucas.ac.cn (X.Z.); ningyq@ciomp.ac.cn (Y.N.)

[2] University of Chinese Academy of Sciences, Beijing 100049, China

* Correspondence: fuxh@ciomp.ac.cn

Abstract: In recent years, optical crystals for 1.3 μm all-solid-state passively Q-switched lasers have been widely studied due to their eye-safe band, atmospheric transmission characteristics, compactness, and low cost. They are widely used in the fields of high-precision laser radar, biomedical applications, and fine processing. In this review, we focus on three types of optical crystals used as the 1.3 μm laser gain media: neodymium-doped vanadate (Nd:YVO$_4$, Nd:GdVO$_4$, Nd:LuVO$_4$), neodymium-doped aluminum-containing garnet (Nd:YAG, Nd:LuAG), and neodymium-doped gallium-containing garnet (Nd:GGG, Nd:GAGG, Nd:LGGG). In addition, other crystals such as Nd:KGW, Nd:YAP, Nd:YLF, and Nd:LLF are also discussed. First, we introduce the properties of the abovementioned 1.3 μm laser crystals. Then, the recent advances in domestic and foreign research on these optical crystals are summarized. Finally, the future challenges and development trend of 1.3 μm laser crystals are proposed. We believe this review will provide a comprehensive understanding of the optical crystals for 1.3 μm all-solid-state passively Q-switched lasers.

Keywords: optical crystals; 1.3 μm laser; passively Q-switched laser; all-solid-state-laser; saturable absorber

1. Introduction

Q-switched technology is used to compress the laser energy to a narrow pulse to improve the peak power of the output laser beam. Passive Q-switched technology uses a saturable absorber (SA) as the Q-switched device to obtain output laser pulses. Since the emergence of the laser diode (LD) in the 1980s, diode-pumped solid-state laser (DPSSL) has developed rapidly owing to the achievement of narrow pulse width, high peak power, compact cavity structure, high efficiency, and low cost.

In recent years, laser radar has been extensively researched for its use in unmanned driving technology. According to the ANSI Z136.1—2014 standard, the allowable power of the 1.34 μm laser is 1.9 times that of the 1.5 μm and 18 times that of the 910 nm laser in the range of Class 1 power. Hence, the 1.3 μm laser radar can output greater power and realize remote eye-safe detection. In addition, due to the low loss and low dispersion characteristics of the 1.3 μm wavelength in the fiber, it has been widely used in the fields of communication and biosensing, for example, in the generation of non-classical optical field [1], spectral detection [2], and remote sensing [3]. Further, the 1.3 μm wavelength laser can be used as a light source to obtain a variety of wavelength lasers through nonlinear changes such as frequency doubling [4], frequency quadrupling [5], sum frequency generation [6], and Raman scattering [7]. Thus, the 1.3 μm passive Q-switched laser has immense application prospects in Figure 1.

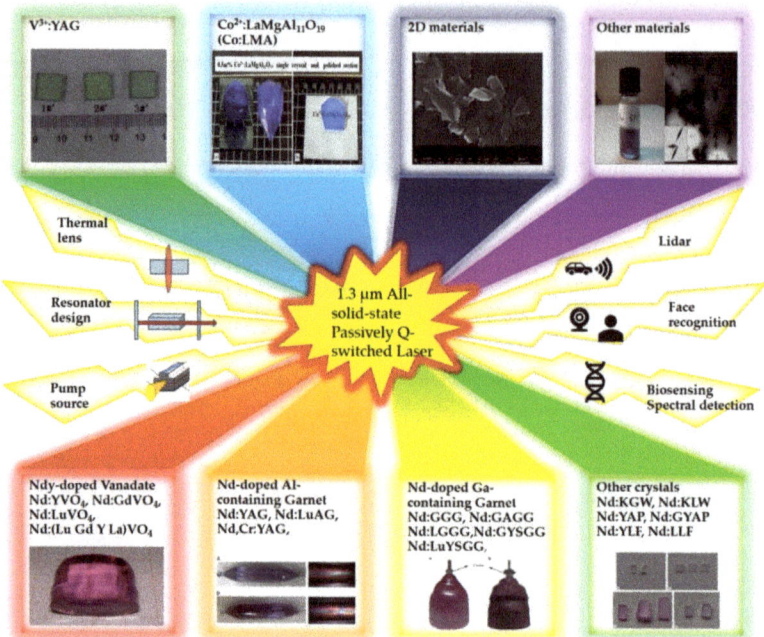

Figure 1. Composition and application of 1.3 µm passively Q-switched laser. V:YAG samples [8], Co^{2+}:LaMgAl$_{11}$O$_{19}$ crystal and polished section [9], SEM image of MoS$_2$-SA [10], Photograph of gold nanobipyramids solution and TEM image of the gold nanobipyramids [11], Nd:YVO$_4$ crystals samples [12], Nd:YAG crystals samples [13], Nd:GGG crystals samples [14], Nd:KGW crystals samples [15].

The laser gain medium as the core component of a solid-state laser is the basis for laser development. Nd^{3+} is the earliest applied doped ion, and its energy level structure is the decisive factor for the spectral characteristics of the gain medium. The substrate significantly affects the mechanical, physical, and chemical properties of the gain medium. Presently, crystal, ceramic, or glass is widely used as the substrate. The central wavelengths of radiation for these materials are generally 0.9 µm, 1.06 µm, and 1.3 µm, which are derived from three energy levels transitions of $^4F_{3/2}$-$^4I_{9/2}$, $^4F_{3/2}$-$^4I_{11/2}$, and $^4F_{3/2}$-$^4I_{13/2}$, respectively. The gain medium materials based on an LD pump must have the following characteristics: wide absorption peak, long fluorescence lifetime, large stimulated emission cross section, good mechanical properties, and high thermal conductivity. In this review, we discuss the 1.3 µm laser crystals, namely Nd:YVO$_4$ [16], Nd:GdVO$_4$ [17], Nd:YAG [18], Nd:GGG [19], Nd:KGW [20], Nd:YAP [21]. Among them, Nd:YVO$_4$, Nd:GdVO$_4$, and Nd:YAG are the major gain medium materials that can obtain high repetition rate and large output power. In 2015 Nikkinen et al. [22] reported a 1.3 µm Nd:YVO$_4$ microchip laser with a dilute nitride GaInNAs/GaAs saturable absorber mirror. The laser produced pulse as narrow as 204 ps with 2.3 MHz repetition rate. In 2015 Wang et al. [23] realized a high-peak-power (64.9 kW), short-pulse-width (6.16 ns) passively Q-switched Nd:YAG/V^{3+}:YAG laser at 1.3 µm. In 2019 Li et al. [24] simultaneously used both V^{3+}:YAG and MoSe$_2$ SA as passively Q-switche device. The pulse duration was 82.4 ns pulse at a repetition rate of 409.3 kHz. During the recent decades, researchers have created new optical crystals such as Nd:(Lu Gd Y La) VO$_4$ mixed crystal [25], Nd,Cr:YAG double-doped crystal [26,27], Nd:GYSGG crystal [28–30] and so on. In 2009 Huang et al. [31] investigated a diode-end-pumped passively Q-switched Nd:Gd$_{0.5}$Y$_{0.5}$VO$_4$ laser at 1.34 µm. For the passive Q-switching operation, the narrowest pulse width was 47.8 ns with 76 kHz repetition rate, with peak power estimated to be

182W, respectively. In 2011 Li et al. [32] realized passively Q-switched laser operation with a mixed c-cut Nd:Gd$_{0.33}$Lu$_{0.33}$Y$_{0.33}$VO$_4$ crystal at 1.34 µm. For passively Q-switched operation, the narrowest pulse width of 26 ns, the highest peak power of 1.8 kW were obtained using V:YAG as Q-switch. In 2016, Lin et al. [33] used Nd,Cr:YAG as gain medium and V^{3+}:YAG as SA to achieve dual-wavelength output (946 nm, 1.3 µm). The maximum average output power of 1.3 µm laser was 0.6 W, the narrowest pulse width was 19.2 ns at the highest repetition rate of 43.25 kHz. In 2017 Lin et al. employed a Co:MgAl$_2$O$_4$ crystal in a Nd:GYSGG passively Q-switched laser. The narrowest pulse width of 20.5 ns was achieved. The highest peak power was 1319 W under a pump power of 7.20 W, respectively. They provide the basis for further improving the output performance of the laser.

SA is considered an important part of a passively Q-switched laser. It utilizes the saturable absorption effect to modulate the loss in the laser cavity for realizing the Q-switching process. V^{3+}:YAG [8] and Co^{2+}:LaMgAl$_{11}$O$_{19}$ (Co:LMA) [9] are the most commonly used in the 1.3 µm band. Their ratios of the excited-state absorption cross section to the ground-state absorption cross section are approximately 0.1 and 0.2, respectively. Moreover, their ground-state recovery time is relatively short; hence, they are easily bleached. When these two materials were used in Q-switched devices, the pulse peak power was above 330 kW [34] and pulse width could reach 1 ns [35]. Further, the output repetition rate of 1820 kHz could be obtained [35]. In recent years, with the rapid development of new materials and nanotechnology, some new SA devices have emerged [36], such as graphene [37–45], black phosphorus [46,47], topological insulators (TI) [48,49], transition metal disulfides (TMDs) [10,24,50–56], gold nanomaterials [11,57], MXene [58–60], and so on. Most novel SA devices have been reported to achieve high-repetition-rate pulse output (>150 kHz) but large pulse width (>60 ns) and low peak power (<30 W). Owing to the development of SA materials, the performance of passively Q-switched lasers is expected to be further improved.

In this review, we first classify the 1.3 µm laser crystals and introduce their properties. Next, we focus on the research progress of different types of 1.3 µm passively Q-switched laser and reveal the development bottleneck for 1.3 µm laser crystals. In addition, we also introduce some new optical crystals and novel SA materials. Finally, we summarize the study and discuss the scope for future development of 1.3 µm laser crystals.

2. Classification of 1.3 µm Laser Crystals

2.1. Neodymium-Doped Vanadate

As the most popular 1.3 µm laser crystals, the Nd-doped vanadate crystals mainly include Nd:YVO$_4$, Nd:GdVO$_4$, Nd:LuVO$_4$, and Nd:Y$_x$Gd$_{1-x}$VO$_4$ (x = 0~1). Among them, Nd:YVO$_4$ and Nd:GdVO$_4$ have been widely researched.

Nd:YVO$_4$ is an excellent laser crystal with mature technology. It was first invented by O'Connor [61] of the MIT Lincoln Laboratory. It is a natural birefringence crystal with thermal conductivity of 5.2 Wm^{-1}K^{-1} and absorption bandwidth of approximately 20 nm. Since the stimulated emission cross section of Nd:YVO$_4$ at 1342 nm (1 at.%, 7.6×10^{-19} cm^2 is 18 times larger than that of Nd:YAG) is smaller than the ground-state absorption cross section of a saturated absorber (V^{3+}:YAG, 7.2×10^{-18} cm^2) and the upper-level lifetime is short (98 µs) [62], the Nd:YVO$_4$ laser can achieve high-repetition-rate pulses output. As early as 1976, Tuker et al. [63] realized a 1.3 µm continuous-wavelength output by end-pumping Nd:YVO$_4$ with an argon-ion laser. However, the slope efficiency was only 7%.

In 1997, Fluck et al. [64] proposed a diode-pumped 1.34-µm-wavelength passively Q-switched microchip laser, and an InGaAsP semiconductor SA mirror was used as a Q-switched device. When pumped at 400 mW, the pulse repetition rate was 53 kHz and pulse width was 230 ps, but the peak power was only 450 mW. In 2005, Lai et al. [65] used the InAs/GaAs quantum dot material as an SA. When the incident pump power was 2.2 W, the output pulse repetition rate of 770 kHz, pulse width of 90 ns, and peak power of above 5 W were obtained. In 2006, Janousek et al. [66] investigated passively synchronous dual-wavelength (1064 and 1342 nm) Q-switched lasers based on V^{3+}:YAG

SA. The schematic of this laser is shown in Figure 2. When the pump power was 3.5 W, the output laser repetition rate was 10 kHz, the pulse width was 70 ns, and the intracavity peak power was 3 kW. In 2020, Kane et al. [35] used a Nd:YVO$_4$ microchip as the gain medium, and employed V^{3+}:YAG and output coupler (OC) mirrors with different transmittances to conduct multiple sets of experiments. In one group of experiments, the repetition rate of the output pulse was 460 kHz, pulse duration was 1.6 ns, and peak power was approximately 500 W. In another group, the repetition rate of the output pulses was 24 kHz, pulse duration was 1.08 ns, and peak power was 2.3 kW. Although only the experimental data were reported by the authors and no detailed experimental results were presented, the study provided the basis for further realizing a 1.3 µm pulse laser with narrow pulse width, high peak power, high repetition rate, and good stability.

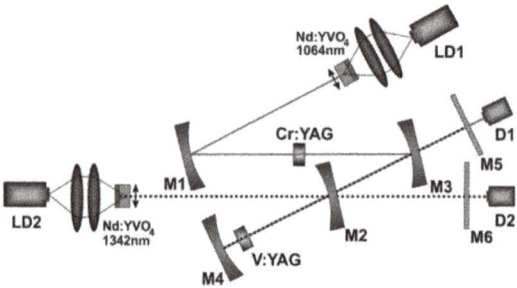

Figure 2. Setup for passively synchronized Q-switched Nd:YVO$_4$ lasers oscillating at 1064 and 1342 nm [66].

Nd:YVO$_4$ is a commonly used crystal, it is usually used to produce 1.3 µm wavelength laser. It has a five times higher absorption efficiency than that of Nd:YAG. The stimulated emission cross section of Nd:YVO$_4$ at 1342 nm is 18 times bigger than that of Nd:YAG at 1.3 µm, resulting in a more compact structure. Table 1 lists the research progress on the 1.3 µm Nd:YVO$_4$ laser. Using Nd:YVO$_4$ as the gain medium not only ensure a high repetition rate of several MHz, but also achieves a high peak power of several kW. In these results, V^{3+}:YAG is considered to be an ideal SA material for Nd:YVO$_4$ at 1.3 µm.

Nd:GdVO$_4$ is also a popular Nd-doped vanadate crystal which has been widely recognized as the gain medium of DPSSLs since its development by Zagumennyĭ et al. [17] in 1992. The structure of the Nd:GdVO$_4$ crystal is the same as that of the Nd:YVO$_4$ crystal, with a zircon structure and tetragonal system. The absorption half-width is 1.6 nm near 808 nm. The branching ratio of the 1.34 µm fluorescence spectrum to 1.06 µm is approximately 0.2, which can emit a 1.3 µm laser. It has a large stimulated emission cross section (c-cut 0.52 at.%, 1.8×10^{-19} cm^2@1342 nm), short upper-level lifetime (100 µs), and high thermal conductivity (11.7 Wm^{-1} K^{-1}), and achieves high doping concentration [25,78]. Therefore, high repetition rate and high-energy pulse output can be realized.

In 2007, Qi et al. [79] investigated an LD-pumped c-cut Nd:GdVO$_4$ crystal with a Co:LMA SA, lasing at 1.34 µm wavelength. The maximum repetition rate of the laser output was 277 kHz, shortest pulse width was 32 ns, output power was 266 mW, and maximum peak power was 187 W. In 2008, Ma et al. [34] compared the output characteristics of a-cut and c-cut Nd:GdVO$_4$ passively Q-switched lasers at 1342 nm. When the pump energy of the flash lamp was 27 J, the corresponding output laser pulse width of the two crystals were 61.72 ns and 53.94 ns, and the single pulse output energies were 15.5 mJ and 17.6 mJ, respectively. The corresponding peak powers were 247 kW and 330 kW. In 2011, Li et al. [80] simultaneously used V^{3+}:YAG and Co:LMA SA in the cavity to obtain a narrower pulse width and higher peak power. The schematic of this laser is shown in Figure 3. The corresponding output laser repetition frequencies were 49.8 kHz and 36 kHz, pulse widths were 16.9 ns and 11.3 ns, maximum average output powers were 0.319 W and 0.268 W, and peak powers were 378.2 W and 659 W.

Table 1. Research progress on 1.3 μm passively Q-switched Nd:YVO$_4$ lasers.

Year	C_{Nd}	SA	T_{OC}	P_{Ave} (W)	Pulse Width (ns)	Peak Power (W)	Repetition Rate (kHz)	Ref.
1997	3 at.%	InGaAsP	8.5%	0.0065	0.23	0.45	53	[64]
2003	1 at.%	PbS (T = 97%)	3%	0.012	110	-	295	[67]
		V^{3+}:YAG (T = 95%)	8%	0.023	200	-	250	
2005	2 at.%	InAs/GaAs	5%	0.013	13	150	7	[65]
2005	1 at.%	V^{3+}:YAG (T = 85%)	6%	0.36	90	>5	770	[62]
2006	1 at.%	V^{3+}:YAG (T = 90%)	7%	0.096	8.8	436	25	[66]
2006	0.5 at.%	InGaAsP	-	-	70	3000 (intra)	10	[68]
2007	0.27 at.%	Co^{2+}:LMA (T = 90%)	6%	0.16	19	220	38	[69]
2011	0.3 at.%	V^{3+}:YAG (T = 94%)	9.7%	0.58	42	346	40	[70]
2011	0.5 at.%	nc-Si/SiN$_x$ film	3%	0.9	54	180	89	[71]
2015	Microchip	GaInNAs/GaAs	8%	0.67	51	~592	22.2	[22]
2017	0.4 at.%	Graphene oxide	5%	0.024	0.204	-	2300	[44]
2018	YVO$_4$/Nd:YVO$_4$/YVO$_4$	MXene Ti$_3$C$_2$T$_x$	5%	0.52	329	7.39	214	[58]
2018	0.4 at.%	Antimonene	4%	0.03	454	0.406	162	[72]
2019	0.1 at.%	Bi:GaAs	5%	0.039	48.33	28.17	28.65	[73]
		GaAs	3.8%	0.435	64	48.7	138	
				0.405	282	~9	158	
2019	YVO$_4$/Nd:YVO$_4$/YVO$_4$ (0.3 at.%)	WS$_2$	12%	0.538	550	10.1	97	[55]
2019	YVO$_4$/Nd:YVO$_4$/Nd:YVO$_4$ 0 at.%/0.1 at.%/0.3 at.%	MoS$_2$	12%	1.1	140	23.8	330	[10]
2020	Microchip	V^{3+}:YAG (T = 97.5%)	4%	-	4.8	89	1820	[35]
		V^{3+}:YAG (T = 97.5%)	4%	-	6.4	144	680	
		V^{3+}:YAG (T = 95%)	4%	-	2.2	344	616	
		V^{3+}:YAG (T = 95%)	4%	-	3.6	383	295	
		V^{3+}:YAG (T = 90%)	4%	-	1.6	500	460	
		V^{3+}:YAG (T = 90%)	14%	-	1.6	2400	93	
		V^{3+}:YAG (T = 79%)	4%	-	1.3	2500	11	
		V^{3+}:YAG (T = 79%)	4%	-	1.08	2300	24	
2020	0.5 at.%	PtSe$_2$	5%	0.209	775	2.61	103.5	[56]
2020	0.2 at.%	GO-FONP	10%	0.306	163	5.98	314	[45]
2020	0.2 at.%	FONP	5%	0.14	767	1.56	116	[74]
2020	0.3 at.%	GaInSn	10%	0.425	32	1622	44	[75]
2020	0.3 at.%	Ti$_3$C$_2$(OH)$_2$/Ti$_3$C$_2$F$_2$	3%	0.48	390	6.25	195	[76]
2020	0.5 at.%	Mo$_2$C	5%	0.236	222	4.5	236	[59]
2020	0.5 at.%	Mo$_2$C	5%	0.293	313	10.04	93	[60]
2022	-	Ti$_2$C Mxene	-	0.215	190	7.75	146	[77]

C_{Nd}, Nd doping concentration; T_{OC}, transmission of output coupler mirror; P_{Ave}, average output power; GO-FONP, graphene oxide and ferroferric-oxide nanoparticle hybrid.

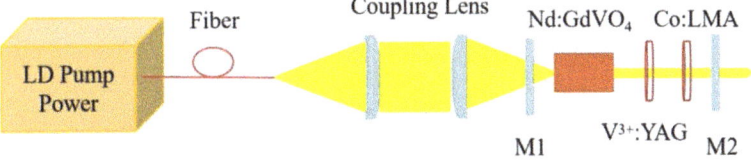

Figure 3. Experimental setup [80].

The Nd:GdVO$_4$ crystal has large stimulated emission cross section and short upper-level lifetime, which ensures high repetition rate, short pulse width, and peak power. Table 2 summarizes the research progress on the 1.3 μm Nd:GdVO$_4$ laser. Compared with the single SA, the pulse width was greatly reduced and peak power was increased by using double SA as the Q-switched device. However, additional losses were introduced, which decreased the output power. Composite crystals with different doping concentrations can enhance the absorption of pump light, thereby increasing the output power. V^{3+}:YAG and Co^{2+}:LMA crystals are ideal SAs for the Nd:GdVO$_4$ crystal. Although the use of two-dimensional materials such as bismuth quantum dots and TMDs as SAs yields high-repetition-rate output (>100 kHz), the pulse width (>80 ns) and peak power (<10 W) are not satisfactory.

Table 2. Research progress on 1.3 µm passively Q-switched Nd:GdVO$_4$ lasers.

Year	C_{Nd}	SA	T_{OC}	P_{Ave} (W)	Pulse Width (ns)	Peak Power (W)	Repetition Rate (kHz)	Ref.
2007	0.52 at.% c-cut	Co^{2+}:LMA (T = 90%)	5.5%	0.266	32	187	277	[79]
2008	0.52 at.% a-cut c-cut	V^{3+}:YAG (T = 54%)	10%	- -	61.72 53.9	247,000 330,000	-	[34]
2009	0.52 at.% a-cut c-cut	V^{3+}:YAG (T = 94%)	3% 10% 3% 10%	0.519 0.441 - -	- - 21.7 22.3	- - 307 316	- - 48.41 53.25	[81]
2010	0.5 at.% a-cut	V^{3+}:YAG (T = 96%)	15%	0.782 *	80	244	76.1 *	[82]
2011	0.5 at.% c-cut	Co^{2+}:LMA +V^{3+}:YAG (T = 94%) (T = 90%) +V^{3+}:YAG (T = 50%)	5%	0.319 0.268	16.9 11.3	378.2 659	49.8 36	[80]
2018	0.5 at.%	Au-NBPs (T = 90%)	4%	0.175	342	3.6	141.8	[11]
2018	0.3 at.% a-cut	Black phosphorus	8%	0.452	77	10.04	625	[47]
2018		1T-TiSe$_2$	15%	0.36	344	4.67	224	[51]
2019	c-cut composite crystal 0.1 at.%/0.3 at.%/0.8 at.%	MoSe$_2$ V^{3+}:YAG+MoSe$_2$ V^{3+}:YAG	3.8%	0.0526 0.1922 0.04	420 82.4 267	0.52565 5.6 -	238 409.3 -	[24]
2019	c-cut composite crystal 0.1 at.%/0.3 at.%/1 at.%	ZIF-67	3.8%	0.109	108	2.43	415	[83]
2019	0.5 at.%	BiQDs	5%	0.125	510	1.8	135	[84]
2020	c-cut composite crystal 0.1 at.%/0.3 at.%/0.8 at.%	BiQDs	3.8%	0.12	155	1.68	457	[85]
2020	-	ITO-NWAs	10%	0.32	296	4.69	230.2	[86]
2021		Co^{2+}:β-Ga$_2$O$_3$	3.8%	0.035	280	-	181	[87]
2021	0.3 at.%	α-Fe$_2$O$_3$ nanosheets	3.8%	0.114	180	1.8	358	[88]
2022	1 at.%	m-BiVO$_4$	3.8%	0.1153	355	1.35	242.6	[89]

C_{Nd}, Nd doping concentration; T_{OC}, transmission of output coupler mirror; P_{Ave}, average output power; Au-NBPs, gold nanobipyramids; ZIF-67, zeolitic imidazolate framework-67; BiQDs, bismuth quantum dots; ITO-NWAs, broadband indium tin oxide nanowire arrays. m-BiVO$_4$, monoclinic bismuth vanadate.

In 2002, Maunier et al. [90] obtained Nd:LuVO$_4$ by replacing yttrium with lutetium. The stimulated emission cross section of the c-cut Nd:LuVO$_4$ at 1.34 µm was 1.5×10^{-19} cm^2 (π polarization) and 1.9×10^{-19} cm^2 (σ polarization), with high thermal conductivity (9.77 Wm^{-1}K^{-1}) and small upper energy lifetime (95 µs) [90–92]. Liu et al. [93] reported a diode-pumped passively Q-switched Nd:LuVO$_4$ laser at 1.34 µm in 2008. The maximum output peak power of 820 W was attained with the pulse repetition rate of 22.4 kHz. In 2010, Liu et al. [94] used Co:LMA as the SA to obtain 534 kHz high-repetition-rate pulse output with an 8% transmission OC mirror.

Usually, a series of Nd:(Lu Gd Y La)VO$_4$ mixed crystals is grown by combining two ions, which is very suitable as the gain medium of Q-switched lasers owing to the long upper-energy-level life and small stimulated emission cross section [25]. For example, Nd:Y$_x$Gd$_{1-x}$VO$_4$ (x = 0~1) crystals were successfully grown by the Czochralski method. The thermal conductivity, specific heat capacity, and stimulated emission cross section of Nd:Y$_x$Gd$_{1-x}$VO$_4$ were different owing to the different crystal composition ratio, Nd ion doping, and cutting direction. For example, when x = 0.37, 0.63, the corresponding specific heat capacities of Nd:Y$_x$Gd$_{1-x}$VO$_4$ are 28.33 and 28.98 cal mol^{-1} K^{-1}, and the thermal conductivities are 4.88 Wm^{-1}K^{-1} and 5.04 Wm^{-1}K^{-1}, respectively [95]. Taking the Nd:Gd$_{0.5}$Y$_{0.5}$VO$_4$/V^{3+}:YAG laser as an example, the stimulated emission cross section of the gain medium is 1.0×10^{-19} cm^2, ground-state absorption cross section is 7.2×10^{-18} cm^2 at 1.3 µm, and V^{3+}:YAG ground-state absorption cross section is 7.2×10^{-18} cm^2; hence, the second threshold conditions are easily achieved [31]. Therefore, the Nd:(Lu Gd Y La) VO$_4$ crystals can achieve 1.3-µm-wavelength high-repetition-rate pulse laser output [31,32,96–99].

In 2010, Omatsu et al. [97] demonstrated an LD side-pumped bounce amplification laser with a slab of Nd:Gd$_{0.6}$Y$_{0.4}$VO$_4$ crystal as the gain medium and V:YAG as the SA. The maximum output power of 6.5 W and peak power of 0.87 kW were obtained at the pump power of 37 W, pulse laser repetition rate of 150 kHz, and pulse width of approximately 50 ns.

In 2011, Li et al. [100] investigated the laser performance with a mixed Nd:Lu$_{0.15}$Y$_{0.85}$VO$_4$ crystal at 1.34 µm wavelength. When V^{3+}:YAG T$_0$ = 89%, pulses with repetition rate of 42.5 kHz, minimum pulse width of 30.6 ns, and peak power of 268 W were obtained. When

$T_0 = 96\%$, the pulse repetition rate was 248 kHz, pulse width was 83.4 ns, and peak power was 21.6 W. In 2022, Cai et al. [101] prepared a tin disulfide saturable absorber. A stable passively Q-switched (PQS) Nd:Lu$_{0.15}$Y$_{0.85}$VO$_4$ 1.3 μm laser was successfully realized. It had a repetition rate of 1.18 MHz, the shortest pulse width of 34 ns, and a peak power of 20.8 W. In another work Cai et al. [102] successfully fabricated a nickel-cobalt layered double hydroxide SA, and it was used as a passively mode-locked modulator for the first time. It could obtain stable pulse sequence with a repetition frequency of 1.18 MHz and a narrowest pulse width of 52 ns, the corresponding peak power was 13.89 W. The experimental device is shown in Figure 4.

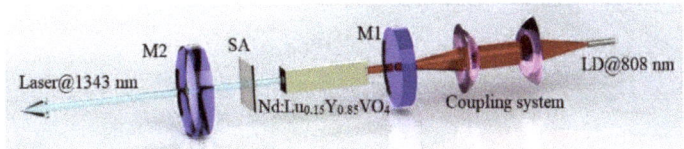

Figure 4. Passively Q-switched 1.34 μm laser experimental device with nickel-cobalt layered double hydroxide SA [102].

In 2013, Han et al. [103] proposed a Nd:La$_{0.05}$Lu$_{0.95}$VO$_4$ crystal as the gain medium to produce pulse laser with repetition rate of 33 kHz, average output power of 0.19 W, pulse width of 41 ns, and peak power of 199 W.

2.2. Neodymium-Doped Aluminum-Containing Garnet

Nd:YAG is a commonly used laser crystal. Geusic et al. [18] fabricated the Nd:YAG crystal output laser for the first time in 1964. The thermal conductivity of Nd:YAG is 14 Wm^{-1}K^{-1}, specific heat is 0.59 J/g K, stimulated emission cross sections are 0.8×10^{-19} cm^2 (1319 nm) and 0.9×10^{-19} cm^2 (1338 nm), Mohs hardness is 8.5, absorption line width is 4 nm, and fluorescence lifetime is 230 μs [104]. Hence, the Nd:YAG laser can achieve high repetition rate and peak power pulse output at 1.3 μm wavelength.

In 2006, Jabczynski et al. [105] used a 600 W diode-stack side-pumped triangular Nd:YAG slab laser to achieve pulse width of 6 ns, peak power of 125 kW, and passively Q-switched pulse output at 1.3 μm wavelength. The schematic is displayed in Figure 5. In 2011, Li et al. [106] realized the passive Q-switched output of Nd:YAG at 1319 nm with V^{3+}:YAG SA. When the transmittance of the OC mirror was 2.8%, the repetition rate of the output pulse laser was 230 kHz and minimum pulse width was 128 ns. In 2016 Lin et al. [107] pumped Nd:YAG with 885 nm LD to obtain dual-wavelength output laser at 1319 nm and 1338 nm. The shortest pulse widths are 20.20 ns and 20.86 ns, respectively. The maximum repetition rate is 64.10 kHz.

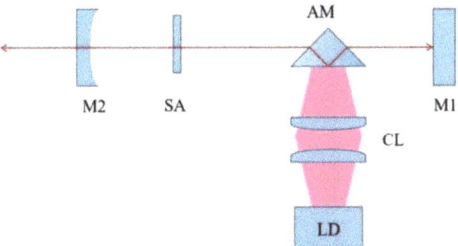

Figure 5. Arrangement of the laser resonator with side-pumped trigonal crystal (LD, fast-collimated pumping laser diode; CL, coupling lens; AM, triangular slab active medium; SA, V:YAG; M2, laser output coupler; M1, laser rear mirror).

In 2018, Zhou et al. [4] experimentally demonstrated a passively Q-switched red laser with an Nd^{3+}:YAG/YAG/V^{3+}:YAG/YAG composite crystal. This work first achieved

1327.6 nm laser and its second harmonic generation from the Nd^{3+}:YAG. In 2019, Lin et al. [52] prepared ReS$_2$ by liquid phase exfoliation method. For the first time, it was used as passively Q-switched devices at 1.3 µm wavelength. The repetition rate of the output pulse laser reached 214 kHz, pulse width was 403 ns, maximum average output power was 78 mW, and pulse peak power was 0.9 W. The schematic is displayed in Figure 6.

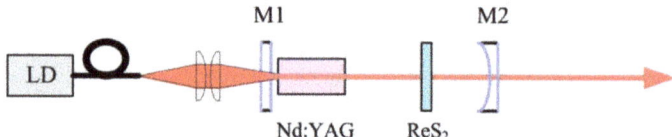

Figure 6. The schematic of Q-switched laser cavity based on ReS$_2$ SA. M1: plane input mirror; M2: concave output mirror with radius of −100 mm and 8% transmittance [52].

Table 3 lists the research progress on the 1.3 µm Nd:YAG laser, in which the combination of Nd:YAG and V^{3+}:YAG demonstrated high repetition rate, peak power, narrow pulse width, and passively Q-switched pulse output at 1.3 µm wavelength. When two-dimensional materials such as graphene and metal disulfide were used as the SA, the pulse width of the high-repetition-frequency pulse output laser was more than 100 ns and peak power was less than 20 W. In the experiment, a composite crystal such as YAG/Nd:YAG/V:YAG could shorten the cavity length and enhance the heat dissipation, thereby shortening the pulse width and improving the average output power and peak power. The Nd:YAG laser could also improve the output power through the dual-wavelength (1319, 1338 nm) output.

Table 3. Research progress on 1.3 µm passively Q-switched Nd:YAG lasers.

	Nd:YAG							
Year	C_{Nd}	SA	T_{OC}	P_{Ave} (W)	Pulse Width (ns)	Peak Power (W)	Repetition Rate (kHz)	Ref.
2003	-	V^{3+}:YAG (T = 95.7%)	15%	1.56	20	1200	60	[108]
		Double V^{3+}:YAG		0.96	20	2400	20	
	YAG/Nd:YAG (4 + 8)mm	V^{3+}:YAG, T = 89%	6%	0.215	21	3100	5.6	
		T = 93%	9%	0.525	30	4300	11	
2004	1 at.%		T = 91%	9%	0.365	25	4800	5.3	[109]
			14% 54 mm L_{cav}	0.5	19.7 ± 0.3	5000	5.2	
	YAG/Nd:YAG/VYAG (4 + 8 + 0.5) mm	V^{3+}:YAG (T = 88%)	9% 42 mm L_{cav}	0.7	14.7 ± 0.5	4600	10.4	
			18% 33 mm L_{cav}	0.43	11.0 ± 0.4	6100	6.4	
2006	Triangular Slab	V^{3+}:YAG (T = 95.7%)	25%	-	6	125,000	-	[105]
2007	1.1 at.%	YAG/Nd:YAG (4 + 8) mm Nd:YAG 4 mm	V^{3+}:YAG (T = 85%)	10%	0.6	6.2 ± 0.2	6000 ± 300	15	[110]
				0.25	1.7 ± 0.1	11,000 ± 2000	11	
2010	2 at.%	T = 89.6% (1319 nm) T = 89.7% (1338 nm) Co^{2+}:LMA	12.5% (1319 nm) 12% (1338 nm)	0.266	15	167	133	[104]
2011	0.6 at.%	V^{3+}:YAG (T = 92%)	2.8%	1.8	128	-	230	[106]
2013	0.6 at.%	Graphene Oxide	10%	0.82	2000	11.7	35	[39]
2015	1 at.%	V^{3+}:YAG (T = 88%)	2.5%	0.628	21	2100	15	[111]
2015	1.1 at%	V^{3+}:YAG, T = 90%	15%	3.3 mJ	-	-	77.5	[23]
		T = 85%	25%	2.4 mJ	6.16	64,900	34.1	
2016	1 at.%	Multilayer Graphene (T = 89%)	2.5%	0.34	380	139	209	[41]
2016	1 at.%	V^{3+}:YAG (T = 90%)	6%	1.97	20.2 1319 nm	-	64.1	[107]
				1.58	20.86 1338 nm	-		
2018	Nd^{3+}:YAG/YAG/V^{3+}:YAG/YAG 1 at.%	V^{3+}:YAG	2.8%	2.41	-	-	-	[4]
2018	1.1 at.%	Co:MgAl$_2$O$_4$ (T = 89.5%)	16%	0.48	18.3	1533	17.5	[112]
2019	1 at.%	ReS$_2$	8%	0.078 W	403	0.9	214	[52]
2019	-	ReS$_2$	-	0.101 W	111	2.95	308.4	[53]
2019	-	SnS$_2$	-	0.136 W	323	1.89	223	[54]

C_{Nd}, Nd doping concentration; T_{OC}, transmission of output coupler mirror; P_{Ave}, average output power; L_{cav}, cavity length.

Nd:Lu$_3$Al$_5$O$_{12}$ (Nd:LuAG) is an isostructure of Nd:YAG, which can be used to grow high-quality single crystals. It has high thermal conductivity (9.6 W·m^{-1}·K^{-1}), large absorption cross section (1.52 × 10^{-20} cm^2), and stimulated radiation cross section at 1.3 µm of 0.5 × 10^{-19} cm^2. The full width at half maximum (FWHM) of the absorption band (5 nm) and fluorescence lifetime (277 µs) are both greater than those of Nd:YAG, but the absorption coefficient is slightly lower, which is suitable for a high-repetition-rate and high-energy laser [113].

In 2015, Liu et al. [114] used V^{3+}:YAG SA to realize Nd:LuAG 1.3 µm passively Q-switched output with minimum pulse width of 17 ns and maximum single pulse energy of 18.9 µJ. In 2017, Wang et al. [50] realized a Nd:LuAG 99 kHz high-repetition-frequency passive Q-switched output based on MoS$_2$ SA.

2.3. Neodymium-Doped Gallium-Containing Garnet

Nd:Gd$_3$Ga$_5$O$_{12}$ (Nd:GGG) was first reported by Geusic et al. [18] in 1964. The stimulated emission cross section of Nd:GGG at 1331 nm is 3.9 × 10^{-20} cm^2, fluorescence lifetime is 240 µs, thermal conductivity is 6.4 W·m^{-1}·K^{-1}, specific heat capacity is 380 J·kg^{-1}·K^{-1}, and absorption linewidth is 4 nm. In addition, it also exhibits the advantage of yielding a large-size crystal [19], which promotes its wide application in solid-state heat capacity lasers.

In 2016, Han et al. [46] performed a passively Q-switched experiment with Nd:GGG crystal and black phosphorus SA. A pulse output with repetition rate of 175 kHz, pulse width of 363 ns, average output power of 157 mW, and peak power of 3 W was obtained.

In 2010, Zhang et al. [115] obtained a 1.33-µm-wavelength pulsed laser by using an LD end-pumped Nd:GAGG crystal. The maximum average output power was 450 mW with a 3% transmittance of the OC mirror. The maximum peak pulse power was 7.1 kW with an 8% transmittance of the OC mirror.

In 2021, Gao et al. [116] investigated passively Q-switched Nd:LGGG lasers at 1.3 µm, In the Q-switching regime, a repetition rate of 8 kHz with a minimum pulse duration of 9.75 ns was obtained, the corresponding peak power was 2.4 kW.

Nd:GGG crystals doped with ions such as Al^{3+}, Y^{3+}, and Lu^{3+} were fabricated, which could grow a variety of new crystals such as Nd:Gd$_3$Al$_x$Ga$_{5-x}$O$_{12}$ (x = 0.94) (Nd:GAGG) [117], Nd:(Lu$_x$Gd$_{1-x}$)$_3$Ga$_5$O$_{12}$ (Nd:LGGG) [118,119], Nd:Gd$_{3x}$Y$_{3(1-x)}$Sc$_2$Ga$_{3(1+\delta)}$O$_{12}$ (Nd:GYSGG) [28], and Nd:Lu$_3$Sc$_{1.5}$Ga$_{3.5}$O$_{12}$ (Nd:LuYSGG) [120]. They can be used as a 1.3 µm solid-state laser gain medium. Since the doped ions replace Gd^{3+} or Ga^{3+} in the crystals, the newly formed crystals have wider non-uniform spectrum broadening, smaller excitation cross section, and greater energy storage capacity. The research progress is detailed in Table 4.

Table 4. Research progress on 1.3 µm passively Q-switched neodymium-doped gallium garnet lasers.

Year	C$_{Nd}$	SA	T$_{OC}$	P$_{Ave}$ (W)	Pulse Width (ns)	Peak Power (W)	Repetition Rate (kHz)	Ref.
				Nd:Gd$_3$Ga$_5$O$_{12}$ (Nd:GGG)				
2009	1 at.%	Co^{2+}:LMA T = 90% T = 81%	8%	0.183 0.131	26.1 16.4	700 1300	- 6.1	[121]
2009	1 at.%	V^{3+}:YAG (T = 94%)	8%	0.46	19	650	39	[122]
2015	0.5 at.%	Graphene	2.2%	0.69	556	7.45	166.7	[40]
2016	0.5 at.%	Black Phosphorus	5%	0.157	363	3	175	[46]
				Nd:Gd$_3$Al$_x$Ga$_{5-x}$O$_{12}$ (Nd:GAGG)				
2010	0.74 at.%	V^{3+}:YAG (T = 94%)	8%	0.29	18.2	2000	8	[115]
2011	0.74 at.%	Co^{2+}:LMA (T = 90%)	8%	0.329	14.6	7100	3	[123]
				Nd:(Lu$_x$Gd$_{1-x}$)$_3$Ga$_5$O$_{12}$ (Nd:LGGG)				
2013	0.96 at.%	V^{3+}:YAG (T = 95%)	8%	0.75	25.9	1700	17.1	[124]
2021	1 at.%	V^{3+}:YAG (T = 90%)	5%	0.176	9.75	2400	8	[116]
				Nd:Gd$_{3x}$Y$_{3(1-x)}$Sc$_2$Ga$_{3(1+\delta)}$O$_{12}$ (Nd:GYSGG)				
2016	2 at.%	V^{3+}:YAG (T = 90%)	8.8%	0.251	23.9	954	11	[29]
2017	1 at.%	Co:MgAl$_2$O$_4$ (T = 82%)	12%	0.225	20.5	1319 *	9.1	[30]
				Nd:Lu$_3$Sc$_{1.5}$Ga$_{3.5}$O$_{12}$ (Nd:LuYSGG)				
2019	1 at.%	V^{3+}:YAG (T = 90%)	5% 10%	0.39 0.34	- 20.8	- 428	41.6 38.2	[125]
2020	1 at.%	Bi$_2$Se$_3$ (T = 75%)	5%	0.36	146	7.05	349.5	[49]

C$_{Nd}$, Nd doping concentration; T$_{OC}$, transmission of output coupler mirror; P$_{Ave}$, average output power.

2.4. The Other Types of Optical Crystals

In 1998, Demidovich et al. [20,126] demonstrated that LD-pumped Nd:KGd(WO$_4$)$_2$ (Nd:KGW) can produce a continuous-wavelength laser output at 1.35 µm. The stimulated emission cross section of Nd:KGW at 1351 nm is 0.9×10^{-19} cm^2, and fluorescence lifetime is 98 µs. In 2003, Savitski et al. [67] used V^{3+}:YAG and PbS-doped glass as the SA and conducted comparative experiments. When the pump power was 47 mW, the output laser repetition rate reached 170 kHz, pulse width was 270 ns, and output power was 1 mW. With the V^{3+}:YAG SA, the laser repetition rate was 100 kHz, pulse width was 250 ns, and output power was 5 mW. However, the output pulse width of this type of gain medium in the 1.3 µm band was large and peak power was low.

Nd:YAP has a large emission cross section (1.8×10^{-19} cm^2) at 1341 nm, long fluorescence lifetime (170 µs), and high thermal conductivity (11 W·m^{-1}·K^{-1}). It is an excellent solid-state laser gain medium. In 2015, Chen et al. [127] demonstrated a diode-side-pumped passively Q-switched Nd:YAP laser operating at 1.34 µm with V^{3+}:YAG SA. When the pump current was 34 A, the repetition rate of 5.51 kHz, pulse width of 197 ns, and maximum peak power of 6.93 kW were obtained. In 2016 Xu et al. [43] demonstrated a single and multi-wavelength Nd:YAP laser. The diode-pumped continuous-wave laser operated at 1364 nm by using a 0.08 mm glass etalon, and dual-wavelength laser was also achieved at 1328 and 1340 nm as well as at 1340 and 1364 nm. Replacing the etalons with Go SA, stable Q-switched laser operated at 1339 nm, the repetition rate of 76.9 kHz, pulse width of 380 ns, and maximum peak power of 14.5 W were obtained. The schematic of this laser is shown in Figure 7.

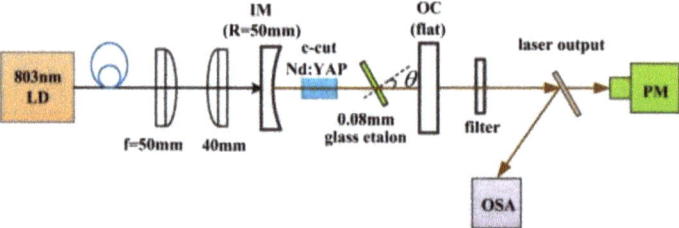

Figure 7. Set-up used for the LD-pumped Nd:YAP continuous-wave laser experiments. OSA, optical spectrum analyzer; PM, power meter; OC, output coupler [43].

Nd:LiYF$_4$(Nd:YLF) is widely used, as it is suitable for lasers with different structures and pumping modes. Nd:YLF is a natural birefringence crystal, which has a long upper-level life ($\sim$ 520 µs) and small emission cross section ($\sim$2–2.5 $\times$ 10^{-20} cm^2; two polarizations). Hence, it possesses large energy storage capacity. In addition, owing to its high thermal conductivity (6 W·m^{-1}·K^{-1}) and negative thermal lens effect dn/dT (-4.3×10^{-6} π polarization, -2×10^{-6} σ polarization), it reduces the effect of the positive thermal lens. Moreover, it also has some advantages such as high crystal quality [128,129]. In 2013, Botha et al. [130] realized the Q-switched output of a Nd:YLF laser with maximum peak power of 6.1 kW at 1314 nm. In 2015, Xu et al. [48] reported a 1.3 µm passively Q-switched Nd:YLF laser by using few-layer TI Bi$_2$Se$_3$ as the SA. They obtained a pulse repetition rate of 161.3 kHz, shortest pulse width of 433 ns, and pulse energy of approximately 1.23 µJ.

Nd:LuLiF$_4$ (Nd:LLF) is an isostructure of Nd:YLF, which can also be used as a solid gain medium at the 1.3 µm band. Compared with Nd:YLF, it has larger emission cross section (5.1 $\times$ 10^{-20} π polarization, 2.2 $\times$ 10^{-20} σ polarization) and similar fluorescence lifetime (489 µs). In 2013, Li et al. [131] reported a dual-wavelength (1314 nm and 1321 nm) output of the Nd:LLF laser. When the repetition rate of the Q-switched pulse was 17.2 kHz, the peak power of 885 W was obtained. In 2019, Qian et al. [57] obtained a 227 kHz high-repetition-rate pulse output by using gold nanorods (GNRs) with aspect ratio of 8 as the SA.

Several other types of optical crystals can also emit a 1.3 μm laser, such as: Nd:Lu$_2$O$_3$ [132,133] and Nd,Cr:YAG [33,134]. Here, we introduce only the main types. Table 5 presents the details of the research on the Nd:KGW, Nd:YAP, Nd:YLF, and Nd:LLF lasers.

Table 5. Research progress on 1.3 μm passively Q-switched Nd:KGW, Nd:YAP, Nd:YLF, and Nd:LLF lasers.

Year	C_{Nd}	SA	T_{OC}	P_{Ave} (W)	Pulse Width (ns)	Peak Power (W)	Repetition Rate (kHz)	Ref.
		Nd:KGd(WO$_4$)$_2$ (Nd:KGW)						
2001	4 at.%	V^{3+}:YAG (T = 96.5%)	2%	0.087	78	40	27.6	[135]
2003	7 at.%	PbS (T = 97%)	4%	0.001	270	-	170	[67]
		V^{3+}:YAG (T = 95%)		0.005	250	-	100	
2006	3 at.%	PbS (T = 98%)	4%	0.012	50	8	31	[136]
		V^{3+}:YAG (T = 98%)		0.042	95	7.6	58	
		Nd:KLu(WO$_4$)$_2$ (Nd:KLW)						
2012	0.5 at.%	Graphene on SiC (T = 80%)	8%	0.89	466	-	135	[38]
		Nd:YAlO$_3$ (Nd:YAP)						
2006	Triangular Slab	V^{3+}:YAG (T = 95.7%)	25%	5.7	77,000	-	-	[105]
2015	0.9 at.%	V^{3+}:YAG (T = 93%)	7%	7.52	197	6930	5.51	[127]
2016	0.8 at.%	Graphene Oxide;	2.3%	0.43	380	14.7	76.9	[43]
		Nd:GdAlO$_3$ (Nd:GYAP)						
2022	1 at.%	Bi nanosheets	5%	0.2	361	1.25	365	[137]
		Nd:LiYF$_4$ (Nd:YLF)						
2013	0.5 at.%	V^{3+}:YAG (T = 97%)	5%	5.2	135	6100	6.3	[130]
2015	1 at.%	Bi$_2$Se$_3$ (T = 95.1%)	2.3%	0.2	433	2.84	161.3	[48]
		Nd:LuLiF$_4$ (Nd:LLF)						
2013	1 at.%	V^{3+}:YAG (T = 80%)	3%	1.87	120	885	17.2	[131]
2016	1 at.%	Monolayer Graphene	3.8%	1.03	133	84.9	91	[42]
			8%	1.33	155	111.6	77	
2017	1 at.%	g-C$_3$N$_4$	3.8%	0.96	275	-	154	[138]
2019	1 at.%	GNRs aspect ratio 5	3.8%	1.432	328	-	200	[57]
			8%	1.173	460	-	205	
		GNRs aspect ratio 8	3.8%	1.209	271	-	218	
			8%	1.247	438	-	227	

C_{Nd}, Nd doping concentration; T_{OC}, transmission of output coupler mirror; P_{Ave}, average output power; g-C$_3$N$_4$, two-dimensional (2D) graphite carbon nitride; GNRs, gold nanorods.

3. Conclusions

3.1. Summary

This review discussed 1.3 μm laser crystals systematically. In recent years, the highest repetition rate of 1.3 μm passively Q-switched laser has exceeded MHz, and highest peak power of 70 kW has been achieved. Although researchers have made great progress in 1.3 μm passively Q-switched laser, there are also some factors limit the laser performance, such as, the low-gain emission line at 1.3 μm, the heat accumulation of crystals, the stability at high repetition rate, and so on. We make a summary of 1.3 μm passively Q-switched laser, and provide some research perspectives.

- The peak value of the gain medium at 1.06 μm is much higher than that at 1.3 μm in the fluorescence spectrum, and the transition probability at $^4F_{3/2}$-$^4I_{11/2}$ is greater than that at $^4F_{3/2}$-$^4I_{13/2}$. Hence, it is necessary to suppress the 1.06 μm wavelength oscillation in the cavity to obtain 1.3 μm output light.
- When the stimulated emission cross section of the gain medium is large, the threshold can be reduced and laser oscillation can be easily realized, whereas a small stimulated emission cross section of the gain medium can improve the energy storage capacity. Long fluorescence lifetime can increase the accumulation of upper-level particles and obtain larger energy storage, whereas short fluorescence lifetime is beneficial in obtaining a stable high-repetition-rate pulse output. Therefore, further in-depth research is required on the gain medium materials.
- Owing to the differences in the band gap, nonlinear absorption, and saturated absorption of the SA materials, the ground-state absorption cross section, excited-state absorption loss, modulation depth, and damage threshold are different.
- The resonator design (flat–flat, plane–concave, Z-cavity, V-cavity) affects the laser performance. The flat–flat cavity has the advantages of good directivity and large

mode volume, and it is easy to obtain single-mode oscillations with this cavity. The Z-shaped cavity can not only adjust the focusing position and mode matching, but also limit the output beam astigmatism with a smaller folding angle. V-cavity can adjust the mode matching of the pump light, prevent the SA from absorbing the residual pump energy. The plane–concave cavity can improve the effective area ratio between of the gain medium and the SA, and achieves a compact structure while meeting the second threshold condition.
- The pump source's power, center wavelength and mode matching influence the output power.
- Pulse fluctuations of the passively Q-switched laser, caused by the thermal lens effect also influence the output power.

3.2. Outlook

In view of the existing problems, researchers have put forward the following solutions from different perspectives. These improvement measures effectively accelerate and promote the development of 1.3 μm passively Q-switched laser. So, they also represent the current research trend.

- New crystals of better quality: researchers have continuously developed new crystals, such as Nd:GYSGG, Nd: (Lu Gd Y La) VO$_4$ mixed crystals, and Nd,Cr:YAG, etc. These crystals not only improve the performance, but can also be applied to some special fields due to their unique properties.
- V^{3+}:YAG and Co:LMA are popular 1.3 μm wavelength Q-switched devices. They have great absorption of the 1.3 μm wavelength and can be easily bleached. Hence, they achieve good experimental results (pulse width of several ns, peak power approaching the order of MW). During the past few decades, many new SA materials have been used as Q-switch devices such as graphene, black phosphorus, gold nanomaterials, and MXene. These materials can obtain high-repetition-rate pulse output (several hundreds of kHz), but their peak power is very low (few tens of W).
- Optimization of resonant cavity: selecting the appropriate cavity type and device can reduce the unsaturated absorption loss in the cavity, thereby improving the output light quality. For example, the combination of SA and coupling output mirror transmittance can affect the repetition frequency, pulse width, and power of the output pulses. The laser output power can be improved through lamp pumping, multi-LD side pumping, slab gain medium, and multi-wavelength output. The pulse width can be compressed by using double SA, composite crystal, and mode-locking.
- Reducing the thermal effect: the thermal effect of high-thermal-conductivity crystals and composite crystals can be reduced by using a thermoelectric cooler for controlling the crystal temperature.
- Reducing timing and amplitude jitter: researchers have proposed various methods to reduce the pulse jitter, such as external modulation, pulsed LD pump source, self-injection seeding, and pre-pumping mechanism.

Author Contributions: Conceptualization, X.F. (Xihong Fu) and X.F. (Xinpeng Fu); methodology, X.F. (Xihong Fu) and X.F. (Xinpeng Fu); investigation, Y.S.; resources, X.F. (Xihong Fu) and Y.N.; data curation, Y.S.; writing—original draft preparation, Y.S.; writing—review and editing, Y.S., X.F. (Xihong Fu) and X.F. (Xinpeng Fu); visualization, C.Y., W.L. and X.Z.; supervision, X.F. (Xihong Fu) and X.F. (Xinpeng Fu); project administration, X.F. (Xihong Fu) and Y.N.; funding acquisition, X.F. (Xihong Fu) and Y.W. All authors have read and agreed to the published version of the manuscript.

Funding: This research was funded by Science and Technology Development Project of Jilin Province (grant numbers 20200401060GX, 20210201028GX, 20200501008GX).

Institutional Review Board Statement: Not applicable.

Informed Consent Statement: Not applicable.

Data Availability Statement: Not applicable.

Conflicts of Interest: The authors declare no conflict of interest.

References

1. Zheng, Y.H.; Wu, Z.Q.; Huo, M.R.; Zhou, H.J. Generation of a continuous-wave squeezed vacuum state at 1.3 μm by employing a home-made all-solid-state laser as pump source. *Chin. Phys.* **2013**, *22*, 431–434. [CrossRef]
2. Zhu, W.; Liu, Q.; Wu, Y. Aerosol absorption measurement at SWIR with water vapor interference using a differential photoacoustic spectrometer. *Opt. Express* **2015**, *23*, 23108–23116. [CrossRef]
3. Sorokin, E.; Naumov, S.; Sorokina, I.T. Ultrabroadband infrared solid-state lasers. *IEEE J. Sel. Top. Quantum Electron.* **2005**, *11*, 690–712. [CrossRef]
4. Zhou, H.Q.; Bi, X.L.; Zhu, S.Q.; Li, Z.; Yin, H.; Zhang, P.X.; Chen, Z.Q.; Lv, Q.T. Multi-wavelength passively Q-switched red lasers with Nd^{3+}: YAG/YAG/V^{3+}: YAG/YAG composite crystal. *Opt. Quantum Electron.* **2018**, *50*, 56. [CrossRef]
5. Tu, W.; Shang, L.Q.; Dai, S.B.; Zong, N.; Wang, Z.M.; Zhang, F.F.; Chen, Y.; Liu, K.; Zhang, S.J.; Yang, F.; et al. 0.95 W high-repetition-rate, picosecond 335 nm laser based on a frequency quadrupled, diode-pumped $Nd:YVO_4$ MOPA system. *Appl. Opt.* **2015**, *54*, 6182–6185. [CrossRef]
6. Li, Y.L.; Yao, J.B.; Zeng, Y.H.; Zhang, Y.L. Sum-frequency yellow light laser. *J. Changchun Univ. Sci. Technol. (Nat. Sci. Ed.)* **2010**, *33*, 51–53.
7. Niu, X. Extracavity all-solid-state eye-safe Raman laser based on $Ba(NO_3)_2$ crystal pumped by 1342nm laser. Master's Thesis, Beijing University of Technology, Beijing, China, 2016.
8. Zhang, D.; Su, L.; Li, H.; Qian, X.; Xu, J. Characteristics and optical spectra of V:YAG crystal grown in reducing atmosphere. *J. Cryst. Growth* **2006**, *294*, 437–441. [CrossRef]
9. Ge, W.; Zhang, H.; Wang, J.; Ran, D.; Sun, S.; Xia, H.; Liu, J.; Xu, X.; Hu, X.; Jiang, M. Growth and thermal properties of Co^{2+}:$LaMgAl_{11}O_{19}$ crystal. *J. Cryst. Growth* **2005**, *282*, 320–329. [CrossRef]
10. Zhang, G.; Wang, Y.; Wang, J.; Jiao, Z. Passively Q-switched laser at 1.34 μm using a molybdenum disulfide saturable absorber. *Infrared Phys. Technol.* **2019**, *96*, 311–315. [CrossRef]
11. Chu, Z.; Zhang, H.; Wu, Y.; Zhang, C.; Liu, J.; Yang, J. Passively Q-switched laser based on gold nanobipyramids as saturable absorbers in the 1.3 μm region. *Opt. Commun.* **2018**, *406*, 209–213. [CrossRef]
12. Zhang, G.; Tu, H.; Liu, Y.; Hu, Z. Heat treatment and optical absorption studies on $Nd:YVO_4$ crystal. *J. Cryst. Growth* **2009**, *311*, 912–915. [CrossRef]
13. Saeedi, H.; Yadegari, M.; Enayati, S.; Asadian, M.; Shojaee, M.; Khodaei, Y.; Mirzaei, N.; Mashayekhi Asl, I. Thermal shocks influence on the growth process and optical quality of Nd: YAG crystal. *J. Cryst. Growth* **2013**, *363*, 171–175. [CrossRef]
14. Asadian, M.; Hajiesmaeilbaigi, F.; Mirzaei, N.; Saeedi, H.; Khodaei, Y.; Enayati, S. Composition and dissociation processes analysis in crystal growth of Nd:GGG by the Czochralski method. *J. Cryst. Growth* **2010**, *312*, 1645–1650. [CrossRef]
15. Senthil Kumaran, A.; Moorthy Babu, S.; Ganesamoorthy, S.; Bhaumik, I.; Karnal, A.K. Crystal growth and characterization of $KY(WO_4)_2$ and $KGd(WO_4)_2$ for laser applications. *J. Cryst. Growth* **2006**, *292*, 368–372. [CrossRef]
16. Bowkett, G.C.; Baxter, G.W.; Booth, D.J.; Taira, T.; Teranishi, H.; Kobayashi, T. Single-mode 1.34-μm $Nd:YVO_4$ microchip laser with cw Ti:sapphire and diode-laser pumping. *Opt. Lett.* **1994**, *19*, 957–959. [CrossRef]
17. Zagumennyĭ, A.I.; Ostroumov, V.G.; Shcherbakov, I.A.; Jensen, T.; Meyen, J.P.; Huber, G. The $Nd:GdVO_4$ crystal: A new material for diode-pumped lasers. *Sov. J. Quantum Electron.* **1992**, *22*, 1071. [CrossRef]
18. Geusic, J.E.; Marcos, H.M.; Van Uitert, L. Laser oscillations in Nd-doped yttrium aluminum, yttrium gallium and gadolinium garnets. *Appl. Phys. Lett.* **1964**, *4*, 182–184. [CrossRef]
19. Yoshida, K.; Yoshida, H.; Kato, Y. Characterization of high average power Nd:GGG slab lasers. *IEEE J. Quantum Electron.* **1988**, *24*, 1188–1192. [CrossRef]
20. Demidovich, A.A.; Kuzmin, A.N.; Ryabtsev, G.I.; Strek, W.; Titov, A.N. A 1.35 μm laser diode pumped continuous wave KGW:Nd laser. *Spectrochim. Acta Part A Mol. Biomol. Spectrosc.* **1998**, *54*, 1711–1713. [CrossRef]
21. Helena, J.; Pavel, C.; Jan, S.; Jan Karol, J.; Krzysztof, K.; Waldemar, Z.; Zygmunt, M.; Mitsunobu, M. In Passively mode-locked Q-switched Nd:YAP 1.34-um/1.08-um laser with efficient hollow-waveguide radiation delivery. In *Solid State Lasers XI: High-Power Lasers and Applications*; SPIE: San Jose, CA, USA, 2002.
22. Nikkinen, J.; Korpijärvi, V.M.; Leino, I.; Härkönen, A.; Guina, M. Microchip laser Q-switched with GaInNAs/GaAs SESAM emitting 204 ps pulses at 1342 nm. *Electron. Lett.* **2015**, *51*, 850–852. [CrossRef]
23. Wang, Z.; Zhang, B.; Ning, J.; Zhang, X.; Su, X.; Zhao, R. High-peak-power passively Q-switched 1.3 μm $Nd:YAG/V^{3+}$:YAG laser pumped by a pulsed laser diode. *Chin. Opt. Lett.* **2015**, *13*, 021403–21406. [CrossRef]
24. Dong, L.; Li, D.; Pan, H.; Li, Y.; Zhao, S.; Li, G.; Chu, H. Pulse characteristics from a $MoSe_2$ Q-switched $Nd:GdVO_4$ laser at 1.3 μm. *Appl. Opt.* **2019**, *58*, 8194–8199. [CrossRef]
25. Li, X. Research on 1.3 μm short pulse laser characteristics of LD-pumped Nd-doped vanadate crystals. Ph.D. Thesis, Shan Dong University, Jinan, China, April 2012.
26. Wang, S.; Li, Q.; Du, S.; Zhang, Q.; Shi, Y.; Xing, J.; Zhang, D.; Feng, B.; Zhang, Z.; Zhang, S. Self-Q-switched and mode-locked Nd,Cr:YAG laser with 6.52-W average output power. *Opt. Commun.* **2007**, *277*, 130–133. [CrossRef]
27. Li, J.; Wu, Y.S.; Pan, Y.B.; Zhu, Y.; Guo, J.K. Spectral properties of Nd,Cr:YAG self-Q-switched laser transparent ceramics. *Chin. J. Lumin.* **2007**, 219–224.

28. Zhong, K.; Sun, C.L.; Yao, J.Q.; Xu, D.G.; Pei, Y.Q.; Zhang, Q.L.; Luo, J.Q.; Sun, D.L.; Yin, S.T. Continuous-wave Nd:GYSGG laser around 1.3 μm. *Laser Phys. Lett.* **2012**, *9*, 491–495. [CrossRef]
29. Song, T.; Li, P.; Chen, X.; Ma, B.; Dun, Y. Passively Q-switched Nd:GYSGG laser operating at 1.3 μm with V:YAG as saturable absorber. *Optik* **2016**, *127*, 10621–10625. [CrossRef]
30. Lin, H.-Y.; Sun, D.; Copner, N.; Zhu, W.-Z. Nd:GYSGG laser at 1331.6 nm passively Q-switched by a Co:MgAl$_2$O$_4$ crystal. *Opt. Mater.* **2017**, *69*, 250–253. [CrossRef]
31. Huang, H.-T.; Zhang, B.-T.; He, J.-L.; Yang, J.-F.; Xu, J.-L.; Yang, X.-Q.; Zuo, C.-H.; Zhao, S. Diode-pumped passively Q-switched Nd:Gd$_{0.5}$Y$_{0.5}$VO$_4$ laser at 1.34 μm with V^{3+}:YAG as the saturable absorber. *Opt. Express* **2009**, *17*, 6946–6951. [CrossRef]
32. Li, X.; Li, G.Q.; Zhao, S.Z.; Zhao, B.; Li, Y.F.; Yin, L. CW and passively Q-switched laser performance of a mixed c-cut Nd:Gd$_{0.33}$Lu$_{0.33}$Y$_{0.33}$VO$_4$ crystal operating at 1.34 μm. *Opt. Mater.* **2011**, *34*, 159–163. [CrossRef]
33. Lin, B.; Zhang, Q.-L.; Zhang, D.-X.; Feng, B.-H.; He, J.-L.; Zhang, J.-Y. Passively Q-Switched Nd,Cr:YAG Laser Simultaneous Dual-Wavelength Operation at 946 nm and 1.3 μm. *Chin. Phys. Lett.* **2016**, *33*, 074203. [CrossRef]
34. Ma, J.S.; Li, Y.F.; Sun, Y.M.; Qi, H.J.; Lan, R.J.; Hou, X.Y. Passively Q-switched 1.34 μm Nd:GdVO$_4$ laser with V:YAG saturable absorber. *Laser Phys. Lett.* **2008**, *5*, 593. [CrossRef]
35. Kane, T.J.; Clarkson, W.A.; Shori, R.K. 1.34 μm Nd:YVO$_4$ laser passively Q-switched by V:YAG and optimized for lidar. In *Solid State Lasers XXIX: Technology and Devices*; SPIE: San Francisco, CA, USA, 2020; Volume 11259.
36. Yang, F.; Li, M.; Zhao, S.; Fu, X.H.; Gao, L.L. Research progress on passively Q-switched lasers based on new saturable absorption devices. *Laser Optoelectron. Prog.* **2020**, *57*, 15.
37. Liang, L.; Lin, Z.H.; Chen, S.; Wang, J.X. Graphene passively Q-switching for dual-wavelength lasers at 1064 nm and 1342 nm in Nd:YVO$_4$ laser. *Chin. J. Laser* **2014**, *41*, 53–56. [CrossRef]
38. Shen, H.; Wang, Q.; Zhang, X.; Liu, Z.; Bai, F.; Cong, Z.; Chen, X.; Gao, L.; Zhang, H.; Xu, X.; et al. Passively-switched Nd:KLu(WO$_4$)$_2$ laser at 1355 nm with graphene on SiC as saturable absorber. *Appl. Phys. Express* **2012**, *5*, 092703. [CrossRef]
39. Zhang, L.; Yu, H.; Yan, S.; Zhao, W.; Sun, W.; Yang, Y.; Wang, L.; Hou, W.; Lin, X.; Wang, Y.; et al. A 1319 nm diode-side-pumped Nd:YAG laser Q-switched with graphene oxide. *J. Mod. Opt.* **2013**, *60*, 1287–1289. [CrossRef]
40. Xu, B.; Wang, Y.; Cheng, Y.; Yang, H.; Xu, H.; Cai, Z. Nanosecond pulse generation in a passively Q-switched Nd:GGG laser at 1331 nm by CVD graphene saturable absorber. *J. Opt.* **2015**, *17*, 105501. [CrossRef]
41. Feng, C.; Zhang, H.; Wang, Q.; Xu, S.; Fang, J. 1357 nm passively Q-switched crystalline ceramic laser based on multilayer graphene. *Laser Phy.* **2016**, *26*, 055802. [CrossRef]
42. Li, S.; Li, T.; Zhao, S.; Li, G.; Hang, Y.; Zhang, P. 1.31 and 1.32 μm dual-wavelength Nd:LuLiF$_4$ laser. *Opt. Laser Technol.* **2016**, *81*, 14–17.
43. Xu, B.; Wang, Y.; Lin, Z.; Peng, J.; Cheng, Y.; Luo, Z.; Xu, H.; Cai, Z.; Weng, J.; Moncorgé, R. Single- and multi-wavelength Nd:YAlO$_3$ lasers at 1328, 1339 and 1364 nm. *Opt. Laser Technol.* **2016**, *81*, 1–6. [CrossRef]
44. Lin, H.Y.; Zhao, M.J.; Lin, H.J.; Wang, Y.P. Graphene-oxide as saturable absorber for a 1342 nm Q-switched Nd:YVO4 laser. *Optik* **2017**, *135*, 129–133. [CrossRef]
45. Zhang, H.; Peng, J.; Yang, X.; Ma, C.; Zhao, Q.; Chen, G.; Su, X.; Li, D.; Zheng, Y. Passively Q-switched Nd:YVO$_4$ laser operating at 1.3 μm with a graphene oxide and ferroferric-oxide nanoparticle hybrid as a saturable absorber. *Appl. Opt.* **2020**, *59*, 1741–1745. [CrossRef]
46. Han, S.; Zhang, F.; Wang, M.; Wang, L.; Zhou, Y.; Wang, Z.; Xu, X. Black phosphorus based saturable absorber for Nd-ion doped pulsed solid state laser operation. *Indian J. Phys.* **2016**, *91*, 439–443. [CrossRef]
47. Sun, X.; Nie, H.; He, J.; Zhao, R.; Su, X.; Wang, Y.; Zhang, B.; Wang, R.; Yang, K. Passively Q-Switched Nd:GdVO$_4$ 1.3 μm laser with few-layered black phosphorus saturable absorber. *IEEE J. Sel. Top. Quantum Electron.* **2018**, *24*, 1600405. [CrossRef]
48. Xu, B.; Wang, Y.; Peng, J.; Luo, Z.; Xu, H.; Cai, Z.; Weng, J. Topological insulator Bi$_2$Se$_3$ based Q-switched Nd:LiYF$_4$ nanosecond laser at 1313 nm. *Opt Express* **2015**, *23*, 7674–7680. [CrossRef]
49. Wang, B. Passively-Q-switched 1.33 μm Nd:LuYSGG laser with the Bi$_2$Se$_3$ topological insulator as a saturable absorber. *J. Russ. Laser Res.* **2020**, *41*, 358–363. [CrossRef]
50. Wang, Y.; Yang, K.; Zhang, Y.; Zhao, S.; Luan, C.; Liu, C.; Wang, J.; Xu, X.; Xu, J. Passively Q-switched laser at 1.3 μm with few-layered MoS$_2$ saturable absorber. *IEEE J. Sel. Top. Quantum Electron.* **2017**, *23*, 71–75. [CrossRef]
51. Yan, B.; Zhang, B.; Nie, H.; Li, G.; Sun, X.; Wang, Y.; Liu, J.; Shi, B.; Liu, S.; He, J. Broadband 1T-titanium selenide-based saturable absorbers for solid-state bulk lasers. *Nanoscale* **2018**, *10*, 20171–20177. [CrossRef]
52. Lin, M.; Peng, Q.; Hou, W.; Fan, X.; Liu, J. 1.3 μm Q-switched solid-state laser based on few-layer ReS$_2$ saturable absorber. *Opt. Laser Technol.* **2019**, *109*, 90–93. [CrossRef]
53. Liu, S.; Wang, M.; Yin, S.; Xie, Z.; Wang, Z.; Zhou, S.; Chen, P. Nonlinear optical properties of few-layer rhenium disulfide nanosheets and their passively Q-switched laser application. *Phys. Status Solidi (A)* **2019**, *216*, 1800837. [CrossRef]
54. Wang, M.; Wang, Z.; Xu, X.; Duan, S.; Du, C. Tin diselenide-based saturable absorbers for eye-safe pulse lasers. *Nanotechnology* **2019**, *30*, 265703. [CrossRef]
55. Zhang, G.; Wang, Y.; Jiao, Z.; Li, D.; Wang, J.; Chen, Z. Tungsten disulfide saturable absorber for passively Q-Switched YVO$_4$/Nd:YVO$_4$/YVO$_4$ laser at 1342.2 nm. *Opt. Mater.* **2019**, *92*, 95–99. [CrossRef]
56. Yang, Z.; Han, L.; Zhang, J.; Zhang, Y.; Zhang, F.; Lin, Z.; Ren, X.; Yang, Q.; Zhang, H. Passively Q-switched laser using PtSe$_2$ as saturable absorber at 1.3 μm. *Infrared Phys. Technol.* **2020**, *104*, 103155. [CrossRef]

57. Qian, Q.; Wang, N.; Zhao, S.; Li, G.; Li, T.; Li, D.; Yang, K.; Zang, J.; Ma, H. Gold nanorods as saturable absorbers for the passively Q-switched Nd:LLF laser at 1.34 μm. *Chin. Opt. Lett.* **2019**, *17*, 041401. [CrossRef]
58. Wang, C.; Peng, Q.-Q.; Fan, X.-W.; Liang, W.-Y.; Zhang, F.; Liu, J.; Zhang, H. MXene $Ti_3C_2T_x$ saturable absorber for pulsed laser at 1.3 μm. *Chin. Phys. B* **2018**, *27*, 094214. [CrossRef]
59. Wang, J.; Liu, S.; Wang, Y.; Wang, T.; Shang, S.; Ren, W. Magnetron-sputtering deposited molybdenum carbide MXene thin films as a saturable absorber for passively Q-switched lasers. *J. Mater. Chem. C* **2020**, *8*, 1608–1613. [CrossRef]
60. Wang, J.; Wang, Y.; Liu, S.; Li, G.; Zhang, G.; Cheng, G. Nonlinear Optical Response of Reflective MXene Molybdenum Carbide Films as Saturable Absorbers. *Nanomaterials (Basel)* **2020**, *10*. [CrossRef]
61. O'Connor, J.R. Unusual crystal-field energy levels and efficient laser properties of YVO_4:Nd. *Appl. Phys. Lett.* **1966**, *9*, 407–409. [CrossRef]
62. Xue, Q.H.; Zheng, Q.; Bu, Y.K.; Qian, L.S. LD-pumped Nd:YVO_4/V:YAG passively Q-switched 1.34 μm laser. *Acta Photonica Sinica* **2005**, *34*, 971–974.
63. Tucker, A.W.; Birnbaum, M.; Fincher, C.L.; DeShazer, L.G. Continuous-wave operation of Nd : YVO_4 at 1.06 and 1.34 μm. *J. Appl. Phys.* **1976**, *47*, 232–234. [CrossRef]
64. Fluck, R.; Braun, B.; Gini, E.; Melchior, H.; Keller, U. Passively Q-switched 1.34-μm Nd:YVO_4 microchip laser with semiconductor saturable-absorber mirrors. *Opt. Lett.* **1997**, *22*, 991–993. [CrossRef]
65. Lai, H.C.; Li, A.; Su, K.W.; Ku, M.L.; Chen, Y.F.; Huang, K.F. InAs/GaAs quantum-dot saturable absorbers for diode-pumped passively Q-switched Nd-doped 1.3-μm lasers. *Opt. Lett.* **2005**, *30*, 480–482. [CrossRef]
66. Janousek, J.; Tidemand-Lichtenberg, P.; Mortensen, J.L.; Buchhave, P. Investigation of passively synchronized dual-wavelength Q-switched lasers based on V:YAG saturable absorber. *Opt. Commun.* **2006**, *265*, 277–282. [CrossRef]
67. Savitski, V.G.; Malyarevich, A.M.; Yumashev, K.V.; Sinclair, B.D.; Lipovskii, A.A. Diode-pumped Nd:YVO_4 and Nd:KGd$(WO_4)_2$ 1.3 μm lasers passively Q-switched with PbS-doped glass. *Appl. Phys. B* **2003**, *76*, 253–256. [CrossRef]
68. Li, A.; Liu, S.C.; Su, K.W.; Liao, Y.L.; Huang, S.C.; Chen, Y.F.; Huang, K.F. InGaAsP quantum-wells saturable absorber for diode-pumped passively Q-switched 1.3-μm lasers. *Appl. Phys. B* **2006**, *84*, 429–431. [CrossRef]
69. Huang, H.T.; He, J.L.; Zuo, C.H.; Zhang, H.J.; Wang, J.Y.; Liu, Y.; Wang, H.T. Co^{2+}: LMA crystal as saturable absorber for a diode-pumped passively Q-switched Nd:YVO_4 laser at 1342 nm. *Appl. Phy. B* **2007**, *89*, 319–321. [CrossRef]
70. Xu, J.L.; Huang, H.T.; He, J.L.; Yang, J.F.; Zhang, B.T.; Yang, X.Q.; Liu, F.Q. Dual-wavelength oscillation at 1064 and 1342 nm in a passively Q-switched Nd:YVO_4 laser with V^{3+}:YAG as saturable absorber. *Appl. Phys. B* **2011**, *103*, 75–82. [CrossRef]
71. Zhai, Y.; Wang, J.X.; Wang, Y.F. Optical characteristic of nc-Si/SiN_x film and its Q-Switching to 1342 nm laser. *Laser J.* **2011**, *32*, 10–11.
72. Wang, M.; Zhang, F.; Wang, Z.; Wu, Z.; Xu, X. Passively Q-switched Nd^{3+} solid-state lasers with antimonene as saturable absorber. *Opt. Express* **2018**, *26*, 4085–4095. [CrossRef]
73. Pan, H.; Chu, H.; Li, Y.; Li, G.; Zhao, S.; Li, D. Bismuth functionalized GaAs as saturable absorber for passive Q-switching at 1.34 μm. *Opt. Mater.* **2019**, *98*, 109457. [CrossRef]
74. Zhang, H.; Peng, J.; Yao, J.; Yang, X.; Li, D.; Zheng, Y. 1.3 μm passively Q-switched mode-locked laser with Fe_3O_4 nanoparticle saturable absorber. *Laser Phys.* **2020**, *30*, 125801. [CrossRef]
75. Zhang, T.; Wang, M.; Xue, Y.; Xu, J.; Xie, Z.; Zhu, S. Liquid metal as a broadband saturable absorber for passively Q-switched lasers. *Chin. Opt. Lett.* **2020**, *18*, 111901. [CrossRef]
76. Cao, L.; Chu, H.; Pan, H.; Wang, R.; Li, Y.; Zhao, S.; Li, D.; Zhang, H.; Li, D. Nonlinear optical absorption features in few-layered hybrid $Ti_3C_2(OH)_2$/$Ti_3C_2F_2$ MXene for optical pulse generation in the NIR region. *Opt. Express* **2020**, *28*, 31499–31509. [CrossRef]
77. Zhang, T.; Chu, H.; Li, Y.; Zhao, S.; Ma, X.; Pan, H.; Li, D. Third-order optical nonlinearity in Ti_2C MXene for Q-switching operation at 1–2 μm. *Opt. Mater.* **2022**, *124*, 112054. [CrossRef]
78. Zhang, H.; Meng, X.; Liu, J.; Zhu, L.; Wang, C.; Shao, Z.; Wang, J.; Liu, Y. Growth of lowly Nd doped $GdVO_4$ single crystal and its laser properties. *J. Cryst. Growth* **2000**, *216*, 367–371. [CrossRef]
79. Qi, H.J.; Liu, X.D.; Hou, X.Y.; Li, Y.F.; Sun, Y.M. A c-cut Nd:$GdVO_4$ solid-state laser passively Q-switched with Co^{2+}:$LaMgAl_{11}O_{19}$ lasing at 1.34 μm. *Laser Phys. Lett.* **2007**, *4*, 576–579. [CrossRef]
80. Li, Y.F.; Zhao, S.Z.; Sun, Y.M.; Qi, H.J.; Cheng, K. Diode-pumped doubly passively Q-switched c-cut Nd:$GdVO_4$ 1.34 μm laser with V^{3+}:YAG and Co:LMA saturable absorbers. *Opt. Laser Technol.* **2011**, *43*, 985–988. [CrossRef]
81. Ma, J.; Li, Y.; Sun, Y.; Xu, J.; He, J. Diode-pumped passively Q-switched Nd:$GdVO_4$ laser at 1342 nm with V:YAG saturable absorber. *Opt. Commun.* **2009**, *282*, 958–961. [CrossRef]
82. Xu, C.; Li, G.; Zhao, S.; Li, X.; Cheng, K.; Zhang, G.; Li, T. LD-pumped passively Q-switched Nd:$GdVO_4$ laser at 1342 nm with high initial transmission V:YAG saturable absorber. *Laser Phys.* **2010**, *20*, 1335–1340. [CrossRef]
83. Pan, H.; Wang, X.; Chu, H.; Li, Y.; Zhao, S.; Li, G.; Li, D. Optical modulation characteristics of zeolitic imidazolate framework-67 (ZIF-67) in the near infrared regime. *Opt. Lett.* **2019**, *44*, 5892–5895. [CrossRef]
84. Su, X.; Wang, Y.; Zhang, B.; Zhang, H.; Yang, K.; Wang, R.; He, J. Bismuth quantum dots as an optical saturable absorber for a 1.3 μm Q-switched solid-state laser. *Appl. Opt.* **2019**, *58*, 1621–1625. [CrossRef]
85. Dong, L.; Huang, W.; Chu, H.; Li, Y.; Wang, Y.; Zhao, S.; Li, G.; Zhang, H.; Li, D. Passively Q-switched near-infrared lasers with bismuthene quantum dots as the saturable absorber. *Opt. Laser Technol.* **2020**, *128*, 106219. [CrossRef]

86. Feng, X.; Liu, J.; Yang, W.; Yu, X.; Jiang, S.; Ning, T.; Liu, J. Broadband indium tin oxide nanowire arrays as saturable absorbers for solid-state lasers. *Opt. Express* **2020**, *28*, 1554–1560. [CrossRef]
87. Zhang, J.; Wang, Y.; Mu, W.; Jia, Z.; Zhang, B.; He, J.; Tao, X. New near-infrared optical modulator of Co^{2+}:β-Ga_2O_3 single crystal. *Opt. Mater. Express* **2021**, *11*. [CrossRef]
88. Dong, L.; Chu, H.; Li, Y.; Zhao, S.; Li, G.; Li, D. Nonlinear optical responses of α-Fe_2O_3 nanosheets and application as a saturable absorber in the wide near-infrared region. *Opt. Laser Technol.* **2021**, *136*, 106812.
89. Zhang, C.; Zheng, L.; Chu, H.; Pan, H.; Hu, Y.; Li, D.; Dong, L.; Zhao, S.; Li, D. Monoclinic bismuth vanadate nanoparticles as saturable absorber for Q-switching operations at 1.3 and 2 μm. *Appl. Phys. Express* **2022**, *15*, 072004. [CrossRef]
90. Maunier, C.; Doualan, J.L.; Moncorgé, R.; Speghini, A.; Bettinelli, M.; Cavalli, E. Growth, spectroscopic characterization, and laser performance of Nd:$LuVO_4$ a new infrared laser material that is suitable for diode pumping. *J. Opt. Soc. Am. B* **2002**, *19*, 1794–1800. [CrossRef]
91. Zhang, H.; Liu, J.; Wang, J.; Xu, X.; Jiang, M. Continuous-wave laser performance of Nd:$LuVO_4$ crystal operating at 1.34 μm. *Appl. Opt.* **2005**, *44*, 7439–7441. [CrossRef]
92. Ran, D.; Xia, H.; Sun, S.; Liu, F.; Ling, Z.; Ge, W.; Zhang, H.; Wang, J. Thermal properties of a Nd:$LuVO_4$ crystal. *Cryst. Res. Technol.* **2007**, *42*, 920–925. [CrossRef]
93. Liu, F.; He, J.; Zhang, B.; Xu, J.; Dong, X.; Yang, K.; Xia, H.; Zhang, H. Diode-pumped passively Q-switched Nd:$LuVO_4$ laser at 1.34 μm with a V^{3+}:YAG saturable absorber. *Opt. Express* **2008**, *16*, 11759–11763. [CrossRef]
94. Liu, F.Q.; He, J.L.; Xu, J.L.; Yang, J.F.; Zhang, B.T.; Huang, H.T.; Gao, C.Y.; Xu, J.Q.; Zhang, H.J. Passive Q-switching performance with Co:LMA crystal in a diode-pumped Nd:$LuVO_4$ laser. *Laser Phys.* **2010**, *20*, 786–789. [CrossRef]
95. Yu, Y.; Wang, J.; Zhang, H.; Yu, H.; Wang, Z.; Jiang, M.; Xia, H.; Boughton, R.I. Growth and characterization of Nd:$Y_xGd_{1-x}VO_4$ series laser crystals. *J. Opt. Soc. Am. B* **2008**, *25*, 995–1001. [CrossRef]
96. Li, P.; Li, Y.; Sun, Y.; Hou, X.; Zhang, H.; Wang, J. Passively Q-switched 1.34 μm Nd:$Y_xGd_{1-x}VO_4$ laser with Co^{2+}:$LaMgAl_{11}O_{19}$ saturable absorber. *Opt. Express* **2006**, *14*, 7730–7736. [CrossRef]
97. Omatsu, T.; Miyamoto, K.; Okida, M.; Minassian, A.; Damzen, M.J. 1.3-μm passive Q-switching of a Nd-doped mixed vanadate bounce laser in combination with a V:YAG saturable absorber. *Appl. Phys. B* **2010**, *101*, 65–70. [CrossRef]
98. Zhang, B.T.; Huang, H.T.; He, J.L.; Yang, J.F.; Xu, J.L.; Zuo, C.H.; Zhao, S. Diode-end-pumped passively Q-switched 1.34 μm Nd:$Gd_{0.5}Y_{0.5}VO_4$ laser with Co^{2+}:LMA saturable absorber. *Opt. Mater.* **2009**, *31*, 1697–1700. [CrossRef]
99. Zhang, S.; Wang, X.; He, J.; Yang, Q.; Li, X.; Zhao, B. Passively Q-switched laser performance of an a-cut Nd:$Lu_{0.33}Y_{0.36}Gd_{0.3}VO_4$ crystal at 1.34 μm with V^{3+}:YAG as the saturable absorber. *Laser Phys.* **2013**, *23*, 095805. [CrossRef]
100. Li, X.; Li, G.; Zhao, S.; Xu, C.; Du, G. Diode-pumped passively Q-switched Nd:$Lu_xY_{1-x}VO_4$ laser at 1.34 μm with two V:YAG saturable absorbers. *Opt.Commun.* **2011**, *284*, 1307–1311. [CrossRef]
101. Cai, E.; Xu, J.; Zhang, S.; Wu, Z. Tin disulfide as saturable absorber for the 1.3 μm nanosecond laser. *Laser Phys. Lett.* **2022**, *19*, 065802. [CrossRef]
102. Cai, E.; Xu, J.; Liu, Y.; Zhang, S.; Fan, X.; Wang, M.; Lou, F.; Lv, H.; Wang, X.; Li, T. Passively Q-switched and Q-switched mode-locked Nd:$Lu_{0.15}Y_{0.85}VO_4$ lasers at 1.34 μm with a nickel-cobalt layered double hydroxide saturable absorber. *Opt. Mater. Express* **2022**, *12*, 931–939. [CrossRef]
103. Han, S.; Xu, H.H.; Zhao, Y.G.; Chen, L.J.; Wang, Z.P.; Yu, H.H.; Zhang, H.J.; Xu, X.G. The $^4F_{3/2} \to ^4I_{13/2}$ transition property of Nd:$La_{0.05}Lu_{0.95}VO_4$ crystal. *Laser Phys.* **2013**, *23*, 105814. [CrossRef]
104. Guo, L.; Lan, R.; Liu, H.; Yu, H.; Zhang, H.; Wang, J.; Hu, D.; Zhuang, S.; Chen, L.; Zhao, Y.; et al. 1319 nm and 1338 nm dual-wavelength operation of LD end-pumped Nd:YAG ceramic laser. *Opt. Express* **2010**, *18*, 9098–9106. [CrossRef]
105. Jabczynski, J.K.; Zendzian, W.; Kwiatkowski, J.; Šulc, J.; Nemec, M.; Jelínková, H. Passively Q-switched neodymium slab lasers at 1.3-μm wavelength side-pumped by a 600-W laser diode stack. *Opt. Eng.* **2006**, *45*, 114204. [CrossRef]
106. Li, P.; Chen, X.H.; Zhang, H.N.; Wang, Q.P. Diode-end-pumped passively Q-switched 1319 nm Nd:YAG ceramic laser with a V^{3+}:YAG saturable absorber. *Laser Phys.* **2011**, *21*, 1708–1711. [CrossRef]
107. Lin, B.; Xiao, K.; Zhang, Q.L.; Zhang, D.X.; Feng, B.H.; Li, Q.N.; He, J.L. Dual-wavelength Nd:YAG laser operation at 1319 and 1338 nm by direct pumping at 885 nm. *Appl. Opt.* **2016**, *55*, 1844–1848. [CrossRef] [PubMed]
108. Podlipensky, A.V.; Yumashev, K.V.; Kuleshov, N.V.; Kretschmann, H.M.; Huber, G. Passive Q-switching of 1.44 μm and 1.34 μm diode-pumped Nd:YAG lasers with a V:YAG saturable absorber. *Appl. Phys. B* **2003**, *76*, 245–247. [CrossRef]
109. Šulc, J.; Jelinkova, H.; Nemec, M.; Nejezchleb, K.; Skoda, V. V:YAG saturable absorber for flash-lamp and diode-pumped solid state lasers. In Proceedings of the Solid State Lasers and Amplifiers: Photonics Europe, Strasbourg, France, 24–29 April 2004; Volume 5460, pp. 292–302.
110. Šulc, J.; Jelínková, H.; Nejezchleb, K.; Škoda, V. Nd:YAG/V:YAG monolithic microchip laser operating at 1.3 μm. *Opt. Mater.* **2007**, *30*, 50–53. [CrossRef]
111. Feng, C.; Zhang, H.; Fang, J.; Wang, Q. Passively Q-switched Nd:YAG ceramic laser with V^{3+}:YAG saturable absorber at 1357 nm. *Appl. Opt.* **2015**, *54*, 9902–9905. [CrossRef]
112. Lin, H.Y.; Liu, H.; Zhang, S.Q. Passively Q-Switched 1319 nm Nd:YAG laser based on Co:$MgAl_2O_4$ crystal. *Laser Optoelectron. Prog.* **2018**, *55*, 121404.
113. Xu, X.D.; Wang, X.D.; Meng, J.Q.; Cheng, Y.; Li, D.Z.; Cheng, S.S.; Wu, F.; Zhao, Z.W.; Xu, J. Crystal growth, spectral and laser properties of Nd:LuAG single crystal. *Laser Phys. Lett.* **2009**, *6*, 678–681. [CrossRef]

114. Liu, C.; Zhao, S.; Li, G.; Yang, K.; Li, D.; Li, T.; Qiao, W.; Feng, T.; Chen, X.; Xu, X.; et al. Experimental and theoretical study of a passively Q-switched Nd:LuAG laser at 1.3 μm with a V^{3+}:YAG saturable absorber. *J. Opt. Soc. Am. B* **2015**, *32*, 1001–1006. [CrossRef]
115. Zhang, B.; Yang, J.; He, J.; Huang, H.; Liu, S.; Xu, J.; Liu, F.; Zhi, Y.; Tao, X. Diode-end-pumped passively Q-switched 1.33 μm Nd:$Gd_3Al_xGa_{5-x}O_{12}$ laser with V^{3+}: YAG saturable absorber. *Opt. Express* **2010**, *18*, 12052–12058. [CrossRef]
116. Gao, S.; Wang, W. Thermal optical properties, Q-switching and frequency tuning of an Nd:LGGG laser based on the $^4F_{3/2} \rightarrow {}^4I_{13/2}$ transition of a neodymium ion. *Quantum Electron.* **2021**, *51*, 149–152. [CrossRef]
117. Kuwano, Y.; Saito, S.; Hase, S. Crystal growth and optical properties of Nd: GGAG. *J. Cryst. Growth* **1988**, *92*, 17–22. [CrossRef]
118. Jia, Z.; Tao, X.; Yu, H.; Dong, C.; Zhang, J.; Zhang, H.; Wang, Z.; Jiang, M. Growth and properties of Nd:$(Lu_xGd_{1-x})_3Ga_5O_{12}$ laser crystal by Czochralski method. *Opt. Mater.* **2008**, *31*, 346–349. [CrossRef]
119. Fu, X.; Jia, Z.; Li, Y.; Yuan, D.; Dong, C.; Tao, X. Crystal growth and characterization of Nd^{3+}:$(La_xGd_{1-x})_3Ga_5O_{12}$ laser crystal. *Opt. Mater. Express* **2012**, *2*, 1242–1253. [CrossRef]
120. Wang, B.; Tian, L.; Yu, H.; Zhang, H.; Wang, J. Energy enhancement of mixed Nd:LuYSGG crystal in passively Q-switched lasers. *Opt. Lett.* **2015**, *40*, 3213–3216. [CrossRef]
121. Zuo, C.H.; Zhang, B.T.; He, J.L.; Dong, X.L.; Yang, J.F.; Huang, H.T.; Xu, J.L.; Zhao, S.; Dong, C.M.; Tao, X.T. CW and passive Q-switching of 1331-nm Nd:GGG laser with Co^{2+}:LMA saturable absorber. *Appl. Phys. B* **2009**, *95*, 75–80. [CrossRef]
122. Zuo, C.H.; Zhang, B.T.; He, J.L.; Dong, X.L.; Yang, K.J.; Huang, H.T.; Xu, J.L.; Zhao, S.; Dong, C.M.; Tao, X.T. CW and passively Q-switching characteristics of a diode-end-pumped Nd:GGG laser at 1331 nm. *Opt. Mater.* **2009**, *31*, 976–979. [CrossRef]
123. Zuo, C.H.; Zhang, B.T.; He, J.L. Passively Q-switched 1.33 μm Nd:GAGG laser with Co^{2+}:LMA saturable absorber. *Laser Phys. Letters* **2011**, *8*, 782–786. [CrossRef]
124. Wang, Z.W.; Fu, X.W.; He, J.L.; Jia, Z.T.; Zhang, B.T.; Yang, H.; Wang, R.H.; Liu, X.M.; Tao, X.T. The performance of 1329 nm CW and passively Q-switched Nd:LGGG laser with low Lu-doping level. *Laser Phys. Lett.* **2013**, *10*, 055005. [CrossRef]
125. Wang, B.; Yu, H.; Zhang, H. Passively Q-switched 1.33 μm Nd:LuYSGG laser with V^{3+}:YAG as the saturable absorber. *Laser Phys. Lett.* **2019**, *16*, 015801. [CrossRef]
126. Demidovich, A.A.; Shkadarevich, A.P.; Danailov, M.B.; Apai, P.; Gasmi, T.; Gribkovskii, V.P.; Kuzmin, A.N.; Ryabtsev, G.I.; Batay, L.E. Comparison of cw laser performance of Nd:KGW, Nd: YAG, Nd:BEL, and Nd:YVO_4 under laser diode pumping. *Appl. Phys. B* **1998**, *67*, 11–15. [CrossRef]
127. Chen, X.; Liu, J.; Yu, Y.; Li, T.; Sun, H.; Jin, G. Diode-side-pumped passively Q-switched Nd:YAP laser operating at 1.34 μm with V^{3+}:YAG saturable absorber. *J. Russ. Laser Res.* **2015**, *36*, 86–91.
128. Ryan, J.R.; Beach, R. Optical absorption and stimulated emission of neodymium in yttrium lithium fluoride. *J. Opt. Soc. Am. B* **1992**, *9*, 1883–1887. [CrossRef]
129. Hardman, P.J.; Clarkson, W.A.; Friel, G.J.; Pollnau, M.; Hanna, D.C. Energy-transfer upconversion and thermal lensing in high-power end-pumped Nd:YLF laser crystals. *IEEE J. Quantum Electron.* **1999**, *35*, 647–655. [CrossRef]
130. Botha, R.C.; Strauss, H.J.; Bollig, C.; Koen, W.; Collett, O.; Kuleshov, N.V.; Esser, M.J.D.; Combrinck, W.L.; von Bergmann, H.M. High average power 1314 nm Nd:YLF laser, passively Q-switched with V:YAG. *Opt. Lett.* **2013**, *38*, 980–982. [CrossRef]
131. Li, H.; Zhang, R.; Tang, Y.; Wang, S.; Xu, J.; Zhang, P.; Zhao, C.; Hang, Y.; Zhang, S. Efficient dual-wavelength Nd:$LuLiF_4$ laser. *Opt. Lett.* **2013**, *38*, 4425–4428. [CrossRef]
132. Li, J.H.; Liu, X.H.; Wu, J.B.; Zhang, X.; Li, Y.L. High-power diode-pumped Nd:Lu_2O_3 crystal continuous wave thin-disk laser at 1359 nm. *Laser Phys. Lett.* **2012**, *9*, 195–198. [CrossRef]
133. Li, J.; Vannini, M.; Wu, L.; Tian, F.; Hu, D.; Chen, X.; Feng, Y.; Patrizi, B.; Pirri, A.; Toci, G.; et al. Fabrication and Optical Property of Nd:Lu_2O_3 Transparent Ceramics for Solid-state Laser Applications. *J. Inorg. Mater.* **2021**, *36*, 210. [CrossRef]
134. Rao, H.; Liu, Z.; Cong, Z.; Huang, Q.; Liu, Y.; Zhang, S.; Zhang, X.; Feng, C.; Wang, Q.; Ge, L.; et al. High power YAG/Nd:YAG/YAG ceramic planar waveguide laser. *Laser Phys. Lett.* **2017**, *14*, 045801. [CrossRef]
135. Grabtchikov, A.S.; Kuzmin, A.N.; Lisinetskii, V.A.; Orlovich, V.A.; Demidovich, A.A.; Yumashev, K.V.; Kuleshov, N.V.; Eichler, H.J.; Danailov, M.B. Passively Q-switched 1.35 μm diode pumped Nd:KGW laser with V:YAG saturable absorber. *Opt. Mater.* **2001**, *16*, 349–352. [CrossRef]
136. Zolotovskaya, S.A.; Savitski, V.G.; Gaponenko, M.S.; Malyarevich, A.M.; Yumashev, K.V.; Demchuk, M.I.; Raaben, H.; Zhilin, A.A.; Nejezchleb, K. Nd:$KGd(WO_4)_2$ laser at 1.35 μm passively Q-switched with V^{3+}:YAG crystal and PbS-doped glass. *Opt. Mater.* **2006**, *28*, 919–924. [CrossRef]
137. Chen, H.; Zhou, M.; Zhang, P.; Yin, H.; Zhu, S.; Li, Z.; Chen, Z. Passively Q-switched Nd:GYAP laser at 1.3 μm with bismuthene nanosheets as a saturable absorber. *Infrared Phys. Technol.* **2022**, *121*, 104023. [CrossRef]
138. Gao, X.; Li, S.; Li, T.; Li, G.; Ma, H. $g-C_3N_4$ as a saturable absorber for the passively Q-switched Nd:LLF laser at 13 μm. *Photonics Res.* **2017**, *5*, 33–36. [CrossRef]

Review

Principles of Selective Area Epitaxy and Applications in III–V Semiconductor Lasers Using MOCVD: A Review

Bin Wang [1,2], Yugang Zeng [1,2,*], Yue Song [1,2], Ye Wang [1,3], Lei Liang [1,2], Li Qin [1,2], Jianwei Zhang [1,2], Peng Jia [1,2], Yuxin Lei [1,2], Cheng Qiu [1,2], Yongqiang Ning [1,2] and Lijun Wang [1,2,4,5]

1. State Key Laboratory of Luminescence and Application, Changchun Institute of Optics, Fine Mechanics and Physics, Chinese Academy of Sciences, Changchun 130033, China; wangbin201@mails.ucas.ac.cn (B.W.); songyue@ciomp.ac.cn (Y.S.); 2020200039@mails.cust.edu.cn (Y.W.); liangl@ciomp.ac.cn (L.L.); qinl@ciomp.ac.cn (L.Q.); zjw1985@ciomp.ac.cn (J.Z.); jiapeng@ciomp.ac.cn (P.J.); leiyuxin@ciomp.ac.cn (Y.L.); qiucheng@ciomp.ac.cn (C.Q.); ningyq@ciomp.ac.cn (Y.N.); wanglj@ciomp.ac.cn (L.W.)
2. Daheng College, University of Chinese Academy of Sciences, Beijing 100049, China
3. School of Opto-Electronic Engineering, Changchun University of Science and Technology, Changchun 130022, China
4. Peng Cheng Laboratory, No. 2, Shenzhen 518000, China
5. Academician Team Innovation Center of Hainan Province, Key Laboratory of Laser Technology and Optoelectronic Functional Materials of Hainan Province, School of Physics and Electronic Engineering, Hainan Normal University, Haikou 570206, China
* Correspondence: zengyg@ciomp.ac.cn

Abstract: Selective area epitaxy (SAE) using metal–organic chemical vapor deposition (MOCVD) is a crucial fabrication technique for lasers and photonic integrated circuits (PICs). A low-cost, reproducible, and simple process for the mass production of semiconductor lasers with specific structures was realized by means of SAE. This paper presents a review of the applications of SAE in semiconductor lasers. Growth rate enhancement and composition variation, which are two unique characteristics of SAE, are attributed to a mask. The design of the mask geometry enables the engineering of a bandgap to achieve lasing wavelength tuning. SAE allows for the reproducible and economical fabrication of buried heterojunction lasers, quantum dot lasers, and heteroepitaxial III–V compound lasers on Si. Moreover, it enables the fabrication of compact photonic integrated devices, including electro-absorption modulated lasers and multi-wavelength array lasers. Results show that SAE is an economical and reproducible method to fabricate lasers with desired structures. The goals for SAE applications in the future are to improve the performance of lasers and PICs, including reducing the defects of the grown material introduced by the SAE mask and achieving precise control of the thickness and composition.

Keywords: selective area epitaxy; MOCVD; semiconductor laser; quantum dot; heteroepitaxy; EML; multi-wavelength laser arrays

1. Introduction

Semiconductor lasers have become crucial light sources because of their advantages, including small size, light weight, high reliability, high modulation speed, and easy monolithic integration with other optoelectronic devices, in the fields of modern optical communication, industry, medical, and military fields [1–3]. The fabrication of semiconductor laser epitaxial structures has been achieved due to the emergence and development of metal–organic chemical vapor deposition (MOCVD) and molecular beam epitaxy (MBE). Selective area epitaxy (SAE) is a method used to controllably grow III–V, II–VI and lead salts lasing materials with specific shapes and geometries. III–V compounds are widely used semiconductor laser materials. Some novel III–V nanostructure materials fabricated with SAE, such as nanowires (NWs), twin-free nanosheets, and low-dislocation nanomembranes at the heterogeneous interface, have been attractive for use in solar cells and lasers due to their good optical and electrical properties [4,5]. Although SAE using MBE has the advantages of providing a controllable thickness and ultra-high clean surface, the MBE process is time-consuming and the required equipment is expensive, which limits the large-scale production of epitaxial structures. In the fabrication of III–V semiconductor lasers, SAE employing MOCVD technologies can realize the desired laser structure on a large scale, such as buried heterostructure (BH) [6] and quantum dot (QD) lasers. SAE can decrease the high defect density due to a lattice being mismatched at the interface between III and V materials and Si, demonstrating the feasibility of high-performance III–V lasers on Si. Over the past few decades, the integration of semiconductor lasers and other optoelectronic devices on monolithic chips has been an inevitable trend to satisfy the demands of high-speed optical communication [7,8]. SAE is an effective method that realizes integrated lasers, such as electro-absorption modulated lasers (EML) [9], multi-wavelength laser arrays (MWLA) [10].

The improvement of laser structure and the design of advanced monolithic integrated devices require a deeper understanding of the principles and applications of SAE. SAE has proven to be a powerful tool for fabricating semiconductor lasers with desired structure and compact photonic integrated circuits (PICs). In this paper, we interpret SAE principles including growth-rate enhancement, composition variation, vapor-phase diffusion model, and bandgap engineering. The SAE applications such as BH laser, QD laser, heteroepitaxial laser on Si, EML, and MWLA are introduced in detail. The problems of defects in the growth of materials introduced by masks needs to be solved urgently. It is also crucial to achieving precise and controllable material growth. SAE can potentially be the primary technology for future optoelectronic devices. It is hoped that this paper can provide a reference for clearly understanding the mechanism of SAE, reducing material defects caused by selective epitaxial growth, and improving the performance of lasers.

2. Principles of SAE

SAE refers to the method of growing materials with different thicknesses and compositions at various locations on the same wafer. It involves the deposition of a dielectric material on the substrate as a mask, most commonly silicon oxide (SiO_2). After designing the desired substrate pattern, the substrate was divided into regions covered by the mask and exposed areas using pattern techniques. The most straightforward and practical mask pattern is a pair of rectangular strips, as shown in Figure 1. The area exposed between the two stripes was the mask-opening region. The dielectric mask can inhibit the precursor deposition and nucleation. As a result, deposition and growth occur only in the opening region and hardly on the mask, realizing the selective growth of the material on the substrate.

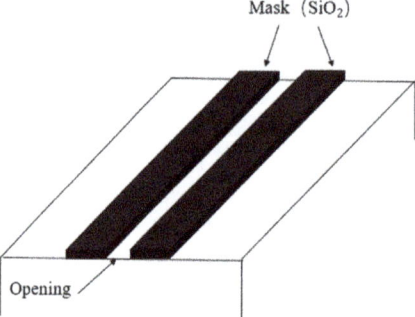

Figure 1. A pair of oxide strips pattern mask for selective area epitaxy.

The mask material should have a low sticking coefficient to the precursor gas [11,12], can withstand high temperatures [13], is insensitive to precursors, and is compatible with subsequent processes, such as inductively coupled plasma (ICP) or chemical wet etching [14]. Common mask materials are amorphous SiN_x and SiO_2.

Pattern techniques realize the design of a patterned mask over the substrate. There are three primary patterning design methods. Deep ultraviolet (DUV) lithography realizes the mask size in the hundreds of nanometers [15]. Electron beam lithography (EBL) can enable the resolution of SAE on a scale of tens to hundreds of nanometers [16]. Nano-imprinting lithography (NIL) can realize the pattern design of wafer-level size.

The geometry and size of the mask affect the epilayer thickness and composition obtained via SAE compared to the uniform epilayer obtained by MOCVD on exposed planar substrates. The successful fabrication of lasers and integrated optoelectronic devices is based on the control of the surface morphology and the composition of selective area growth.

2.1. Growth-Rate Enhancement and Composition Variation

The growth-rate enhancement (or "growth-enhancement effect", GRE/GEE) effect is one of the essential properties of SAE. During the MOCVD process, the precursor gas-phase molecules were selectively adsorbed at the substrate opening, whereas few molecules were attached to the mask surface. This occurred because the sticking coefficients of precursors on the mask were much lower than that of the surface of the epitaxial region in the high temperature of MOCVD. The mask suppressed the nucleation of the crystals. Material growth occurred selectively in the opening region. According to the conservation of mass, the concentration of the precursor above the mask was larger than that of the opening area. The precursor gas was transported from the mask to the opening driven by the concentration gradient. The material grown on the opening area of the masked substrate was thicker than that grown on the unmasked planar substrate at the same time. The deposition rate near the edge of the stripe mask was higher than that of the opening area far from the mask. The lateral gas diffusion over the mask area to the opening area leads to this phenomenon [17–19]. The phenomenon mentioned above is the GRE. GRE is expressed as the ratio of the thickness of the film grown at the opening area on the masked substrate to the thickness of the film grown on the unmasked wafer. Figure 2 shows a schematic of the cross-section of the GRE cross-section.

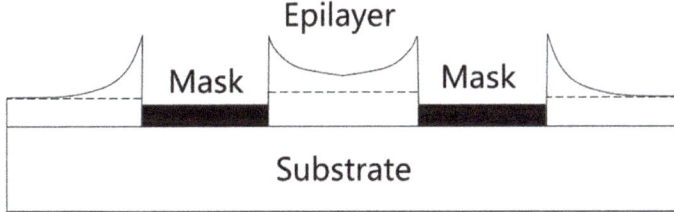

Figure 2. Schematic of the cross-section of the growth-rate enhancement process. The area below the dashed line is the layer thickness grown on the unmasked substrate, and the excess thickness due to GRE is the area above the dashed line.

Another critical property of SAE is the compositional variation with the GRE. The MOCVD growth of III–V alloys on wafers is dominated by the absorption and vapor-phase diffusion of III species. The absorption, surface migration, and re-evaporation capabilities of III-group molecules on the mask are different. These processes affect the flux of III-species gas absorbed on the mask reaching the opening area.

A compositional variation appears in the opening area, especially in the sidewalls of the growth layers near the edges of the strip masks, as diffusion from the mask to the unmasked wafer occurred [20,21].

The typical opening width between the mask was 5–20 μm, and the stripe width was 5–50 μm; the material growth in the openings was determined by the widths of both the oxide mask and the opening [22]. The thickness and GRE of the growth layer increased as the width of the mask increased, which was the exact opposite of the case noted considering the opening area [23]. The GRE was still clearly visible in the opening area with a 100 μm width [24]. The GRE disappeared at positions very far from the edge of the mask stripes, which does not differ from the case involving growth on a planar substrate. The controllable growth of layers with different thicknesses and compositions was realized by changing the spacing and width of the mask stripes.

2.2. Vapor-Phase Diffusion Model of SAE

Several studies have proposed models for SAE-grown materials that quantify the diffusion process. The GRE and composition shift were considered functions of the mask geometry. The GRE mechanism comprises two parts: surface migration and vapor-phase diffusion [23]. Surface migration was thought to occur because the gas molecules absorbed on the dielectric mask migrated along the surface of the mask and the sidewalls of the epilayer. Vapor-phase diffusion refers to the flux to the unmasked region due to the concentration gradient of the undeposited precursor above the mask. The mask width is typically on the order of tens of micrometers. Surface migration in the masked area is ignored as it only occurs within a few microns of the mask surface [25].

The three-dimensional vapor-phase diffusion model is established by solving the state diffusion equation in an assumed region that is similar to the experimental conditions, as shown in Figure 3. The thickness and composition profiles of various compounds were predicted using vapor-phase diffusion calculations.

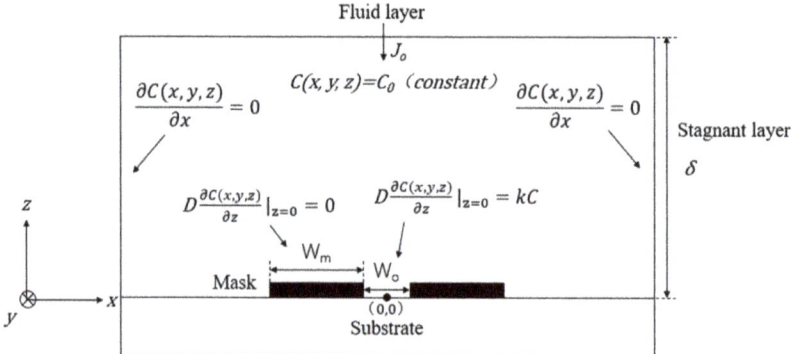

Figure 3. Schematic diagram of the 3D vapor-phase diffusion model.

The model assumed a stagnant area above the wafer. The height of the stagnant area is "δ"(~500 μm) [22]. The fluid layer above the stagnant layer provided a constant flow. In the equations below, D and k are the diffusion length and reaction rate of the reactant precursor, respectively. W_o and W_m are the widths of the opening area and masks, respectively. C is the concentration of the reactant gas molecules, usually group-III species. The growth direction was z, and the length along the mask was y. The horizontal direction, x, is perpendicular to the z- and y-axes.

Because the spatial gas concentration is constant, the substance diffusion equation in three-dimensional space is

$$\frac{\partial^2 C}{\partial x^2} + \frac{\partial^2 C}{\partial y^2} + \frac{\partial^2 C}{\partial z^2} = 0. \tag{1}$$

Above the stagnant area, the concentration of precursors is constant.

$$C(x, y, z = \delta) = C_0 \tag{2}$$

The symmetry of the computational domain defines the left and right boundaries is expressed as

$$D\frac{\partial C(x,y,z)}{\partial x} = 0. \tag{3}$$

There is no growth occurring on the mask surfaces.

$$D\frac{\partial C(x,y,z)}{\partial z}\Big|_{z=0} = 0 \tag{4}$$

The growth process in the unmasked region is given by

$$D\frac{\partial C(x,y,z)}{\partial z}\Big|_{z=0} = kC. \tag{5}$$

In this model, D/k is the effective diffusion length of the species. The relationship between the effective diffusion length and mask length determines the dominant process in GRE. Surface migration plays a major role when the mask geometry is close to the surface-diffusion length. On the contrary, the vapor-phase diffusion model dominates this process because the mask size is larger than the diffusion length [26]. A decrease in the effective diffusion length increased the growth rate next to the stripe mask. The vapor-phase diffusion equations for different III-species precursors were obtained to obtain the GRE and composition changes.

The GRE factor R is expressed as

$$R = \frac{C(x,y,z=0)}{C(0)}, \tag{6}$$

where $C(x, y, z = 0)$ is the concentration of group-III vapor-phase molecules on the $z = 0$ plane, and $C(0)$ is the analytic solution of Laplace's equation for the unmasked wafer. $C(0)$ is expressed as [27]

$$C(0) = \frac{C_0}{\frac{k}{D}\delta + 1} \tag{7}$$

Model calculations were performed without considering the interactions between the different III species [28]. Dupuis et al. [25,29] assumed that the growth-rate enhancement of ternary or quaternary alloys was linearly related to group-III precursors. They gave the expressions for the growth-rate enhancement factor of $Al_xGa_yIn_{1-x-y}As$ system film and III–element composition.

$$R(Al_{x0}Ga_{y0}In_{1-x0-y0}As) = x_0 \cdot R_{Al} + y_0 \cdot R_{Ga} + (1 - x_0 - y_0) \cdot R_{In} \tag{8}$$

$$Al\% = \frac{x_0 \cdot R_{Al}}{R(Al_{x0}Ga_{y0}In_{1-x0-y0}As)} \tag{9}$$

$$Ga\% = \frac{y_0 \cdot R_{Ga}}{R(Al_{x0}Ga_{y0}In_{1-x0-y0}As)} \tag{10}$$

$$In\% = \frac{(1-x_0-y_0) \cdot R_{In}}{R(Al_{x0}Ga_{y0}In_{1-x0-y0}As)} \tag{11}$$

The reactivity of V precursors is very low at the typical deposition temperatures of MOCVD. The V precursors account for the majority of the gas-phase components, far exceeding the stoichiometric number in the film compound composition, such that it is therefore difficult to obtain an expression of the V composition of the semiconductor compound. Therefore, the deposition rate of III–V compounds is usually determined by the incorporation of III precursors. The GRE becomes more pronounced with the increase of the III–V ratio. At higher temperatures, the desorption of gas molecules on the surface is enhanced, resulting in weaker selective growth. Building an accurate growth-rate-enhancement computational model for these equations helps predict and control the specific growth, including thickness and composition, enabling the design of strain and bandgap energy variations at different locations on a single wafer. The accuracy and efficiency of a diffusion model are vital for the applications of future SAE-integrated optical devices.

2.3. Bandgap Engineering of SAE

The spatial variation of the energy band and strain for a given patterned mask can be deduced by varying the width of the mask and the opening area. For bulk materials, changes in composition cause a wavelength shift. In quantum wells (QWs), the GRE causes an additional wavelength shift in addition to the composition. The magnitude of the wavelength shifts depends on the width of the mask and the opening area [30]. The effective energy gap, Eg, is sensitive to the thickness of the QW and increases as the layer thickness decreases. The dependence of Eg on the layer thickness enables the tuning of emission wavelength.

Sasaki et al. [31] fabricated a 2 μm wide InGaAs/InGaAsP multiple quantum well (MQW) ridge structure, which was surrounded by stripe masks with various widths on both sides. They observed a 50 nm PL peak wavelength shift by varying the mask width from 4 μm to 10 μm, as shown in Figure 4. A series of studies have demonstrated the relationship between QW emission wavelength tuning and mask geometry [8,28,29].

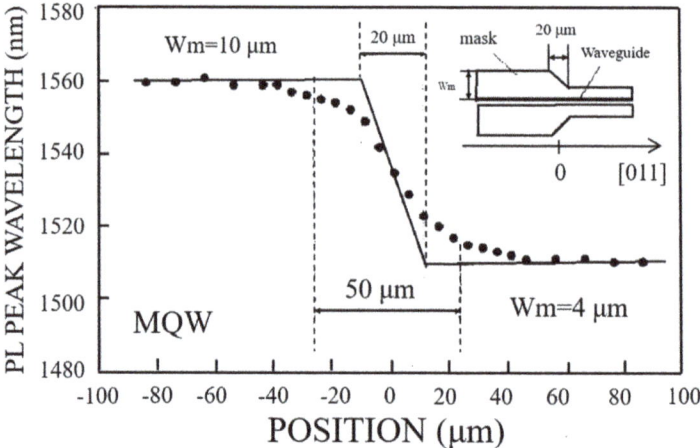

Figure 4. PL peak wavelength profile for MQW structure along [011] direction. A 20 μm tapered transition region formed between the 10 μm and 4 μm wide mask. The inset is mask configuration. Reprinted with permission from ref. [31] © Elsevier. Copyright 1993 Journal of Crystal Growth.

The gradual gradient wavelength profiles were due to GRE along the edge of the mask. Due to the GRE, there is a thick bulge region at the interface between the mask and the epitaxial layer (Figure 2). Flat interfaces are essential in photonic devices. Adjusting MOCVD parameters, such as increasing the growth temperature [32] and reactor pressure [33], can eliminate convexity near the edge of the mask. The in situ etching process can improve the planarity between the opening area and the masked region [34,35].

Nonplanar growth on a substrate is also an indispensable research topic for SAE. In specific applications, such as butt-coupled waveguides and buried heterojunction lasers, wafers are etched to create grooves with well-defined inclined facets. The orientation of the inclined plane of the groove was different from that of the planar wafer. Distinction of the growth rate in different crystal orientations leads to enhanced diffusion on a nonplanar surface. Different growth rates on different crystal planes also enable the fabrication of high-performance laser structures, such as BH, which will be further discussed in the following sections.

3. Applications of SAE in Semiconductor Lasers

3.1. Buried Heterostructure (BH) Lasers

The active region in a buried heterostructure (BH) laser is completely buried in a wide-bandgap hetero-material. The BH structure has a strong lateral refractive index confinement and current confinement due to the lateral heterostructure [36]. The BH laser has a low threshold current, stable waveguide mode, and nearly symmetrical beam distribution, making it a perfect structure for semiconductor lasers operating with a low threshold and high efficiency. However, conventional BH lasers require multiple MOCVD steps, easily introducing contamination and oxidation to the surface. The SAE can complete the BH structure in a single growth step, eliminating the related defects.

Galeuchet et al. [37] presented a GaInAs/InP heterostructure grown in one SAE step. The GaInAs active layer growth occurred on the (100) facet rather than {111}, where the InP layer growth took place. Different sticking coefficients or diffusion constants of InP and GaInAs may be responsible for the different growth in the (100) and {111} facets. The nucleation sites of the {111}A facets were occupied with stable molecules that did not react to a large extent, while the {111}B facets were active and provided reactants to the (100) facets. The width of the mesa decreased as the growth proceeded. Finally, the lateral size of the buried layers became zero, and the sides of the sloping {111}B facets merged into one. A BH laser was fabricated in one growth step because the GaInAs layers were buried in the InP. The catastrophic optical damage (COD) due to the absorption in the active region near the output facets usually occurs in high-power semiconductor lasers. This limits the high-power output of lasers. Lammert et al. [38] presented a BH laser with a nonabsorbing mirror region, where a fusiform patterned mask was used, as shown in Figure 5. The output power of the BH laser with nonabsorbing mirrors exhibited increments of 40%.

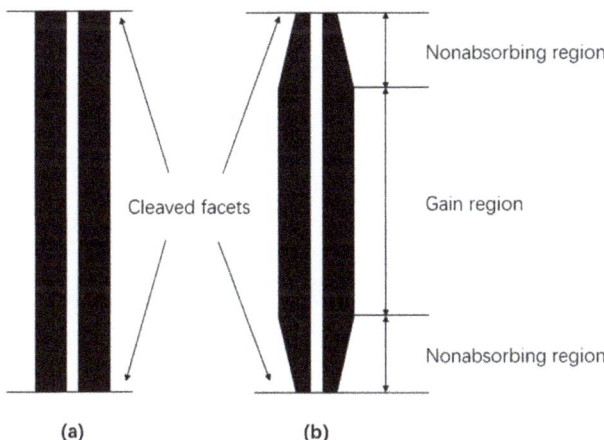

Figure 5. Schematic diagram of a stripe mask for two types of BH laser. For (**a**) Conventional stripe mask for a BH laser. For (**b**), a fusiform mask for nonabsorbing mirrors in a BH laser.

Epitaxial structures with different thicknesses and compositions were obtained by designing the mask geometry. Semiconductor lasers are integrated with other waveguides on the same wafer to form optoelectronic devices for various purposes. Kobayashi et al. [39] proposed an MQW BH laser integrated with a reduced-thickness waveguide layer. Both the gain and waveguide regions were simultaneously grown by SAE on a (100) InP-patterned substrate. The far-field FWHM of the device was 11.8° vertically and 8° laterally. The threshold current was 19 mA, with a high slope efficiency of 0.25 mW/mA.

Takemase et al. [40] reported an AlGaInAs BH laser with a mode profile converter (MPC) for the first time. The device consisted of a constant thickness portion as the gain region and a vertically tapered thickness portion to expand the beam output. The gain region and thickness-tapered portion were grown via SAE. The width of the opening area between the two masks was 20 μm. AlGaInAs MQW was grown between the mask stripes with a width of 100 μm. The gain layer was approximately two times thicker than that of the end of the tapered region. The SiO_2 mask was removed using HF after the current blocking layers were established. The characteristic temperature was 43 K at 60 °C. The poor temperature characteristic was attributed to the poor crystal quality of the AlGaInAs grown by SAE.

Bour et al. [41] described a self-aligned BH AlGaInAs QW laser using a micro-SAE. The general process of fabricating a self-aligned BH laser with a single growth step is shown in Figure 6. The 50 nm SiO_2 mask stripes were formed along [011] on a 2 μm InP n-cladding layer. The widths of opening region and masks were 1.5 and 7 μm, respectively. The sample was returned to the MOCVD reactor for growing AlGaInAs MQW and a separate confinement heterostructure (SCH) under high temperature conditions, accompanied by the {111} B sidewall formation, as shown in Figure 6b. The InP layer nucleated on the {111} sidewall when the temperature was lowered, because of a decrease in the surface mobility, and therefore the InP p-cladding layer was developed and encapsulated in the active region (Figure 6c). The SiO_2 mask was used as the current blocking layer without removal in the finished device, whose performance was improved compared to the conventional reverse-biased p–n junction.

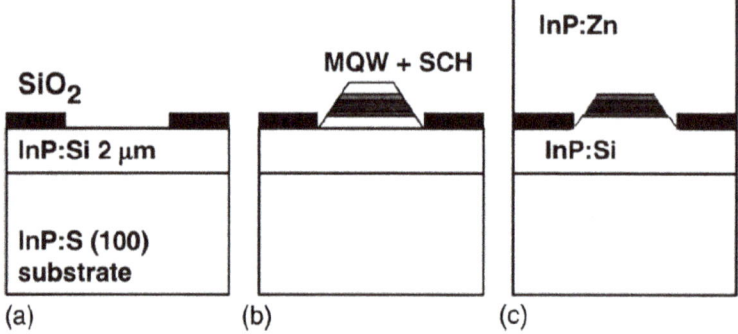

Figure 6. Process for a self-aligned BH laser. (**a**) n-type InP cladding layer and SiO_2 mask; (**b**) the active region and SCH under conditions wherein a {111} no-growth sidewall develops; (**c**) the upper p-type InP cladding layer. Reprinted with permission from ref. [41] © AIP Publishing. Copyright 2004 Applied Physics Letters.

Cai et al. [42] obtained an InGaAs/InGaAsP MQW BH laser in a single growth step. InP cladding layer was grown on {111} B sidewalls. The threshold current of the BH laser is 2.7 mA at room temperature. After a burn-in test of 600 h, the total degradation of the outpower was less than 6%, and the BH laser had excellent reliability. The SAE method has great potential for low-cost high-performance BH laser fabrications and associated device integrations due to its excellent performance and inherent manufacturing simplicity.

3.2. Quantum Dots (QD) Lasers

3.2.1. QDs Lasers by EBL and Lithography

Semiconductor quantum dots (QDs) that achieve full three-dimensional confinement are sphere-like nanostructures with strong carrier confinement effects. However, in MOCVD and MBE, pyramidal or cylindrical QDs are obtained due to the existence of a wetting layer. They are typically embedded in another material with a large bandgap. Since QDs have a lower threshold current density, narrower gain spectrum, higher optical gain, higher temperature stability, and better dynamic characteristics than QWs, they are an ideal gain medium for semiconductor lasers. However, it is challenging to grow QDs with uniform size and controlled position. The growth of high-quality QDs is a popular research topic.

The typical QD growth mode is Stranski–Krastanov (SK) growth, relying on strain-driven self-assembly between different material layers. If the strain is too high to grow continuous layers, the later-grown layers split into nanoscale islands when they reach a thickness of several single atoms. Subsequently, the island structures were embedded in an extensive–bandgap material to form QDs. This process is known as SK growth (Figure 7). The QD obtained using this method is called self-assembled QDs (SAQDs). The precisely controlled diameter and thickness of the QDs can stabilize the energy-level distribution without broadening the laser gain spectrum. The gain increased as density of the QDs increased. The operating threshold and gain spectrum characteristics of QD lasers were better than those of QW lasers. However, SAQDs have two disadvantages: (1) the presence of a wetting layer causes incomplete three-dimensional confinement and (2) the randomness of the nucleation position and the variation in size (diameter, thickness) also limit the advantages of QDs. Both lead to a shift in the quantized energy level and the non-uniform broadening of the gain spectrum, which is broader than that of QWs.

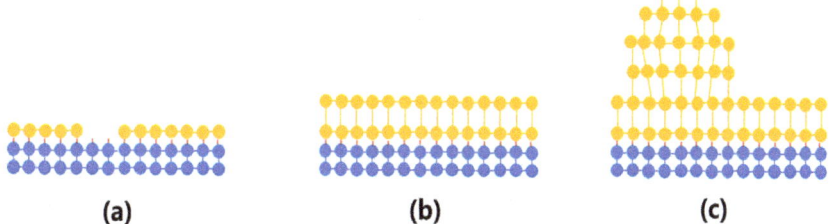

Figure 7. Schematic of SK growth. (**a**) Original molecule deposition. (**b**) Wetting layer formation. (**c**) QD formation and strain relaxation. The atoms of substrate and epitaxial layer were represented with blue and yellow, respectively.

The SAE is a method used to overcome the shortcomings of SAQDs' growth. The patterned grown QDs have no wetting layers and can control their position and diameter to obtain uniform QDs. The most widely used method of fabricating QDs by SAE is to first pattern a mask on the substrate with high-resolution lithography, ensuring that the size of the opening region is small enough to grow QDs.

Elarde et al. [43] combined EBL and SAE to prepare uniformly distributed QDs. The EBL was used to define the exact location of QDs' nucleation, and the SAE realized the control of QDs' geometries by patterned SiO_2 masks. They deposited 10 nm SiO_2 and patterned 4, 6, and 8 μm mask stripes using optical lithography and wet etching on n-type $Al_{0.75}Ga_{0.25}As$ cladding and GaAs barrier layers. Their samples were coated with polymethylmethacrylate to obtain arrays of 30 to 40 nm circular features in diameter using EBL. The arrays were centered over the SiO_2 mask stripes. After the array pattern was transferred to the SiO_2 masks by wet etching, 6.9 nm-thick $In_{0.35}Ga_{0.65}As$ QD layers were formed and encapsulated by 10 nm GaAs. The p-type cladding and contact layer were then grown during a later stage. The density of the fabricated QDs with a center spacing of 100 nm was 1.2×10^{10} cm^{-2}, and the diameter of QDs was an average of 80 nm. This report combined EBL and SAE to manufacture the first QD laser.

Mokkapati et al. [44] presented the InGaAs QD laser with different lasing wavelengths. When the width of the stripe was as small as 5 μm, QD intensity could provide insufficient gain for lasing. Correspondingly, an excessive thickness of InGaAs was obtained when the stripe width reached 20 μm, introducing defects and increasing losses. Akaishi et al. [45] fabricate a series of SiO_2 stripe masks uniformly arrayed on the InP substrate. They utilized a SiO_2 mask array consisting of a wide mask on the side and a series of relatively narrow mask stripes. By changing the width of the wide mask while maintaining the narrow stripe masks at 3 μm widths, the sizes of the InAs QDs varied in each opening region. The double-capped layer enabled the uniform height of QDs. When the width of the wide stripe mask was 200 μm, the emission wavelength range of the QD array grown between the narrow masks exceeded 120 nm.

3.2.2. QDs Lasers by Block Copolymer Lithography

Morever, patterned QDs can also be prepared by block copolymer (BCP) lithography. A typical diblock copolymer consists of alternately polymerizing two polymers, one of which is miscible in small amounts in another polymer. This immiscibility leads to the phase separation of the two materials and maintains the properties of each material. The interaction between the two polymers produces a QD pattern. The absence of a wetting layer resulted in perfect three-dimensional confinement. Figure 8 presents the process of SAE QD fabrication using diblock copolymer lithography.

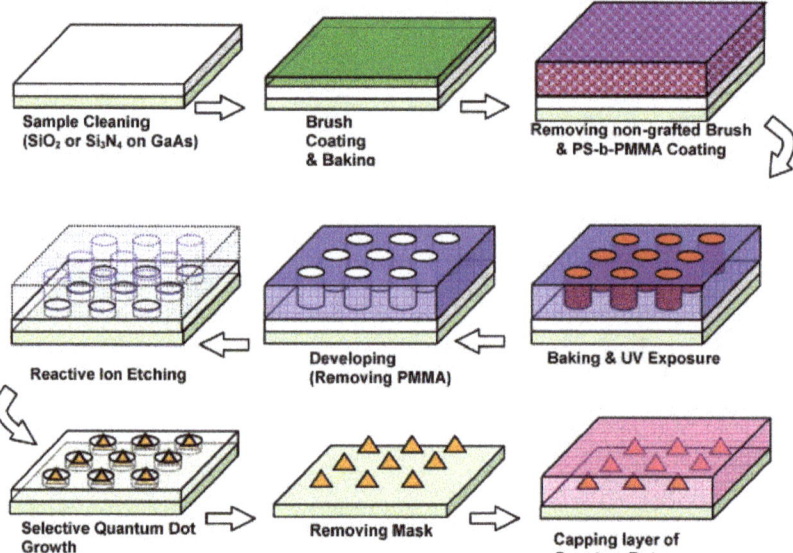

Figure 8. Selective growth of QDs by diblock copolymer lithography. Reprinted with permission from ref. [46] © Elsevier. Copyright 2006 Journal of Crystal Growth.

Polystyrene-block-poly (PS-b-PMMA) was the diblock copolymer in this process. First, the amorphous mask materials were deposited onto the substrate using plasma chemical vapor deposition (PECVD). Then the PS-b-PMMA brush was coated above the mask. After spinning the PS-b-PMMA on wafers, the entire structure was exposed to UV such that the PS area was exposed as the pattern template upon the removal of PMMA. ICP transferred the pattern into the mask layer, followed by the removal of the PS region via etching. Finally, the selective area growth of QD was accomplished in the cylinder hole.

The uniform size distribution and spatial position of QDs fabricated by block copolymer lithography provide more significant advantages than SAQDs grown in SK mode. Li et al. [47] obtained uniform GaAs QDs with a density of approximately 10^{11} cm^{-2} by selective growth in the narrow openings of SiN_x patterned masks using block copolymer lithography. Kim et al. [48] fabricated GaAs/InGaAs/GaAs compressive–strain QDs using SiN_x patterns defined by block copolymer lithography. Figure 9 shows the InGaAs QD laser structure. The CBr_4 in situ etching time was optimized to reduce the processing damage before growing the QDs, leading to a decrease of the nonradiative recombination centers of the QDs. Compared with the devices grown using non-optimized in situ etching times, the QD lasers grown using optimal etching times significantly reduced the threshold and transparency current. Furthermore, Kim et al. [49] fabricated an InAs QDs laser with an $In_{0.1}Ga_{0.9}As$ QW carrier collection layer to increase the carrier injection into the QDs. The QD laser achieved lasing at RT.

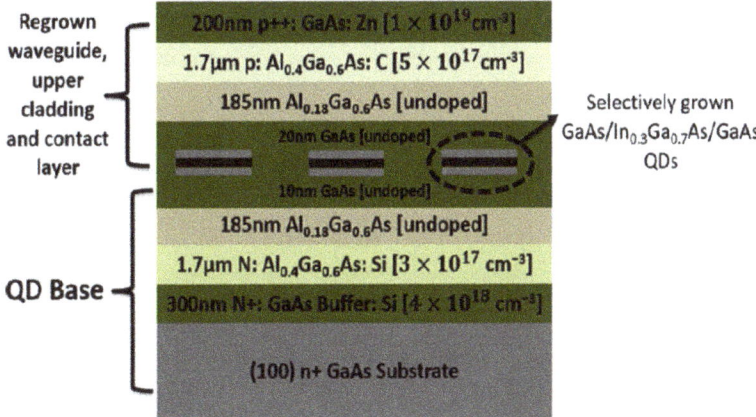

Figure 9. Structure of a InGaAs QD laser. The lower QD base was grown at 700 °C, and the upper regrown layers were grown at 625 °C. Reprinted with permission from ref. [48] © Elsevier. Copyright 2017 Journal of Crystal Growth.

3.3. Heteroepitaxy Lasers on Si or SOI

The natural characteristics of indirect bandgaps limit the wide applications of silicon photonic devices. This inherent property results in weak light emission and poor optoelectronic performance. Direct bandgap materials exhibit good luminescence properties. The direct bandgap III–V semiconductor lasers are high-performance light sources that may be utilized for silicon photonic integrated circuits. However, a high density of threading dislocations (TDs), planar defects, and anti-phase boundaries (APB) arises at the interface of III–V and silicon [50–53], owing to the mismatch of lattice constants, thermal expansion coefficient, and polarity between III–V materials and silicon. There is great research interest in finding methods for growing lattice constant-matched or low-defect-density III–V light sources directly on silicon. Defects introduced by the lattice and polarity mismatches between the III–V materials and Si have been substantially eliminated in a previous study using a template in conjunction with bonding techniques [54]. However, it is expensive and time-consuming to prepare such templates. Therefore, it is more desirable to directly epitaxially grow high-quality III–V materials on Si.

3.3.1. Aspect Ratio Trapping (ART)

Bonding technology is the most common III–V/Si integration technique, combining a known good III–V epitaxial layer with a silicon platform. The main shortcomings of the bonding technique are the strict requirements for ultra-clean and highly flat surfaces, the high thermal resistance introduced by the oxide during bonding, and the wafer size mismatched between the Si substrate and the III–V (InP and GaAs) material substrate. These shortcomings limit the high-density integration of laser on silicon [55]. The local bonding process is technically challenging and expensive, not conducive to mass production [22]. SAE has become an economical and reproductive approach for relieving the defects resulting from mismatched crystal growth and realizing the local integration of III–V lasers on a silicon platform. Figure 10 presents the principle of SAE to decrease dislocation densities in the epitaxial layer at the interface of III–V materials and silicon.

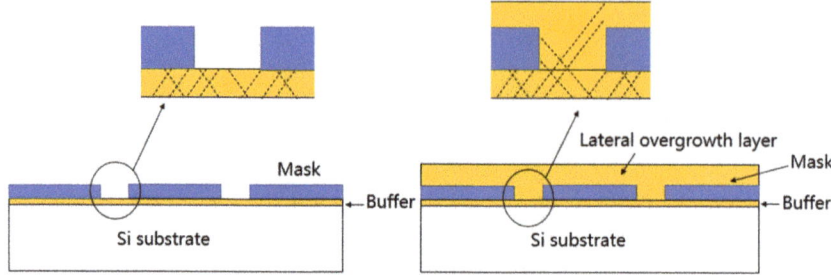

Figure 10. Principle of ART and epitaxial lateral overgrowth (ELO). The circle is a partially enlarged view of the dislocation propagation. The dislocations, marked as dotted lines, only propagate to the lateral overgrowth layer through openings in the patterned mask.

The growth begins in narrow openings on the masked substrate. The dislocations that occur at the interface of III–V and silicon because of the heavily lattice-mismatched can only propagate along a certain angle inclined to the substrate plane from the opening area to the epitaxial layer. For Si (0 0 1) substrates, the dislocations propagate along with the <1 1 1> directions, forming a 54.7° angle with (0 0 1). The thickness of the materials exceeded that of the mask as the growth process proceeded. The epitaxial layer grows laterally along the mask surface, which is referred to as epitaxial lateral overgrowth (ELO or ELOG). Dislocations cannot propagate into the mask and vanish in the epitaxial layers because the interface lattice between the crystalline material and the amorphous mask is discontinuous. If the growth time is sufficiently long, the epitaxial layer grown in the opening region merges with the layer grown in the neighboring opening region to form a complete epitaxial layer above the masks. The mask effectively prevented dislocation propagation, and the dislocation density distributed in the lateral overgrowth partial of the epitaxial layer was several orders of magnitude lower than that grown on the standard unmasked substrate. Figure 11 shows the cross-sectional TEM image of the ELOG InP with SiO_2 masks on silicon. The region above the mask was dislocation-free. Figure 11a shows that the dislocations from the buffer layer could penetrate the region above the mask if the openings were wider than the thickness of the mask. Figure 11b shows that the dislocations were filtered even above the openings if the opening width was smaller than or equal to the mask thickness. There were no coalescence defects above the mask, which refers to defects formed at the junction of two laterally grown InP layers from adjacent openings [56]. This approach, called the aspect ratio trap (ART), has been used in various materials grown on Si [57–59].

Moreover, III–V materials grown by ART are on the sub-micrometer scale, which is not conducive to micrometer-scale ridge waveguide fabrication. Han et al. [60] proposed a lateral ART scheme to grow a micrometer-scale InP sandwiched by oxides on the silicon-on-insulator (SOI). A schematic of lateral ART is shown in Figure 12. Compared to the conventional ART grown on (001) Si, lateral ART started at the {111} Si surface, which would not form APBs. The dimensions of the III–V ridge width were limited by the trench width in conventional ART. The size of III–V materials grown in lateral ART is controlled by the thickness of the Si layer, realizing precise micrometer- or nanometer-scale control.

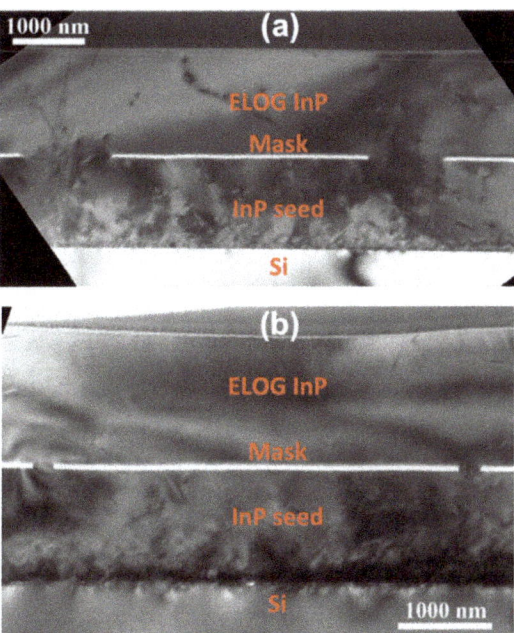

Figure 11. TEM cross-sectional view of ELOG InP on Si; (**a**) mask opening width was larger than the thickness of the mask; (**b**) mask opening width was small to filter dislocations. Reprinted with permission from ref. [56] © Elsevier. Copyright 2012 Current Opinion in Solid State & Materials Science.

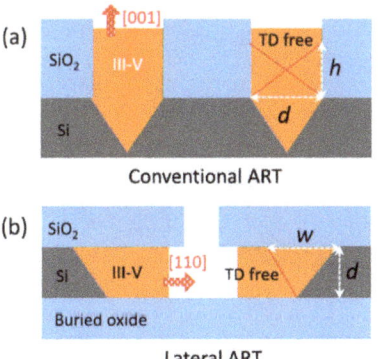

Figure 12. (**a**) Schematic of conventional ART technique. The epitaxy proceeds along [001] direction. (**b**) Schematic of defect trapping and growth mechanism of lateral ART technique. The growth was along the [110] direction. Reprinted with permission from ref. [60] © AIP Publishing. Copyright 2019 Applied Physics Letters.

The ART method has been widely used in shallow trench isolation (STI) structures, where a thin Ge layer is typically grown in situ and covers the bottom of the trench as a buffer layer. The lattice mismatch between InP and Ge is only half that between InP and Si, leading to the easier nucleation of InP on the Ge surface [61]. By designing the Ge surface profiles and setting an aspect ratio larger than two, TDs were confined at the bottom of the trench, and APB formation was suppressed [62,63]. GaAs [64,65], InP [66], and InGaAs [67] have been grown on (001) Si with low defect density using SAE technology. After obtaining defect-less high-quality epitaxial layers, III–V materials were grown to fabricate lasers. Staudinger et al. [68] fabricated wurtzite InP microdisks using SAE. They deposited 300 nm-thick SiO$_2$ and patterned approximately 50 to 100 nm line openings. The wurtzite InP was nucleated in the trenches and grew vertically along the [111] A-direction. The zipper-points at the center of the line openings induced ELO after the wurtzite InP extended out of the SiO$_2$ mask, and {1−100} or {11−20} wurtzite facets were formed. The micro-photoluminescence of this system demonstrated that the wurtzite InP microdisk enabled optically driven lasing at room temperature with a threshold of 365 µJ cm^{-2}. Wong et al. [69] deposited 200 nm-thick SiO$_2$ onto InP (111) A substrate using PECVD. They patterned 120 nm ring-shaped mask openings using EBL and ICP. A micro-ring InP laser was obtained via SAE using a two-stage growth process. In the first stage, the temperature and V/III were 750 °C and 300, respectively. The opening areas were completely filled with high-quality wurtzite InP, and ELO occurred. The growth temperature was then lowered to 730 °C, and the V/III content was increased to suppress vertical growth and enhance lateral growth. The micro-ring laser was operated at room temperature with a low threshold.

3.3.2. Lasers on Si/SOI

Wang et al. [70] demonstrated an optically pumped InP-based DFB laser array integrated on (001) Si, operating at room temperature using the SAE technique. InP was selectively grown in the trenches of the STI structure. ART limits the defective layer to a thickness of approximately 20 nm, instead of several micrometers. The emitted light can couple with the optical waveguide defined in the same horizontal plane. Figure 13 shows the structure of the device. Each DFB laser in the array contained an InP waveguide with a λ/4-phase shift first-order grating, and a second-order grating was defined 30 µm away from the DFB cavities. The Si substrate below the InP laser was intentionally etched away to avoid pump light absorption. Silicon substrates supported suspended cavities at both ends.

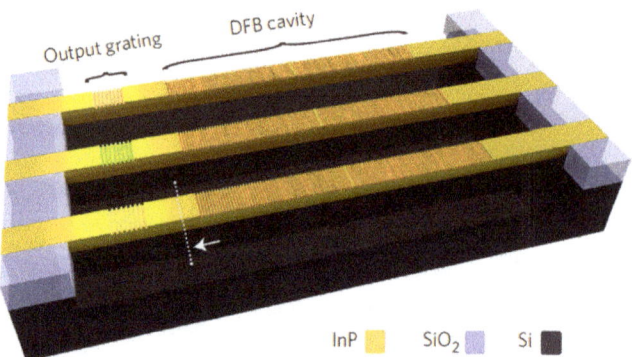

Figure 13. Schematic diagram of the InP-on-Si DFB laser array structure. Reprinted with permission from ref. [70] © Springer Nature. Copyright 2015 Nature Photonics.

The suspended cavity exhibited thermal, electrical injection, and mechanical stability issues. These problems can be addressed if the laser structure is grown on a SOI. Megalini et al. [71] successfully fabricated InGaAsP MQWs in InP nanowires. SEM and TEM images of the InGaAsP MQWs are shown in Figure 14. The 500 nm-thick SiO$_2$ was deposited on the Si surface and patterned into 200 nm-wide stripes with a spacing of 800 nm. The Si layer was etched by KOH to form V-grooves. The inhomogeneous spectrum shape broadening and wavelength peak emission observed in PL tests were attributed to the poor uniformity of the MQWs, as shown in Figure 14a. The MQWs of {001} were thicker than those of {111}, as shown in Figure 14c,d, which caused a spectrum of inhomogeneous broadening and decreased the lasing material gain. The growth was completed during the V-groove patterning of the SOI to trap most of the defects at the Si/SiO$_2$ interface. The PL peak wavelength of the MQW was 1567 nm.

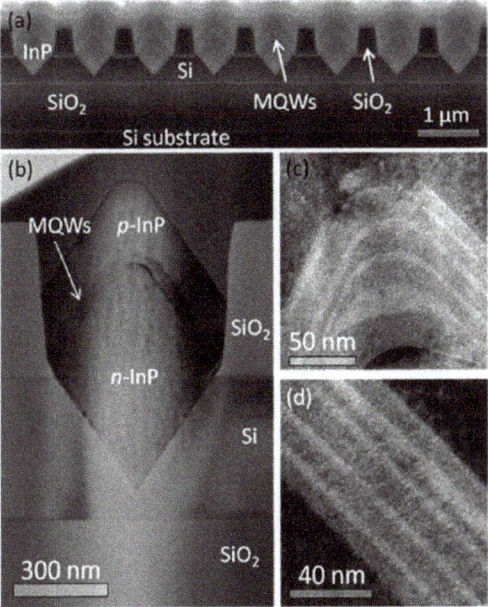

Figure 14. (a) SEM cross-section images of the InP nano-ridge. (b) TEM image of a single nano-ridge with a good symmetry. TEM images of the QWs and the barrier layers in (c) the (001) surface and (d) the {11−1} B facet. The dark-colored layer is the barrier; the light-colored is the QW. Reprinted with permission from ref. [71] © AIP Publishing. Copyright 2017 Applied Physics Letters.

The mask prevents dislocation propagation and the growth of the ~µm buffer layer using the ART method. Compared to optically pumped lasers, electrically pumped lasers are more challenging to manufacture. There are two main difficulties: acquiring high-quality ternary or quaternary compounds and the fabrication of micrometer-scale ridge waveguides with the constraints of sub-micrometer selective masks. Kunert et al. [72,73] integrated an InGaAs/GaAs heterostructure into box-shaped GaAs ridges with (001) flat surfaces outside the trenches, as shown in Figure 15. An apparent PL of the QW was observed for different ridge sizes. A sufficient III–V volume to fabricate a micrometer-order ridge waveguide can be realized by extending the ridge width.

Several studies have reported on electrically pumped lasers. Shi et al. [74] utilized InGaAs/InP strained layer superlattices (SLSs) as filter layers to suppress the propagating TDs. III–V materials were grown in V-grooved (001) Si to realize ART. They fabricated an electrically pumped InP-based laser at 1550 nm on (001) Si. Figure 16 shows the laser structure. The threshold current density was 2.05 kA/cm^2, and the slope efficiency was 0.07 W/A. The maximum output power was 18 mW without facet coating. The continuous-wave (CW) operation temperature reached 65 °C, and the pulsed lasing operation temperature was up to 105 °C.

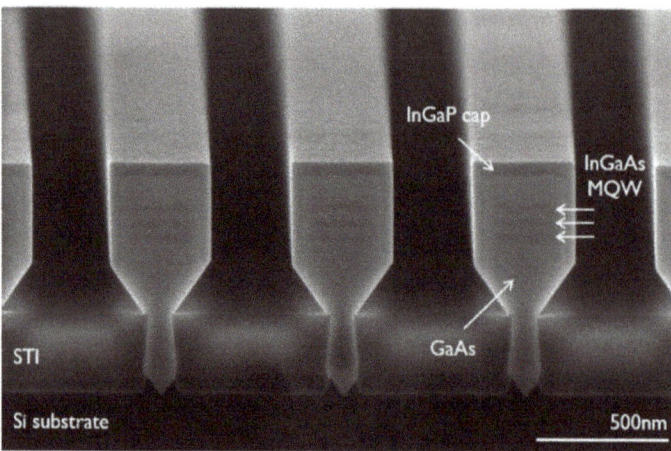

Figure 15. SEM image of the InGaAs/GaAs MQWs in box-shaped GaAs nano ridge structure; the InGaP cap layer was easily observed by the darker contrast. The three MQWs were slightly lighter in color. Reprinted with permission from ref. [72] © IOP Publishing. Copyright 2016 ECS Transactions.

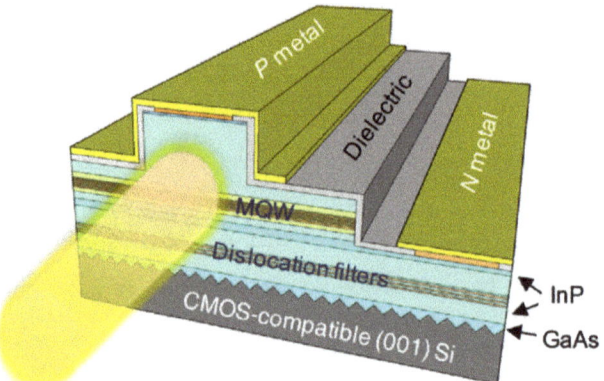

Figure 16. Schematic representation of an InP-based laser on (001) Si. Reprinted with permission from ref. [74] © The Optical Society. Copyright 2019 Optica.

Wei et al. [75] fabricated a 1.3 μm InAs/InAlGaAs quantum dash (QDash) laser on V-grooved (001) Si. The two sets of eight-period $In_{0.16}Ga_{0.84}As$/GaAs films were grown above the V-grooved Si as a buffer layer to trap dislocations at the GaAs/Si interface. TDs were filtered in three sets of ten-period $In_{0.61}Ga_{0.39}As$/InP SLSs. After the n-type InP contact and cladding layers were grown, the three layers of InAs/InAlGaAs QDashes were formed. Finally, the p-type InP cladding and contact layers were deposited. A schematic of the device structure is shown in Figure 17. The threshold current density was 1.05 kA/cm^2, and the output power per facet was 22 mW. The laser realized CW operation at 70 °C.

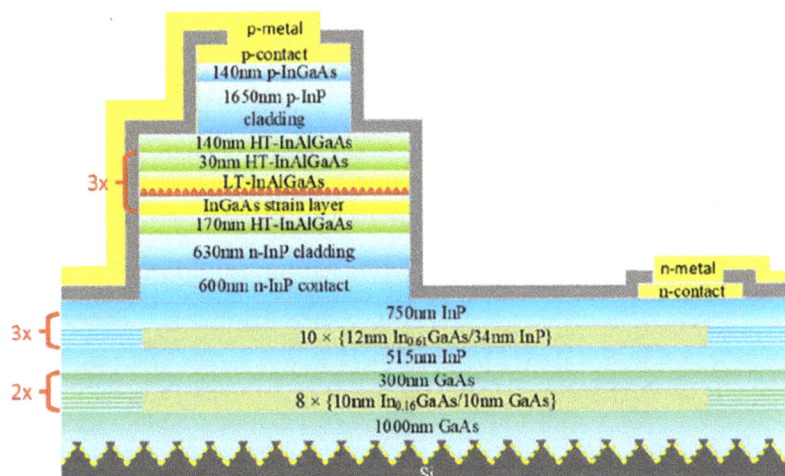

Figure 17. Cross-section schematic of QDash laser on V-grooved (001) Si, including the structural parameters of each epitaxial layer. Reprinted with permission from ref. [75] © AIP Publishing. Copyright 2020 Applied Physics Letters.

Fujii et al. [76] designed an InP-on-SOI template consisting of an InP buffer layer, InGaAs etch stop layer, and InP layer on a (001) InP substrate. The template was fabricated using a bonding technique. A SiO_2 mask was deposited on the template and patterned using photolithography. They successfully fabricated an electrically pumped eight-channel membrane DFB laser array with wavelengths ranging from 1272.3 to 1310.5 nm by adjusting the geometries of masks to optimize the InGaAlAs MQWs on InP-on-SOI. The active regions were buried by n-doped and p-doped InP, which also formed lateral p-i-n structures. A schematic of the membrane laser array is shown in Figure 18. The fiber-coupled output power was greater than 1.5 mW in each channel at 25 °C. The actual lasing wavelength deviated from the designed lasing wavelength by less than 2 nm, and the average channel spacing was 860 GHz.

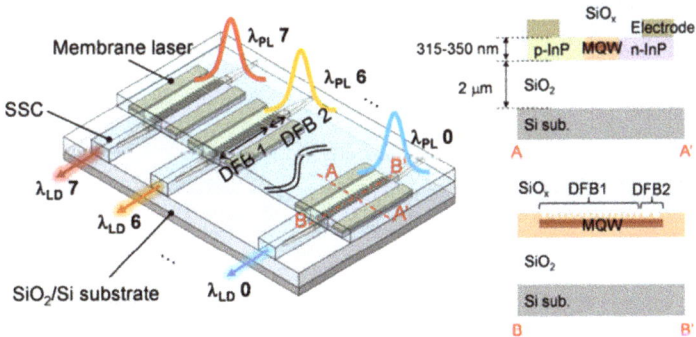

Figure 18. Schematic diagram of eight-channel laser array and membrane laser cross–sections. The output light for each channel was coupled with single–mode fiber by butt–coupling the facets of the high–numerical–aperture fiber and spot-size converter (SSC). Reprinted with permission from ref. [76] © The Optical Society. Copyright 2020 Optica.

Several studies on the SAE growth of III–V materials on SOI or Si have been conducted. The aim is to obtain epitaxial structures with a low defect density on Si. The main method is ART, which confines most of the dislocations in a nanometer–thick buffer layer, preventing defects from reaching the surface of the epitaxial layer, thus deteriorating the performance of the optoelectronic device. However, the problem of obtaining high-quality ternary or quaternary III–V compounds limits the epitaxial laser structure on Si.

3.4. Integrated Semiconductor Laser

Monolithic PIC has been developed to meet the increasing data flow transmission demand of the Internet. Advanced material growth methods have promoted the development of the PIC. The PIC refers to the integration of numerous optoelectronic components on the same chip [77]. The various device structures in PICs require different energy gaps for each portion of the wafer, and the device must be integrated in a reproducible and cost-effective manner. SAE enables the spatially localized epitaxial growth of the desired material structures for specific chip functions. This is a low-cost and reproducible method to achieve device integration. Recently, integrated semiconductor lasers have included EML and MWLA, which are suitable for wavelength division multiplexing (WDM).

3.4.1. EML

Monolithically integrated EML arrays are a promising light source for modern WDM systems. EML have attracted extensive attention owing to their small size, low packaging cost, low driving voltage, and good stability [78]. In the long waveband of thelong-distance optical communication system, such as 1.31 µm and 1.55 µm, the chirp phenomenon will appear, which is not conducive to the high-speed transmission of information. The modulator is integrated with the laser to form the modulated optical signal source because the modulator can avoid the large-wavelength chirp observed in directly modulated lasers. Using the SAE, the EML is fabricated by combining an electro-absorption modulator (EAM) and a laser on the same chip. Compared with discrete EAM and lasers, EML without fiber coupling reduces loss and cost and improves device reliability. At present, there are two main methods to realize the integration of EAM and laser: butt-joint (BJ) coupling (the laser and EAM are grown separately) and SAE [79].

Figure 19 shows the QWs achieved by BJ, wherein the MQWs of the EAM and the laser are grown in separate epitaxy steps. The MQW laser was grown on the entire wafer in the first step. Then, wet or dry etching was used to selectively etch the region, where the EAM MQW would grow in the second epitaxy. Two epitaxy steps increase the manufacturing cost, and the etched interface became rough and easily formed defects, deteriorating the output performance of the EML. BJ enables the optimization of the laser and EAM.

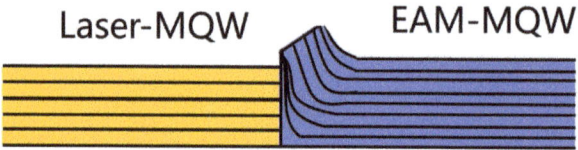

Figure 19. Schematic of the EAM and laser MQWs achieved using the butt-joint technique.

The SAE technique avoids poor interfaces and multiple epitaxy, as shown in Figure 20. Dielectric oxide masks were deposited and patterned in the laser region. The mask width was changed such that the laser region between the masks had a higher growth rate than that of the EAM region without mask coverage. The lower bandgaps of the MQWs of the laser were formed in one epitaxy step because of the growth-rate enhancement, which avoided the absorption of light due to the rough interface introduced by the etching process between multiple epitaxial growth layers and increased the output power of this system [80]. The disadvantage of the SAE is that the optimal parameters of the active region for the laser and modulator cannot be obtained simultaneously [81]. The bandgap of the MQWs in the EAM region is typically designed to be larger than that of the laser MQWs. The MQWs were grown between and outside of the mask stripes via SAE. The thickness and composition of the MQWs in the two regions differed in one-step epitaxy. The MQWs located between the mask stripes could be well controlled to obtain the desired gain region, while it was difficult to simultaneously obtain perfect EAM MQWs outside of the mask region. For example, the laser preferably exhibited few QWs at the lower threshold current, while the EAM region required more QWs to obtain a high extinction ratio.

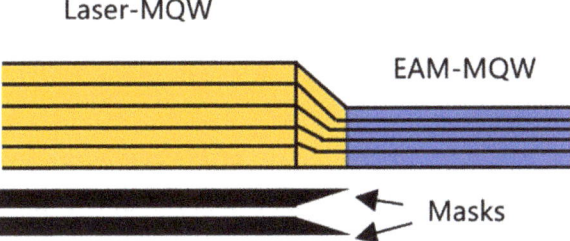

Figure 20. Schematic of the EAM and laser MQWs achieved using the SAE technique.

Zhao et al. [82] integrated a high-mesa DFB laser and EAM using SAE. The absorption and active regions consisted of an InGaAsP/InGaAsP MQW structure and an optical confinement layer with different thicknesses caused by the fusiform mask, as shown in Figure 5b. The spacing of the two SiO$_2$ mask stripes was fixed at 15 µm, and the mask width varied from 30 to 15 µm in the tapered region. Figure 21 presents a schematic of this device. The EML CW threshold current was 26 mA without the modulation bias for the uncoated laser. The output power of the modulation was 5.5 mW when the current was 100 mA at the laser, in single–mode operation with a side–mode suppression ratio (SMSR) > 40 dB at 1.552 µm. The on/off ratio was 15 dB at the biased voltage of −5 V on the EAM.

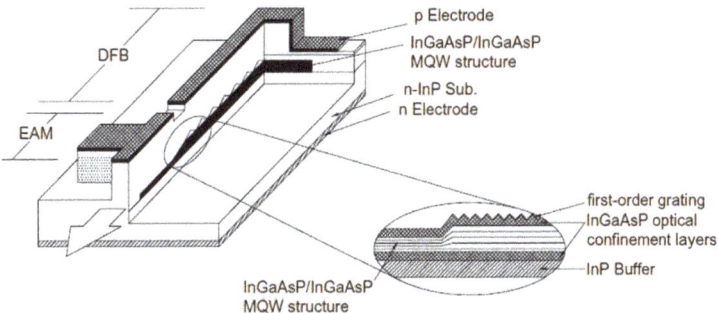

Figure 21. Ridge-waveguide EML consisting of a DFB laser and EAM. The enlarged portion shows the MQWs and interface of the EAM and laser regions. Reprinted with permission from ref. [82] © IOP Publishing. Copyright 2005 Semiconductor Science and Technology.

Kim et al. [83] designed and fabricated an EML consisting of a distributed Bragg reflector (DBR) laser and an EAM. The mask geometry and schematic of the device are shown in Figure 22. The threshold current was 5.7 mA. The output power was 5 mW when the active region injection current was 65 mA. The SMSR was >45 dB at a laser current of 60 mA. A static extinction ratio greater than 20 dB at a biased voltage of −3 V was applied to the EAM.

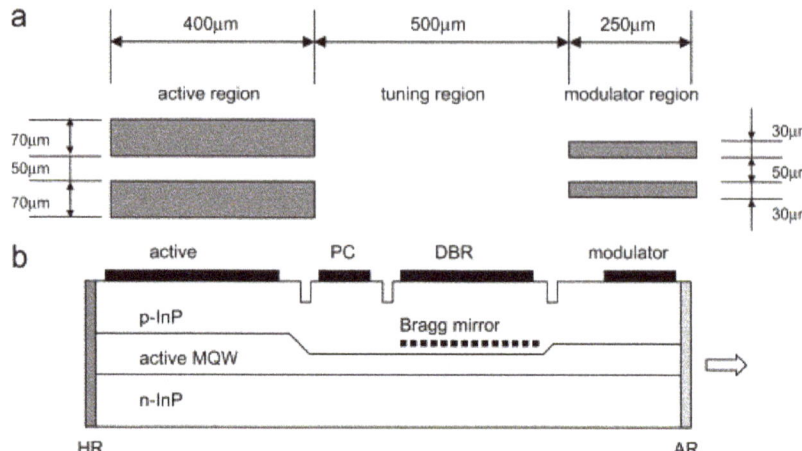

Figure 22. (a) Geometry of the SiN$_x$ mask and (b) schematic diagram of an EAM integrated with a tunable DBR laser. Reprinted with permission from ref. [83] © Elsevier. Copyright 2007 Journal of Crystal Growth.

There is a transition zone of a few microns between the EAM and the laser. The energy bandgap in this region changes slowly because of the bandgap engineering for the SAE (shown in Figure 4), leading to light absorption losses. The quantum well intermixing (QWI) technique enables bandgap changes within a few microns. QWI refers to the generation of defects on the surface using a special method involving QWs. Defects can become incorporated into the QW active region under external actions, such as thermal annealing, thereby changing the QW bandgaps. Combining the QWI and SAE techniques further reduce the coupling losses of the EML. After undergoing SAE, the laser gain and EAM regions were covered with a thermal silicon oxide layer. The QW bandgap of the interface between the EAM and laser was abruptly changed within several micrometers through ion implantation and rapid thermal annealing, reducing the absorption loss of light.

Liu et al. [84] integrated a sampled-grating distributed Bragg reflector (SG-DBR) laser with an EAM by combining SAE and QWI, as shown in Figure 23. The mask width was 20 µm, and the gap of SiO$_2$ stripes was 16 µm. When the current was 200 mA, the threshold current was 62 mA and the output power was 3.6 mW. The current of the front and rear mirrors varied from 0 to 70 mA, with increments of 2 mA; the gain part current was 100 mA unchanged, and the phase current was kept at 0 mA. The lasing wavelength ranged from 1552 to 1582 nm, and all SMSRs were larger than 30 dB. The extinction ratio was 17 dB at a bias voltage of −5 V in the EAM.

In addition to the conventional SAE technology that fabricates EML, Zhu et al. [85] used a modified double-stack active layer (DSAL) SAE technology, which has the advantages of both BJ and conventional SAE. Contrary to Figure 20, SiO$_2$ was deposited in the EAM region, followed by the growth of the EAM and laser MQWs sequentially in a single-step epitaxy process. The SAE-DSAL technique enabled the optimization of the EAM and laser MQWs separately because the growth proceeded at different times. The energy-gap difference became larger between the upper and lower MQWs in the laser region, which reduced the absorption of the EAM. The SAE-DSAL technique can reduce the threshold current of the EML. The threshold current of the EML was 20 mA. The output power was 10 mW with an injection current of 100 mA. The lasing wavelength was 1550.5 nm with an SMSR of more than 41 dB. The extinction ratio was 12 dB when the −3 V bias voltage was applied to the EAM. Zhu et al. [86] also fabricated an EML that combined SAE and DSAL. The threshold current was further decreased to 16 mA, and the output power was larger than 10 mW when the injection current was 64 mA. The wavelength was 1552.28 nm with an SMSR larger than 53 dB. A 30 dB static extinction ratio over 30 was obtained when the bias voltage in the EAM was −5 V.

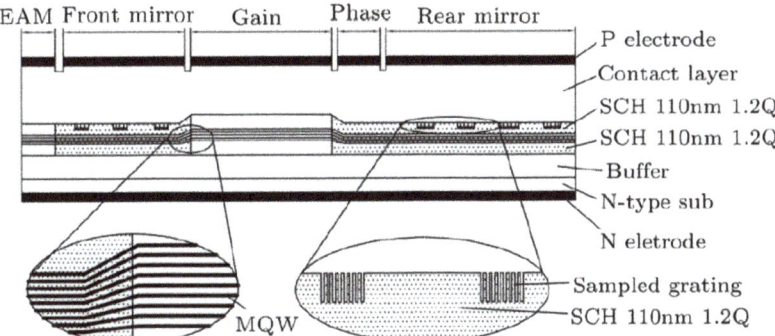

Figure 23. Schematic of EML consisting of a SG-DBR laser and EAM by combining SAE and QWI. The active region consisted of eight compressively strained InGaAsP MQWs. The 1.2 Q InGaAsP was used as the SCH structure. The grating mask and contact layer were InP layers. Reprinted with permission from ref. [84] © Chinese Physical Society. Copyright 2008 Chinese Physics Letters.

Monolithically integrated EML arrays are key light sources in modern dense wavelength division multiplexing (DWDM) systems. DWDM systems require high-speed, high-coupling output power, single-mode, and low-chirp multi-wavelength light sources in long-distance optical communications [87]. SAE reduces the complexity of EML array fabrication. Cheng et al. [88] designed a four-channel EML array and completed the device using SAE. The threshold current was approximately 18 mA, and the output power at 100 mA was 9 mW. The lasing wavelength ranged from 1551.8 nm to 1554 nm, and the average channel spacing was approximately 0.8 nm. The average value of the single-mode SMSR was up to 45 dB. The extinction ratio was 15 dB with −5 V voltage applied to the modulator.

Xu et al. [89] reported a ten-channel EML array using SAE. The arrayed waveguide grating (AWG) combiner was integrated with an EML array using BJ. The Ti film heaters integrated into the device achieved thermal tuning. The fabricated device is shown in Figure 24. The emission of the ten-channel EML spacing was 1.8 nm. The PL peak wavelength of the ten-channel in the laser region ranged from 1530 nm to 1580 nm due to the increasing width of the stripe mask. The threshold currents of each laser in the array was between 30 and 60 mA. The output power ranged from 8 to 13 mW at an injected current of 200 mA injected current. All channels had high single-mode light emission with an SMSR > 40 dB. The device had a static extinction ratio greater than 11 dB and a modulation bandwidth larger than 8 GHz.

3.4.2. MWLA

Multi-wavelength DFB laser arrays (MWLAs) can realize a wide range of wavelength tuning in WDM optical communications. The advantages of monolithic integrated laser arrays are miniaturization, cost-effectiveness, high reliability, and low consumption [90,91]. Lasers with different emission wavelengths and uniform channel spacing are realized in MWLAs by adjusting the grating pitch and the ridge waveguide width [92–95]. The use of EBL to fabricate MWLA is time–consuming and expensive, and the limitation of EBL resolution makes it difficult to fabricate MWLA with a channel spacing of 1 nm. In the SAE, the same channel spacing can be achieved with oxide stripe mask widths of a few micrometers. The reproducibility and processing simplicity of SAE are much better than those of EBL. In SAE, the lasing wavelength of the active region and the effective refractive index of the waveguide are modulated by the thickness and composition variation of the materials, which is achieved through the control of the dielectric mask geometry and material growth conditions.

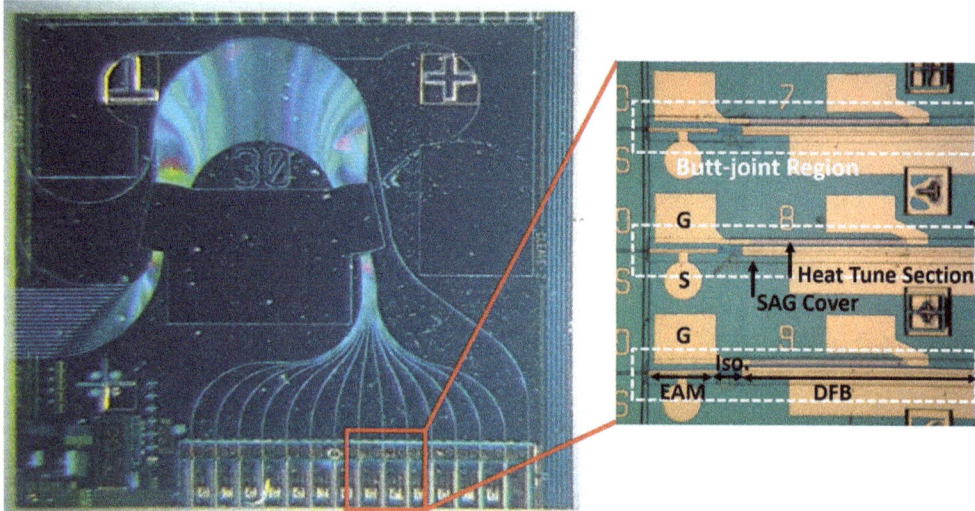

Figure 24. Optical graph of the fabricated device. The size of the integrated chip was 5800 × 5800 µm^2. The lengths of the DFB laser and EAM were 500 and 150 µm, respectively. A 50 µm isolation region was located between the EAM and DFB laser. Reprinted with permission from ref. [89] © Elsevier. Copyright 2017 Optics & Laser Technology.

Darja et al. [96] reported a four–channel DFB laser array with multi-mode interference (MMI) for 1.55 µm coarse wavelength division multiplexing (CWDM) systems. The width of the opening region between the mask stripes was 15 µm. The device fabrication process included: (1) the formation of the InGaAsP DFB grating; (2) SiO$_2$ mask pattern fabrication; and (3) the growth of the InGaAsP MQW, InP cladding layers, and InGaAs capping layer. The average threshold current of the four-channel was an average of 70 mA. The lasing wavelengths of the four-channel DFB laser arrays were 1521.2, 1541.4, 1564, and 1580.6 nm, with SMSR > 30 dB.

The lasing wavelength and channel spacing of the MWLA must be consistent with the desired wavelength. It is not easy to guarantee that the actual situation of growing materials between different elements in MWLA matches the ideal design. Zhang et al. [97] modified the SAE method as shown in Figure 25. The conventional SAE-deposited and patterned masks on the buffer layer realized the selective growth of the laser arrays. The thicknesses of the SCH and MQW layers increased due to growth-rate enhancement, as shown in Figure 25a. The lower SCH and MQW active regions were first formed in the buffer layer. Then, the patterned masks were deposited above the MQW layer, followed by the completion of the remainder of the laser structure, as shown in Figure 25b. The gap of each mask stripe was fixed at 20 µm. The widths of the stripe masks varied from 0 µm by 1.5 µm steps to adjust the thickness of the upper SCH layer to different Brag wavelengths. The threshold current of each channel was approximately 18 mA in MWLA, with a wavelength spacing of 0.42 nm. The lower SCH and MQW layers achieved excellent control of the lasing wavelength and channel spacing because they were insensitive to the growth-rate enhancement and composition variation of SAE. The wavelength matching of the device was better than that of the MWLA fabricated using EBL.

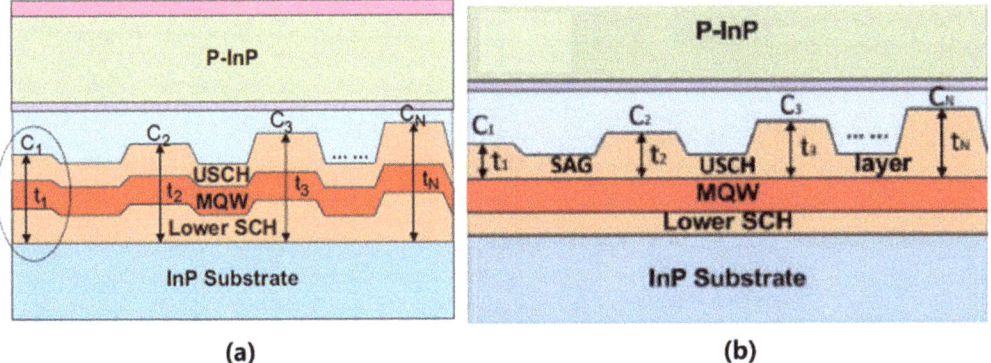

Figure 25. (a) Schematic of the MWLA structure achieved via conventional SAE. The upper and lower SCH and MQW were thicker due to the growth-rate enhancement generated by the masks. Reprinted with permission from ref. [98] © Elsevier. Copyright 2013 Optics Communications. (b) Modified SAE. The lower SCH and MQW were hardly affected by the growth-rate enhancement, enabling the realization of precise emission wavelengths. Reprinted with permission from ref. [95] © Springer Nature. Copyright 2018 Science China Information Sciences.

Zhang et al. [99] subsequently combined the modified SAE and bundle integrated guide (BIG) to fabricate a four-channel DFB laser array integrated with an MMI and a semiconductor optical amplifier (SOA). The output power of each channel in the array was 17 mW at a current of 200 mA, and the slope efficiency was 0.11 W/A. The optical spectra of the lasers in the array showed SMSRs larger than 42 dB, with an average channel spacing of 0.6 nm.

Zhang et al. [100] integrated a ten-channel EML array with MMI to fabricate a transmitter by SAE and BJ. The average output power of each channel was 0.25 W with an injected current of 200 mA. The threshold current of the ten channels was between 20 and 40 mA. The SMSRs of the entire channel were greater than 40 dB. The extinction ratio ranged from 15 to 27 dB at a biased voltage of -6 V.

Guo et al. [101] conducted a 1.3 μm six-channel DFB laser based on an SAE for the first time. The width between the SiO_2 mask stripes was fixed at 30 μm, and the mask widths in the array were 20, 30, 40, 50, 55, and 65 μm. The threshold current was approximately 20 mA, and the output power of all channels was greater than 10 mW at an injection current of 100 mA. The lasing wavelengths of the six-channel DFB laser were 1301.8, 1305.1, 1308.1, 1311.2, 1313, and 1315.7 nm, respectively. The threshold current was approximately 20 mA, and the output power of all channels was greater than 10 mW at an injection current 100 mA.

Kwon et al. [102] developed a ten-channel DFB laser array operating at 1.55 μm for the transmission of a 100 Gbit/s Ethernet system. The average SMSR was larger than 50 dB, and the average channel spacing was 8.2 nm at an injection current of 50 mA. The average threshold current in the ten-channel DFB laser arrays was 25 mA. Subsequently, Kwon et al. [103] fabricated a planar-buried heterostructure (PBH) eight-channel DFB laser array to obtain a higher energy efficiency and better output beam quality attributed to fine current confinement than that of the ridge waveguide structure. The threshold current of the lasers ranged from 8.5 to 11 mA, which is lower than that of the ridge waveguide structure DFB laser array described in a previous study [102]. The output spectra showed that the lasing wavelength changed from 1528.4 to 1584 nm, with an interval of 8 nm. The SMSR of each channel was larger than 50 dB, and the power loss was less than 2 dB after a 2 km transmission.

4. Conclusions

SAE is a crucial design and integration tool for fabricating semiconductor lasers. The thickness and composition of the active layer depend on the geometry of the patterned mask. Knowing the characteristics of growth-rate enhancement and composition variation in SAE is significant in realizing wavelength-tunable lasers. SAE has completed the fabrication of the BH laser with a low threshold and nearly circulated light spot, owing to its cost-effectiveness and inherent manufacturing simplicity. The SAE achieved uniform distribution and size of QDs. The main challenges in obtaining QDs using SAE are increasing the density of QDs and reducing the defects at the interfaces between the QDs and the surrounding embedded materials to improve the laser gain. Although the realization of electrically driven lasers on Si is limited by the availability of micrometer-scale materials and the large number of defects created by direct epitaxy on Si, novel growth schemes, including lateral ART, enable the material dimension to be micrometer-scale and block material defects within an appreciable size, which is beneficial for achieving electrically pumped lasers on Si.

SAE integrates lasers with other optoelectronic devices that are repeatable and cost-effective, typically EML and MWLA. The EML and MWLA will continue to move towards high-speed arrays with low power consumption and cost-effectiveness. The emission intensity of the active region of the EML array channel decreases as the wavelength increases because of the accumulation of SAE mask defects, deteriorating long-distance optical transmission. The SAE should improve the precisely controllable lasing wavelength and channel spacing in the MWLA. Although remarkable progress has been made in the material growth and device fabrication in SAE, specific methods that can realize the precise control of emission wavelength and the position of materials in the active region are still unknown. Defects caused by the presence of the mask more or less deteriorate the output characteristics of lasers. The potential of SAE to reduce material defects, lower device consumption, achieve controllable lasing wavelengths, and improve modulation performance is yet to be explored and will be investigated in future studies regarding the SAE technique.

Author Contributions: Conceptualization, Y.Z. and Y.S.; methodology, Y.Z. and L.Q.; software, Y.W., L.L. and Y.S.; validation, B.W. and Y.Z.; formal analysis, J.Z., P.J., Y.L. and C.Q.; investigation, Y.Z., L.L., Y.S., J.Z., P.J., Y.L. and C.Q.; resources, Y.Z. and L.Q.; data curation, B.W. and Y.Z.; writing—original draft preparation, B.W.; writing—review and editing, Y.Z.; visualization, B.W.; supervision, Y.Z.; project administration, L.Q. and Y.N.; funding acquisition, L.Q. and L.W. All authors have read and agreed to the published version of the manuscript.

Funding: This work was funded by National Science and Technology Major Project of China (2018YFB2200300); National Natural Science Foundation of China (NSFC) (62090051, 62090052, 62090054, 11874353, 62121005, 62090061, 61935009, 61934003, 61904179, 62004194); Science and Technology Development Project of Jilin Province (20200401069GX, 20200401062GX, 20200501006GX, 20200501007GX, 20200501008GX); Key R&D Program of Changchun (21ZGG13, 21ZGN23); Innovation and entrepreneurship Talent Project of Jilin Province (2021Y008); Special Scientific Research Project of Academician Innovation Platform in Hainan Province (YSPTZX202034); Lingyan Research Program of Zhejiang Province (2022C01108); Dawn Talent Training Program of CIOMP.

Data Availability Statement: Not applicable.

Acknowledgments: The authors would like to thank Yugang Zeng, Yue Song, Li Qin, Lei Liang, Jianwei Zhang, Peng Jia, Yuxin Lei, Cheng Qiu, Yongqiang Ning, and Lijun Wang for helping with this article.

Conflicts of Interest: The authors declare no conflict of interest.

References

1. Suematsu, Y.; Iga, K. Semiconductor Lasers in Photonics. *J. Lightwave Technol.* **2008**, *26*, 1132–1144. [CrossRef]
2. Xiong, G.X.; Li, P. Study of the effects of semiconductor laser irradiation on peripheral nerve injury. *Laser Phys.* **2012**, *22*, 1752–1754. [CrossRef]

3. Nasim, H.; Jamil, Y. Diode lasers: From laboratory to industry. *Opt. Laser Technol.* **2014**, *56*, 211–222. [CrossRef]
4. Chi, C.-Y.; Chang, C.-C.; Hu, S.; Yeh, T.-W.; Cronin, S.B.; Dapkus, P.D. Twin-Free GaAs Nanosheets by Selective Area Growth: Implications for Defect-Free Nanostructures. *Nano Lett.* **2013**, *13*, 2506–2515. [CrossRef]
5. Bollani, M.; Fedorov, A.; Albani, M.; Bietti, S.; Bergamaschini, R.; Montalenti, F.; Ballabio, A.; Miglio, L.; Sanguinetti, S. Selective Area Epitaxy of GaAs/Ge/Si Nanomembranes: A Morphological Study. *Crystals* **2020**, *10*, 57. [CrossRef]
6. Cockerill, T.M.; Forbes, D.V.; Dantzig, J.A.; Coleman, J.J. Strained-layer InGaAs-GaAs-AlGaAs buried-heterostructure quantum-well lasers by three-step selective-area metalorganic chemical vapor deposition. *IEEE J. Quantum Electron.* **1994**, *30*, 441–445. [CrossRef]
7. Dai, D.; Bowers, J.E. Silicon-based on-chip multiplexing technologies and devices for Peta-bit optical interconnects. *Nanophotonics* **2014**, *3*, 283–311. [CrossRef]
8. Heck, M.J.R. Highly integrated optical phased arrays: Photonic integrated circuits for optical beam shaping and beam steering. *Nanophotonics* **2017**, *6*, 93–107. [CrossRef]
9. Aoki, M.; Sano, H.; Suzuki, M.; Takahashi, M.; Uomi, K.; Takai, A. Novel structure MQW electroabsorption modulator/DFB-laser integrated device fabricated by selective area MOCVD growth. *Electron. Lett.* **1991**, *27*, 2138–2140. [CrossRef]
10. Cockerill, T.; Lammert, R.; Forbes, D.; Osowski, M.; Coleman, J. Twelve-channel strained-layer InGaAs-GaAs-AlGaAs buried heterostructure quantum well laser array for WDM applications by selective-area MOCVD. *IEEE Photonics Technol. Lett.* **1994**, *6*, 786–788. [CrossRef]
11. Hiruma, K.; Haga, T.; Miyazaki, M. Surface migration and reaction mechanism during selective growth of GaAs and AlAs by metalorganic chemical vapor deposition. *J. Cryst. Growth* **1990**, *102*, 717–724. [CrossRef]
12. Takahashi, Y.; Sakai, S.; Umeno, M. Selective MOCVD growth of GaAlAs on partly masked substrates and its application to optoelectronic devices. *J. Cryst. Growth* **1984**, *68*, 206–213. [CrossRef]
13. Coleman, J.J. Metalorganic chemical vapor deposition for optoelectronic devices. *Proc. IEEE* **1997**, *85*, 1715–1729. [CrossRef]
14. Yuan, X.; Pan, D.; Zhou, Y.; Zhang, X.; Peng, K.; Zhao, B.; Deng, M.; He, J.; Tan, H.H.; Jagadish, C. Selective area epitaxy of III–V nanostructure arrays and networks: Growth, applications, and future directions. *Appl. Phys. Rev.* **2021**, *8*, 021302. [CrossRef]
15. Karker, O.; Bange, R.; Bano, E.; Stambouli, V. Optimizing interferences of DUV lithography on SOI substrates for the rapid fabrication of sub-wavelength features. *Nanotechnology* **2021**, *32*, 235301. [CrossRef]
16. Verma, V.B.; Elarde, V.C. Nanoscale selective area epitaxy: From semiconductor lasers to single-photon sources. *Prog. Quantum Electron.* **2021**, *75*, 100305. [CrossRef]
17. Zybura, M.F.; Jones, S.H.; Duva, J.M.; Durgavich, J. A Simplified Model Describing Enhanced growth-rates during vapor-phase selective epitaxy. *J. Electron. Mater.* **1994**, *23*, 1055–1059. [CrossRef]
18. Korgel, B.; Hicks, R.F. A diffusion model for selective-area epitaxy by metalorganic chemical vapor deposition. *J. Cryst. Growth* **1995**, *151*, 204–212. [CrossRef]
19. Mircea, A.; Jahan, D.; Ougazzaden, A.; Delprat, D.; Silvestre, L.; Zimmermann, G.; Manolescu, A.; Manolescu, A.M. Computer modelling of selective area epitaxy with organometallics. In Proceedings of the 1996 International Semiconductor Conference, Sinaia, Romania, 12 October 1996; Volume 622, pp. 625–628.
20. Chang, J.S.C.; Carey, K.W.; Turner, J.E.; Hodge, L.A. Compositional non-uniformities in selective area growth of GaInAs on InP grown by OMVPE. *J. Electron. Mater.* **1990**, *19*, 345–348. [CrossRef]
21. Shamakhov, V.; Nikolaev, D.; Slipchenko, S.; Fomin, E.; Smirnov, A.; Eliseyev, I.; Pikhtin, N.; Kop'ev, P. Surface Nanostructuring during Selective Area Epitaxy of Heterostructures with InGaAs QWs in the Ultra-Wide Windows. *Nanomaterials* **2021**, *11*, 11. [CrossRef]
22. Dapkus, P.D.; Chi, C.Y.; Choi, S.J.; Chu, H.J.; Dreiske, M.; Li, R.; Lin, Y.; Nakajima, Y.; Ren, D.; Stevenson, R.; et al. Selective area epitaxy by metalorganic chemical vapor deposition—A tool for photonic and novel nanostructure integration. *Prog. Quantum Electron.* **2021**, *75*, 100304. [CrossRef]
23. Kayser, O.; Westphalen, R.; Opitz, B.; Balk, P. Control of selective area growth of InP. *J. Cryst. Growth* **1991**, *112*, 111–122. [CrossRef]
24. Slipchenko, S.; Shamakhov, V.; Nikolaev, D.; Fomin, E.; ya Soshnikov, I.; Bondarev, A.; Mitrofanov, M.; Pikhtin, N.; Kop'ev, P. Basics of surface reconstruction during selective area metalorganic chemical vapour-phase epitaxy of GaAs films in the stripe-type ultra-wide window. *Appl. Surf. Sci.* **2022**, *588*, 152991. [CrossRef]
25. Dupuis, N.; Decobert, J.; Lagrée, P.Y.; Lagay, N.; Cuisin, C.; Poingt, F.; Ramdane, A.; Kazmierski, C. AlGaInAs selective area growth by LP-MOVPE: Experimental characterisation and predictive modelling. *IEE Proc. Optoelectron.* **2006**, *153*, 276–279. [CrossRef]
26. Ujihara, T.; Yoshida, Y.; Sik Lee, W.; Takeda, Y. Pattern size effect on source supply process for sub-micrometer scale selective area growth by organometallic vapor phase epitaxy. *J. Cryst. Growth* **2006**, *289*, 89–95. [CrossRef]
27. Greenspan, J.E. Alloy composition dependence in selective area epitaxy on InP substrates. *J. Cryst. Growth* **2002**, *236*, 273–280. [CrossRef]
28. Shioda, T.; Sugiyama, M.; Shimogaki, Y.; Nakano, Y. Prediction method for the bandgap profiles of InGaAsP multiple quantum well structures fabricated by Selective Area Metal-Organic Vapor Phase Epitaxy. In Proceedings of the 17th International Conference on Indium Phosphide and Related Materials, Glasgow, UK, 8–12 May 2005; pp. 464–467.

29. Decobert, J.; Dupuis, N.; Lagree, P.Y.; Lagay, N.; Ramdane, A.; Ougazzaden, A.; Poingt, F.; Cuisin, C.; Kazmierski, C. Modeling and characterization of AlGaInAs and related materials using selective area growth by metal-organic vapor-phase epitaxy. *J. Cryst. Growth* **2007**, *298*, 28–31. [CrossRef]
30. Ougazzaden, A.; Silvestre, L.; Mircea, A.; Bouadma, N.; Patriarche, G.; Juhel, M. Designing the relative impact of thickness/composition changes in selective area organometallic epitaxy for monolithic integration applications. In Proceedings of the 1997 International Conference on Indium Phosphide and Related Materials, Cape Cod, MA, USA, 11–15 May 1997; pp. 598–601.
31. Sasaki, T.; Kitamura, M.; Mito, I. Selective metalorganic vapor phase epitaxial growth of InGaAsP/InP layers with bandgap energy control in InGaAs/InGaAsP multiple-quantum well structures. *J. Cryst. Growth* **1993**, *132*, 435–443. [CrossRef]
32. Tsuchiya, T.; Shimizu, J.; Shirai, M.; Aoki, M. InGaAlAs selective-area growth on an InP substrate by metalorganic vapor-phase epitaxy. *J. Cryst. Growth* **2005**, *276*, 439–445. [CrossRef]
33. Forbes, D.V.; Corbett, P.B.; Hansen, D.M.; Goodnough, T.J.; Zhang, L.; Myli, K.; Yeh, J.Y.; Mawst, L. The effect of reactor pressure on selective area epitaxy of GaAs in a close-coupled showerhead reactor. *J. Cryst. Growth* **2004**, *261*, 427–432. [CrossRef]
34. Ebert, C.; Bond, A.; Cao, H.; Levkoff, J.; Roberts, J. Selective area etching of InP with PCl3 in MOVPE. *J. Cryst. Growth* **2007**, *307*, 92–96. [CrossRef]
35. Ebert, C.; Levkoff, J.; Roberts, J.; Seiler, J.; Wanamaker, C.; Pinnington, T. Selective area etching of InP with CBr4 in MOVPE. *J. Cryst. Growth* **2007**, *298*, 94–97. [CrossRef]
36. Lammert, R.M.; Cockerill, T.M.; Forbes, D.V.; Smith, G.M.; Coleman, J.J. Submilliampere threshold buried-heterostructure InGaAs/GaAs single quantum well lasers grown by selective-area epitaxy. *IEEE Photonics Technol. Lett.* **1994**, *6*, 1073–1075. [CrossRef]
37. Galeuchet, Y.D.; Roentgen, P.; Graf, V. GaInAs/InP selective area metalorganic vapor phase epitaxy for one-step-grown buried low-dimensional structures. *J. Appl. Phys.* **1990**, *68*, 560–568. [CrossRef]
38. Lammert, R.M.; Smith, G.M.; Forbes, D.V.; Osowski, M.L.; Coleman, J.J. Strained-layer InGaAs-GaAs-AlGaAs buried-heterostructure lasers with nonabsorbing mirrors by selective-area MOCVD. *Electron. Lett.* **1995**, *31*, 1070–1072. [CrossRef]
39. Kobayashi, H.; Ekawa, M.; Okazaki, N.; Aoki, O.; Ogita, S.; Soda, H. Tapered thickness MQW waveguide BH MQW lasers. *IEEE Photonics Technol. Lett.* **1994**, *6*, 1080–1081. [CrossRef]
40. Takemasa, K.; Kubota, M.; Wada, H. 1.3-μm AlGaInAs-InP buried-heterostructure lasers with mode profile converter. *IEEE Photonics Technol. Lett.* **2000**, *12*, 471–473. [CrossRef]
41. Bour, D.; Corzine, S.; Perez, W.; Zhu, J.; Tandon, A.; Ranganath, R.; Lin, C.; Twist, R.; Martinez, L.; Höfler, G.; et al. Self-aligned, buried heterostructure AlInGaAs laser diodesby micro-selective-area epitaxy. *Appl. Phys. Lett.* **2004**, *85*, 2184–2186. [CrossRef]
42. Cai, J.; Choa, F.-S.; Gu, Y.; Ji, X.; Yan, J.; Ru, G.; Cheng, L.; Fan, J. Very low threshold, carrier-confined diode lasers by a single selective area growth. *Appl. Phys. Lett.* **2006**, *88*, 171110. [CrossRef]
43. Elarde, V.C.; Rangarajan, R.; Borchardt, J.J.; Coleman, J.J. Room-temperature operation of patterned quantum-dot lasers fabricated by electron beam lithography and selective area metal-organic chemical vapor deposition. *IEEE Photonics Technol. Lett.* **2005**, *17*, 935–937. [CrossRef]
44. Mokkapati, S.; Tan, H.H.; Jagadish, C. Multiple wavelength InGaAs quantum dot lasers using selective area epitaxy. *Appl. Phys. Lett.* **2007**, *90*, 171104. [CrossRef]
45. Akaishi, M.; Okawa, T.; Saito, Y.; Shimomura, K. Wide emission wavelength InAs/InP quantum dots grown by double-capped procedure using MOVPE selective area growth. *IEEE J. Sel. Top. Quantum Electron.* **2008**, *14*, 1197–1203. [CrossRef]
46. Park, J.H.; Khandekar, A.A.; Park, S.M.; Mawst, L.J.; Kuech, T.F.; Nealey, P.F. Selective MOCVD growth of single-crystal dense GaAs quantum dot array using cylinder-forming diblock copolymers. *J. Cryst. Growth* **2006**, *297*, 283–288. [CrossRef]
47. Li, R.R.; Dapkus, P.D.; Thompson, M.E.; Jeong, W.G.; Harrison, C.; Chaikin, P.M.; Register, R.A.; Adamson, D.H. Dense arrays of ordered GaAs nanostructures by selective area growth on substrates patterned by block copolymer lithography. *Appl. Phys. Lett.* **2000**, *76*, 1689–1691. [CrossRef]
48. Kim, H.; Choi, J.; Lingley, Z.; Brodie, M.; Sin, Y.; Kuech, T.F.; Gopalan, P.; Mawst, L.J. Selective growth of strained (In)GaAs quantum dots on GaAs substrates employing diblock copolymer lithography nanopatterning. *J. Cryst. Growth* **2017**, *465*, 48–54. [CrossRef]
49. Kim, H.; Wei, W.; Kuech, T.F.; Gopalan, P.; Mawst, L.J. Impact of InGaAs carrier collection quantum well on the performance of InAs QD active region lasers fabricated by diblock copolymer lithography and selective area epitaxy. *Semicond. Sci. Technol.* **2019**, *34*, 025012. [CrossRef]
50. Ishida, K.; Akiyama, M.; Nishi, S. Misfit and Threading Dislocations in GaAs Layers Grown on Si Substrates by MOCVD. *Jpn. J. Appl. Phys.* **1987**, *26*, L163–L165. [CrossRef]
51. Georgakilas, A.; Stoemenos, J.; Tsagaraki, K.; Komninou, P.; Flevaris, N.; Panayotatos, P.; Christou, A. Generation and annihilation of antiphase domain boundaries in GaAs on Si grown by molecular beam epitaxy. *J. Mater. Res.* **1993**, *8*, 1908–1921. [CrossRef]
52. Kakinuma, H.; Ueda, T.; Gotoh, S.; Yamagishi, C. Reduction of threading dislocations in GaAs on Si by the use of intermediate GaAs buffer layers prepared under high V–III ratios. *J. Cryst. Growth* **1999**, *205*, 25–30. [CrossRef]
53. Junesand, C.; Kataria, H.; Metaferia, W.; Julian, N.; Wang, Z.; Sun, Y.-T.; Bowers, J.; Pozina, G.; Hultman, L.; Lourdudoss, S. Study of planar defect filtering in InP grown on Si by epitaxial lateral overgrowth. *Opt. Mater. Express* **2013**, *3*, 1960–1973. [CrossRef]
54. Hu, Y.; Liang, D.; Beausoleil, R.G. An advanced III-V-on-silicon photonic integration platform. *Opto-Electron. Adv.* **2021**, *4*, 200094. [CrossRef]

55. Guo, X.; He, A.; Su, Y. Recent advances of heterogeneously integrated III–V laser on Si. *J. Semicond.* **2019**, *40*, 101304. [CrossRef]
56. Lourdudoss, S. Heteroepitaxy and selective area heteroepitaxy for silicon photonics. *Curr. Opin. Solid State Mater. Sci.* **2012**, *16*, 91–99. [CrossRef]
57. Li, J.Z.; Bai, J.; Park, J.-S.; Adekore, B.; Fox, K.; Carroll, M.; Lochtefeld, A.; Shellenbarger, Z. Defect reduction of GaAs epitaxy on Si (001) using selective aspect ratio trapping. *Appl. Phys. Lett.* **2007**, *91*, 021114. [CrossRef]
58. Lee, S.-M.; Cho, Y.J.; Park, J.-B.; Shin, K.W.; Hwang, E.; Lee, S.; Lee, M.-J.; Cho, S.-H.; Su Shin, D.; Park, J.; et al. Effects of growth temperature on surface morphology of InP grown on patterned Si(001) substrates. *J. Cryst. Growth* **2015**, *416*, 113–117. [CrossRef]
59. Wang, G.; Leys, M.R.; Loo, R.; Richard, O.; Bender, H.; Waldron, N.; Brammertz, G.; Dekoster, J.; Wang, W.; Seefeldt, M.; et al. Selective area growth of high quality InP on Si (001) substrates. *Appl. Phys. Lett.* **2010**, *97*, 121913. [CrossRef]
60. Han, Y.; Xue, Y.; Lau, K.M. Selective lateral epitaxy of dislocation-free InP on silicon-on-insulator. *Appl. Phys. Lett.* **2019**, *114*, 192105. [CrossRef]
61. Wang, G.; Rosseel, E.; Loo, R.; Favia, P.; Bender, H.; Caymax, M.; Heyns, M.M.; Vandervorst, W. High quality Ge epitaxial layers in narrow channels on Si (001) substrates. *Appl. Phys. Lett.* **2010**, *96*, 111903. [CrossRef]
62. Cantoro, M.; Merckling, C.; Jiang, S.; Guo, W.; Waldron, N.; Bender, H.; Moussa, A.; Douhard, B.; Vandervorst, W.; Heyns, M.M.; et al. Towards the Monolithic Integration of III-V Compound Semiconductors on Si: Selective Area Growth in High Aspect Ratio Structures vs. Strain Relaxed Buffer-Mediated Epitaxy. In Proceedings of the 2012 IEEE Compound Semiconductor Integrated Circuit Symposium (CSICS), La Jolla, CA, USA, 14–17 October 2012; pp. 1–4.
63. Loo, R.; Wang, G.; Orzali, T.; Waldron, N.; Merckling, C.; Leys, M.R.; Richard, O.; Bender, H.; Eyben, P.; Vandervorst, W.; et al. Selective Area Growth of InP on On-Axis Si(001) Substrates with Low Antiphase Boundary Formation. *J. Electrochem. Soc.* **2012**, *159*, H260–H265. [CrossRef]
64. Guo, W.; Date, L.; Pena, V.; Bao, X.; Merckling, C.; Waldron, N.; Collaert, N.; Caymax, M.; Sanchez, E.; Vancoille, E.; et al. Selective metal-organic chemical vapor deposition growth of high quality GaAs on Si(001). *Appl. Phys. Lett.* **2014**, *105*, 062101. [CrossRef]
65. Li, S.-Y.; Zhou, X.-L.; Kong, X.-T.; Li, M.-K.; Mi, J.-P.; Bian, J.; Wang, W.; Pan, J.-Q. Selective Area Growth of GaAs in V-Grooved Trenches on Si(001) Substrates by Aspect-Ratio Trapping. *Chin. Phys. Lett.* **2015**, *32*, 028101. [CrossRef]
66. Merckling, C.; Waldron, N.; Jiang, S.; Guo, W.; Collaert, N.; Caymax, M.; Vancoille, E.; Barla, K.; Thean, A.; Heyns, M.; et al. Heteroepitaxy of InP on Si(001) by selective-area metal organic vapor-phase epitaxy in sub-50 nm width trenches: The role of the nucleation layer and the recess engineering. *J. Appl. Phys.* **2014**, *115*, 023710. [CrossRef]
67. Waldron, N.; Merckling, C.; Teugels, L.; Ong, P.; Ibrahim, S.A.U.; Sebaai, F.; Pourghaderi, A.; Barla, K.; Collaert, N.; Thean, A.V.Y. InGaAs Gate-All-Around Nanowire Devices on 300 mm Si Substrates. *IEEE Electron Device Lett.* **2014**, *35*, 1097–1099. [CrossRef]
68. Staudinger, P.; Mauthe, S.; Triviño, N.V.; Reidt, S.; Moselund, K.E.; Schmid, H. Wurtzite InP microdisks: From epitaxy to room-temperature lasing. *Nanotechnology* **2020**, *32*, 075605. [CrossRef]
69. Wong, W.W.; Su, Z.; Wang, N.; Jagadish, C.; Tan, H.H. Epitaxially Grown InP Micro-Ring Lasers. *Nano Lett.* **2021**, *21*, 5681–5688. [CrossRef]
70. Wang, Z.; Tian, B.; Pantouvaki, M.; Guo, W.; Absil, P.; Van Campenhout, J.; Merckling, C.; Van Thourhout, D. Room-temperature InP distributed feedback laser array directly grown on silicon. *Nat. Photonics* **2015**, *9*, 837–842. [CrossRef]
71. Megalini, L.; Bonef, B.; Cabinian, B.C.; Zhao, H.; Klamkin, J. 1550-nm InGaAsP multi-quantum-well structures selectively grown on v-groove-patterned SOI substrates. *Appl. Phys. Lett.* **2017**, *111*, 032105. [CrossRef]
72. Kunert, B.; Guo, W.; Mols, Y.; Langer, R.; Barla, K. (Invited) Integration of III/V Hetero-Structures By Selective Area Growth on Si for Nano- and Optoelectronics. *ECS Trans.* **2016**, *75*, 409–419. [CrossRef]
73. Kunert, B.; Guo, W.; Mols, Y.; Tian, B.; Wang, Z.; Shi, Y.; Thourhout, D.V.; Pantouvaki, M.; Campenhout, J.V.; Langer, R.; et al. III/V nano ridge structures for optical applications on patterned 300 mm silicon substrate. *Appl. Phys. Lett.* **2016**, *109*, 091101. [CrossRef]
74. Shi, B.; Zhao, H.; Wang, L.; Song, B.; Klamkin, J. Continuous-wave electrically pumped 1550 nm lasers epitaxially grown on on-axis (001) silicon. *Optica* **2019**, *6*, 1507. [CrossRef]
75. Luo, W.; Xue, Y.; Shi, B.; Zhu, S.; Dong, X.; Lau, K.M. MOCVD growth of InP-based 1.3 μm quantum dash lasers on (001) Si. *Appl. Phys. Lett.* **2020**, *116*, 142106. [CrossRef]
76. Fujii, T.; Takeda, K.; Nishi, H.; Diamantopoulos, N.-P.; Sato, T.; Kakitsuka, T.; Tsuchizawa, T.; Matsuo, S. Multiwavelength membrane laser array using selective area growth on directly bonded InP on SiO_2/Si. *Optica* **2020**, *7*, 838–846. [CrossRef]
77. Kish, F.; Lal, V.; Evans, P.; Corzine, S.W.; Ziari, M.; Butrie, T.; Reffle, M.; Tsai, H.S.; Dentai, A.; Pleumeekers, J.; et al. System-on-Chip Photonic Integrated Circuits. *IEEE J. Sel. Top. Quantum Electron.* **2018**, *24*, 6100120. [CrossRef]
78. Kobayashi, W.; Fujisawa, T.; Ito, T.; Kanazawa, S.; Ueda, Y.; Sanjoh, H. Advantages of EADFB laser for 25 Gbaud/s 4-PAM (50 Gbit/s) modulation and 10 km single-mode fibre transmission. *Electron. Lett.* **2014**, *50*, 683–685. [CrossRef]
79. Yun, H.; Choi, K.; Kwon, Y.; Choe, J.; Moon, J. Fabrication and Characteristics of 40-Gb/s Traveling-Wave Electroabsorption Modulator-Integrated DFB Laser Modules. *IEEE Trans. Adv. Packag.* **2008**, *31*, 351–356. [CrossRef]
80. Zhu, J.T.; Billia, L.; Bour, D.; Corzine, S.; Höfler, G. Performance comparison between integrated 40 Gb/s EAM devices grown by selective area growth and butt-joint overgrowth. *J. Cryst. Growth* **2004**, *272*, 576–581. [CrossRef]
81. Cheng, Y.; Pan, J.; Wang, Y.; Zhou, F.; Wang, B.; Zhao, L.; Zhu, H.; Wang, W. 40-Gb/s Low Chirp Electroabsorption Modulator Integrated With DFB Laser. *IEEE Photonics Technol. Lett.* **2009**, *21*, 356–358. [CrossRef]

82. Zhao, Q.; Pan, J.Q.; Zhou, F.; Wang, B.J.; Wang, L.F.; Wang, W. Monolithic integration of an InGaAsP–InP strained DFB laser and an electroabsorption modulator by ultra-low-pressure selective-area-growth MOCVD. *Semicond. Sci. Technol.* **2005**, *20*, 544–547. [CrossRef]
83. Kim, S.-B.; Sim, J.-S.; Kim, K.S.; Sim, E.-D.; Ryu, S.-W.; Park, H.L. Selective-area MOVPE growth for 10 Gbit/s electroabsorption modulator integrated with a tunable DBR laser. *J. Cryst. Growth* **2007**, *298*, 672–675. [CrossRef]
84. Hong-Bo, L.; Ling-Juan, Z.; Jiao-Qing, P.; Hong-Liang, Z.; Fan, Z.; Bao-Jun, W.; Wei, W. Monolithic Integration of Sampled Grating DBR with Electroabsorption Modulator by Combining Selective-Area-Growth MOCVD and Quantum-Well Intermixing. *Chin. Phys. Lett.* **2008**, *25*, 3670–3672. [CrossRef]
85. Zhu, H.; Liang, S.; Zhao, L.; Kong, D.; Zhu, N.; Wang, W. A selective area growth double stack active layer electroabsorption modulator integrated with a distributed feedback laser. *Chin. Sci. Bull.* **2009**, *54*, 3627. [CrossRef]
86. Deng, Q.; Zhu, H.; Xie, X.; Guo, L.; Sun, S.; Liang, S.; Wang, W. Low chirp EMLs fabricated by combining SAG and double stack active layer techniques. *IEEE Photonics J.* **2018**, *10*, 7902007. [CrossRef]
87. Tanbun-Ek, T.; Fang, W.-C.; Bethea, C.; Sciortino, P.; Sergent, A.; Wisk, P.; People, R.; Chu, S.-N.; Pawelek, R.; Tsang, W.-T.; et al. *Wavelength Division Multiplexed (WDM) Electroabsorption Modulated Laser Fabricated by Selective Area Growth MOVPE Techniques*; SPIE: Bellingham, WA, USA, 1997; Volume 3006.
88. Cheng, Y.; Wang, Q.J.; Pan, J. 1.55 µm high speed low chirp electroabsorption modulated laser arrays based on SAG scheme. *Opt. Express* **2014**, *22*, 31286–31292. [CrossRef]
89. Xu, J.; Liang, S.; Zhang, Z.; An, J.; Zhu, H.; Wang, W. EML Array fabricated by SAG technique monolithically integrated with a buried ridge AWG multiplexer. *Opt. Laser Technol.* **2017**, *91*, 46–50. [CrossRef]
90. Corzine, S.W.; Evans, P.; Fisher, M.; Gheorma, J.; Kato, M.; Dominic, V.; Samra, P.; Nilsson, A.; Rahn, J.; Lyubomirsky, I.; et al. Large-Scale InP Transmitter PICs for PM-DQPSK Fiber Transmission Systems. *IEEE Photonics Technol. Lett.* **2010**, *22*, 1015–1017. [CrossRef]
91. Fujisawa, T.; Kanazawa, S.; Takahata, K.; Kobayashi, W.; Tadokoro, T.; Ishii, H.; Kano, F. 1.3-µm, 4 × 25-Gbit/s, EADFB laser array module with large-output-power and low-driving-voltage for energy-efficient 100 GbE transmitter. *Opt. Express* **2012**, *20*, 614–620. [CrossRef]
92. Miller, L.M.; Beernink, K.J.; Hughes, J.S.; Bishop, S.G.; Coleman, J.J. Four wavelength distributed feedback ridge waveguide quantum-well heterostructure laser array. *Appl. Phys. Lett.* **1992**, *61*, 2964–2966. [CrossRef]
93. Vieu, C.; Carcenac, F.; Pépin, A.; Chen, Y.; Mejias, M.; Lebib, A.; Manin-Ferlazzo, L.; Couraud, L.; Launois, H. Electron beam lithography: Resolution limits and applications. *Appl. Surf. Sci.* **2000**, *164*, 111–117. [CrossRef]
94. Li, G.P.; Makino, T.; Sarangan, A.; Huang, W. 16-wavelength gain-coupled DFB laser array with fine tunability. *IEEE Photonics Technol. Lett.* **1996**, *8*, 22–24. [CrossRef]
95. Liang, S.; Lu, D.; Zhao, L.; Zhu, H.; Wang, B.; Zhou, D.; Wang, W. Fabrication of InP-based monolithically integrated laser transmitters. *Sci. China Inf. Sci.* **2018**, *61*, 080405. [CrossRef]
96. Darja, J.; Chan, M.; Wang, S.-R.; Sugiyama, M.; Nakano, Y. Four channel ridge DFB laser array for 1.55 mu m CWDM systems by wide-stripe selective area MOVPE. *IEICE Trans. Electron.* **2007**, *E90-C*, 1111–1117. [CrossRef]
97. Zhang, C.; Liang, S.; Zhu, H.; Wang, B.; Wang, W. A modified SAG technique for the fabrication of DWDM DFB laser arrays with highly uniform wavelength spacings. *Opt. Express* **2012**, *20*, 29620–29625. [CrossRef] [PubMed]
98. Zhang, C.; Liang, S.; Zhu, H.; Han, L.; Lu, D.; Ji, C.; Zhao, L.; Wang, W. The fabrication of 10-channel DFB laser array by SAG technology. *Opt. Commun.* **2013**, *311*, 6–10. [CrossRef]
99. Han, L.; Liang, S.; Wang, H.; Xu, J.; Qiao, L.; Zhu, H.; Wang, W. Fabrication of Low-Cost Multiwavelength Laser Arrays for OLTs in WDM-PONs by Combining the SAG and BIG Techniques. *IEEE Photonics J.* **2015**, *7*, 1502807. [CrossRef]
100. Zhang, C.; Zhu, H.; Liang, S.; Cui, X.; Wang, H.; Zhao, L.; Wang, W. Ten-channel InP-based large-scale photonic integrated transmitter fabricated by SAG technology. *Opt. Laser Technol.* **2014**, *64*, 17–22. [CrossRef]
101. Guo, F.; Zhang, R.; Lu, D.; Wang, W.; Ji, C. 1.3-µm multi-wavelength DFB laser array fabricated by mocvd selective area growth. *Opt. Commun.* **2014**, *331*, 165–168. [CrossRef]
102. Kwon, O.K.; Leem, Y.A.; Han, Y.T.; Lee, C.W.; Kim, K.S.; Oh, S.H. A 10 × 10 Gb/s DFB laser diode array fabricated using a SAG technique. *Opt. Express* **2014**, *22*, 9073–9080. [CrossRef]
103. Oh, S.H.; Kwon, O.K.; Kim, K.S.; Han, Y.T.; Lee, C.W.; Leem, Y.A.; Shin, J.W.; Nam, E.S. A Multi-Channel Etched-Mesa PBH DFB Laser Array Using an SAG Technique. *IEEE Photonics Technol. Lett.* **2015**, *27*, 2567–2570. [CrossRef]

Review

Research Progress of Monolithic Integrated DFB Laser Arrays for Optical Communication

Shen Niu [1,2], Yue Song [1,2,*], Ligong Zhang [1,2,*], Yongyi Chen [1,2,3,*], Lei Liang [1,2], Ye Wang [1,2,4], Li Qin [1,2], Peng Jia [1,2], Cheng Qiu [1,2], Yuxin Lei [1,2], Yubing Wang [1,2], Yongqiang Ning [1,2] and Lijun Wang [1,2,5,6]

1. State Key Laboratory of Luminescence and Applications, Changchun Institute of Optics, Fine Mechanics and Physics, Chinese Academy of Sciences, Changchun 130033, China; niushen20@mails.ucas.ac.cn (S.N.); lianglei@ciomp.ac.cn (L.L.); wangye@ciomp.ac.cn (Y.W.); qinl@ciomp.ac.cn (L.Q.); jiapeng@ciomp.ac.cn (P.J.); qiucheng@ciomp.ac.cn (C.Q.); leiyuxin@ciomp.ac.cn (Y.L.); wangyubing@ciomp.ac.cn (Y.W.); ningyq@ciomp.ac.cn (Y.N.); wanglj@ciomp.ac.cn (L.W.)
2. Daheng College, University of Chinese Academy of Sciences, Beijing 100049, China
3. Jlight Semiconductor Technology Co., Ltd., Changchun 130033, China
4. College of Opto-Electronic Engineering, Changchun University of Science and Technology, Changchun 130022, China
5. Peng Cheng Laboratory, No.2, Xingke 1st Street, Nanshan, Shenzhen 518000, China
6. Academician Team Innovation Center of Hainan Province, Key Laboratory of Laser Technology and Optoelectronic Functional Materials of Hainan Province, School of Physics and Electronic Engineering, Hainan Normal University, Haikou 570206, China
* Correspondence: songyue@ciomp.ac.cn (Y.S.); zhanglg@ciomp.ac.cn (L.Z.); chenyy@ciomp.ac.cn (Y.C.)

Abstract: Photonic integrated circuits (PICs) play a leading role in modern information and communications technology. Among the core devices in PICs is the distributed feedback (DFB) multi-wavelength semiconductor laser array. Multi-wavelength semiconductor laser arrays can be integrated on a single chip and have the advantages of high stability, good single-mode performance, and narrow line width. The wavelength tuning range has been expanded through the design of the DFB laser array, which is an ideal light source for wavelength-division multiplexing systems. The preparation of DFB laser arrays with a large number of channels, ease of mass production, and accurate emission wavelengths has become an important field of research. The connection methods of lasers in DFB laser arrays are introduced systematically and the current methods of manufacturing multi-wavelength DFB laser arrays covering the perspective of technical principles, technical advantages and disadvantages, main research progress, and research status are summarized.

Keywords: DFB laser array; multi-wavelength; series and parallel; optical communication; photonic integrated circuits

Citation: Niu, S.; Song, Y.; Zhang, L.; Chen, Y.; Liang, L.; Wang, Y.; Qin, L.; Jia, P.; Qiu, C.; Lei, Y.; et al. Research Progress of Monolithic Integrated DFB Laser Arrays for Optical Communication. *Crystals* **2022**, *12*, 1006. https://doi.org/10.3390/cryst12071006

Academic Editor: Alessandro Chiasera

Received: 16 June 2022
Accepted: 18 July 2022
Published: 21 July 2022

Publisher's Note: MDPI stays neutral with regard to jurisdictional claims in published maps and institutional affiliations.

Copyright: © 2022 by the authors. Licensee MDPI, Basel, Switzerland. This article is an open access article distributed under the terms and conditions of the Creative Commons Attribution (CC BY) license (https://creativecommons.org/licenses/by/4.0/).

1. Introduction

The rapid development of optical and wireless networks has driven the proportion of fiber broadband users in China from 56% at the end of 2015 up to 94% currently. The gigabit optical network covers more than 120 million households, and the end-to-end user experiences speeds of 51.2 Mbps. The continuous expansion of optical communication networks and the rapid development of big data, cloud computing, and other fields increase the requirement for sources of light. Semiconductor lasers are small in size, lightweight, have long lifespans, high efficiency, and integration, and have become an important light source for optical communication systems.

The speed of development of the optical communication industry relies on breakthroughs made in the optoelectronic device industry, such as semiconductor lasers and optical amplifiers. The current developmental trends in optical device technology are integration, intelligence, and tunability. Miniaturization, environmental protection, low power consumption, and low cost are also important areas of research for future light sources used in communication.

In the NG-PON2 standard, a fiber capacity of 40 Gbit/s is achieved by utilizing multiple wavelengths and user terminal tunable transceiver technology at dense wavelength-division multiplexing (WDM) channel spacing. In terms of current actual products, the maturity of tunable receivers is lower than that of tunable transmitters. Therefore, tunable lasers are still the focus of this technology [1].

Tunable semiconductor lasers are an important component of future WDM systems, including external cavity diode lasers (ECDLs), vertical-cavity surface-emitting lasers (VCSELs), quantum cascade lasers (QCLs), sampling grating-distributed Bragg reflectors (SG-DBR), and adjustable distributed feedback multi-wavelength lasers and arrays (DFB-MLAs) [2].

The ECDL is usually based on a gain chip and an external mode-selection component and is used to extend the resonant cavity outside of the laser chip. By adjusting the external mode-selection component, ECDL realizes the wavelength-tuning function, allowing the line width to be significantly narrowed and a large tuning range to be obtained. However, this type of laser is large, the optical path is difficult to align, and the mechanical structure lags and wears out during the adjustment process, making it difficult to use in optical communication networks.

The unipolar light source of the QCL, based on the principle of electron transition between quantum-well sub-bands, is different from that of conventional semiconductor lasers. In the case of QCL, the lasing wavelength is limited by the forbidden bandwidth of the material and determined by the conduction and neutron bands. The energy level spacing between sub-bands can be changed by adjusting the thickness of the quantum well, thereby changing the lasing wavelength, and as a result, the QCL is widely used in the mid-and far-infrared bands. At present, QCL is mainly used in free-space communication, gas detection, and other fields, and the 1.3–1.5 µm band of optical communication does not need to use this scheme to change the wavelength.

Semiconductor lasers, such as VCSEL, SG-DBR, a vertical grating-assisted codirectional coupler laser with a sampled Bragg reflector (GCSR), DFB, and DFB-MLA, are available for use in photonic integrated circuits (PICs), which play a key role in overcoming the bottlenecks of transmission capacity and energy consumption in future broadband networks.

Tunable VCSEL lasers usually introduce air gaps in the laser manufacturing process and adjust their position using micro-electromechanical systems (MEMS) to change the size of the air gap, thereby changing the equivalent cavity length to achieve the desired wavelength tuning. However, the mechanical structure of wavelength-tunable VCSELs using MEMS results in wider line width, slower response, and lower output power.

The SG-DBR laser and GCSR laser have a front and rear grating, gain, and phase sections. Periodic modulation is performed on the uniform grating for the wavelength-tuning function, and the vernier effect of the sampling grating is used to expand the tuning range. Its tuning speed is very fast [3]. The mode stability difference between the SG–DBR laser and GCSR laser is its key weakness, which limits its application in optical communication.

Compared to the above tunable lasers, DFB lasers are advantageous because of their small sizes, easy monolithic integration, simple tuning schemes, high wavelength stability, and good single-mode performance. These characteristics promote their wide application in optical communication networks.

For DFB lasers, the output wavelength is related to the refractive index of the active region and Bragg grating period. For tunable DFB lasers, by changing the period of the laser Bragg grating and the injection current or temperature of the laser, the refractive index of the active region is changed accordingly to achieve the desired wavelength. The DFB laser is the most widely used as a mature semiconductor light source for optical communication networks. It has an excellent dynamic single mode and stability, avoids the influence of multi-longitudinal mode dispersion, and is suitable for multi-channel optical fiber communication systems. The tunable DFB laser can be set to any channel of the dense wavelength-division multiplexing system, and the device can be easily integrated into a single chip to reduce the power consumption of wavelength switching and simplify the system, thereby reducing the cost of the coupling, packaging, and energy consumption. In addition, in a definable optical network with a wavelength-selective switch, the tunable DFB laser provides a wavelength routing function, which effectively increases the reconfiguration capability of the optical network.

However, for a single DFB laser, the tuning wavelength range is generally below 10 nm, which does not meet the needs of large-scale tuning [4]. To overcome this problem, a series of DFB lasers can be assembled into arrays to expand the wavelength-tuning range. The first proposed scheme was by Japan's NEC Company [5].

The tuning mechanism of the DFB laser array can be divided into wavelength tuning of a single DFB laser and switching between different DFB lasers. Among them, the wavelength tuning of a single DFB laser, in principle, is to change the grating structure and refractive index of the laser by a certain amount by heating or adding current, thereby changing the emission wavelength. In addition, through the design of some special grating structures or active region and waveguide layer structures, wavelength tuning under different conditions can also be achieved, such as reconstruction equivalent chirp (REC) technology, selective area growth (SAG) technology, asymmetric periodic structure, and periodic gain coupling; switching between different DFB lasers is realized through a personal computer (PC) and a microprocessor (MCU). For the tuning of the DFB laser array, due to the limited space, only the REC technology and the SAG technology are mainly introduced in this paper.

The DFB laser array has good single-mode stability and a simple wavelength-tuning scheme that can be roughly tuned by switching different lasers and fine-tuned by adjusting the laser temperature or injection current. With careful design, DFB laser arrays for coarse and fine WDM systems (20 nm and 0.8 nm channel pitch, respectively) can be fabricated.

Compared with hybrid integrated laser array modules consisting of discrete DFB lasers, a well-designed monolithic integrated DFB laser array can potentially reduce system costs by simplifying the optical alignment and packaging process. The DFB laser array with accurate multichannel spacing can achieve an accurate wavelength spacing to align with the wavelength standard of all WDM channels. Furthermore, it can be tuned significantly faster than that more mature thermally tuned semiconductor lasers; therefore, it has a wide range of applications in WDM.

The time classification of wavelength channel tuning is also specified in the NG-PON standard. For class 1, the switching time is generally required to be less than 10 µs, which is difficult for ordinary electrically tuned and thermally tuned lasers to achieve, while multi-wavelength DFB laser arrays can be achieved at this point [1].

For DFB laser arrays used in optical communication, in addition to the conventional properties such as wavelength, linewidth, power, and side-mode suppression ratio (SMSR), we also need to pay attention to issues such as wavelength stability, wavelength accuracy, and wavelength locking [6]. For wavelength stability, it is usually defined by the maximum spectral excursion (MSE). In NG-PON2, when the channel spacing is 100 GHz, the MSE should be within the range of ±20 GHz, and when the channel spacing is 50 GHz, the MSE should be within the range of ±12.5 GHz [7]. For DFB laser arrays with channel switching, the MSE when the laser is on/off must also be compliant. Due to the use of tunable lasers in the array, its inherent wavelength accuracy and wavelength stability can be moderately relaxed to reduce cost, and the channel is aligned by tuning, but must be compensated by employing a wavelength locking mechanism, usually relying on optical line terminal feedback.

Multi-wavelength DFB laser arrays have been widely investigated and exploited by the researchers in institutes and corporations.

At present, the main difficulties facing multi-wavelength DFB laser arrays are that the emission wavelength needs to be highly accurate, the channels must not affect each other, the number of integrations must be as large as possible, and it is necessary to find a feasible solution suitable for mass production.

In the introduction, we consider the properties of the required light source from the perspective of practical application and compare various lasers. On this basis, the properties and working principles of the DFB laser array are briefly introduced.

DFB laser arrays can be divided into series, parallel, and series–parallel combination structures, depending on the connection mode; design and manufacturing methods of these arrays include electron beam lithography, nano-imprint lithography, reconstruction equivalent chirp, ridge width change, and selective area growth.

This paper reviews the principles, advantages, and disadvantages, research progress, and application status of the above structures and methods. It is hoped that this study can provide a reference for research on the monolithic integrated DFB laser array for optical communication, as well as the promotion of their development.

2. Connection Method
2.1. Parallel DFB Laser Array
2.1.1. Technical Principle, Advantages, and Disadvantages

A parallel DFB laser array is an array in which each laser has an independent waveguide and is coupled together by an S waveguide and a coupler, as shown in Figure 1.

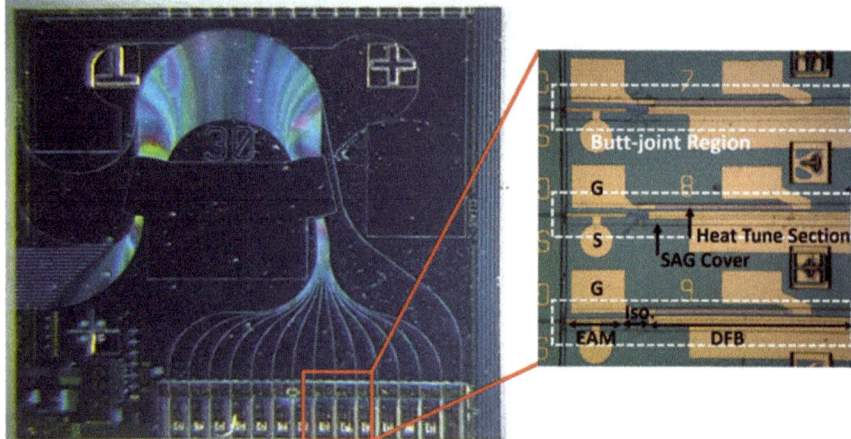

Figure 1. Integrated parallel DFB laser array [8] © Springer link. Copyright 2018 Science China Information Sciences.

The light emitted by the parallel DFB laser array needs to be coupled to a waveguide through photosynthetic wave devices to output. Various photosynthetic wave devices can be used such as a star coupler, directional coupler, multi-mode interference coupler (MMI), array waveguide grating (AWG), and micro-electro-mechanical system (MEMS). A star coupler can be provided with a high wavelength ratio, but its output power between channels is not uniform and produces high power loss. A directional coupler meets the very low insertion loss and reverse reflection requirements; however, the gap between waveguides is sub-micron in size, resulting in a fabrication process that is difficult to control accurately. Currently, directional couplers are commonly used in MMIs and AWGs, as shown in Figure 2.

Based on the principle of self-imaging, the MMI is not sensitive to wavelength and is suitable for the combined output of the multi-wavelength laser array. The process is relatively simple, but it has a low output power and large insertion loss, which increases with the increase in the number of channels. To solve these limitations, the MMI can be etched in both deep and shallow ridges, where the deep ridge can reduce insertion loss and crosstalk, and the shallow ridge has less end reflection.

The AWG is large and has both deep and shallow etching structures. The deep etching structure is small in size but is accompanied by a high insertion loss of greater than 5 dB in general. In contrast, the shallow etching structure has a relatively low insertion loss, generally less than 2 dB, but it requires a large bending radius, and the device is large. In InP-based chips, the AWG has wavelength sensitivity and as the wavelength of the laser changes, its insertion loss further increases.

Additionally, couplers are no longer required when using MEMS. In this configuration, a micromechanical mirror is added to the DFB laser array to achieve coupling. As MEMS tilt mirror achieves precise optical-mechanical alignment electronically, the tolerance is relaxed, and any deviation in the package can be corrected to reduce the package cost [9]. This method increases the device size compared to on-chip integration.

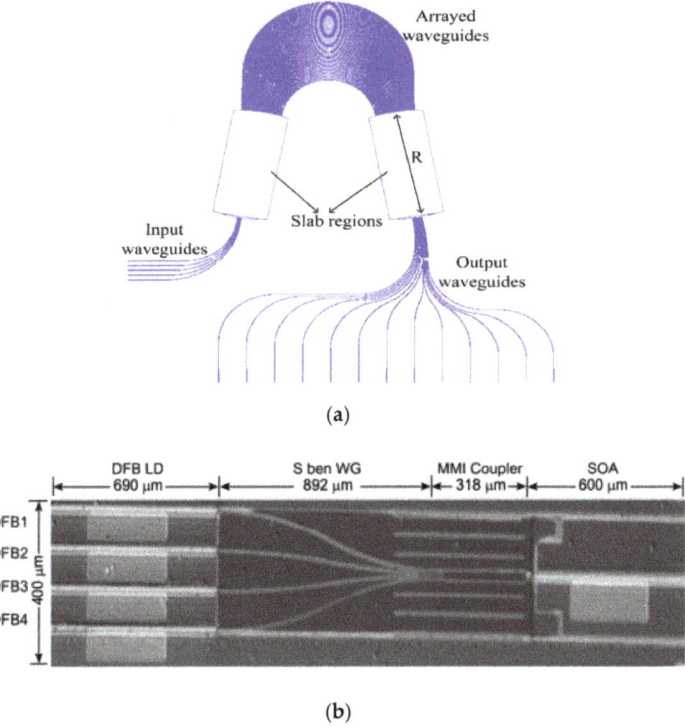

Figure 2. Schematic diagram of different couplers, (a) AWG [10] and (b) MMI [11]. © (a) Elsevier. Copyright 2015 optics communications. (b) Springer link. Copyright 2013 Science China Information Sciences.

In terms of coupling stability, consider a system of N coupled semiconductor lasers described in dimensionless form by the following coupled-mode equations [12]:

$$\frac{dY_j}{dt} = (1 - i\alpha)Z_j Y_j + i\eta \tag{1}$$

$$\frac{TdZ_j}{dt} = P - Z_j - (1 + 2Z_j)|Y_j|^2 \tag{2}$$

where Y_j and Z_j are defined as the normalized electrical field amplitude and normalized excess carrier density in the jth laser in sequence, respectively. t is measured in units of the photon lifetime τ_p. T is the ratio between the carrier recombination time and photon lifetime, and is typically large. P is the normalized excess pumping current for a single laser above the threshold, α is defined as the linewidth enhancement factor, and η is the coupling constant between lasers, which represents the coupling difficulty and coupling strength required to obtain a stable in-phase solution.

If the lasers are coupled in parallel mode, the coupling term $\sum Y_k$ in Equation (1) includes all the elements in the array except $k = j$. Then,

$$\sum Y_k = \sum_{K=1}^{n} Y_k, k \neq j \tag{3}$$

It is convenient to reformulate Equations (1) and (2) in terms of the amplitude and phase of the electrical field. When $Y_j = E_j \exp(i\varphi_j)$ is introduced, Equations (1) and (2) become

$$\frac{dE_j}{dt} = Z_j E_j - \eta \sum [E_k \sin(\varphi_k - \varphi_j)] \tag{4}$$

$$\frac{T dZ_j}{dt} = P - Z_j - (1 + 2Z_j) E_j^2 \tag{5}$$

$$\frac{d\varphi_j}{dt} = -\alpha Z_j + \eta E_j^{-1} \sum [E_k \cos(\varphi_k - \varphi_j)] \tag{6}$$

For parallel coupling, from the characteristic equation [12], use

$$\eta > \eta_p \equiv \frac{2\alpha P}{2(1+2P)N} \tag{7}$$

We will discuss this equation in detail in the next section when η is derived in a series array.

Technical advantage: the current DFB laser arrays are connected in parallel, meaning multiple DFB lasers are designed to be arranged in parallel and are coupled to a multimode interference combiner through a passive S-type waveguide. The advantage of this design is that its fabrication is relatively simple, only the fabrication of a single DFB laser and its couplings are considered, the mutual influence between the lasers is small, and the single-mode stability is high, and the tuning scheme is simple.

Technical disadvantage: during the coupling process, the output optical power encounters large losses, which is particularly critical for DFB lasers with an output wavelength of 1550 nm; therefore, an integrated optical amplifier (SOA) is often required. Moreover, as the number of lasers increases, the complexity and loss of parallel DFB laser arrays gradually increase when combining waves, the area occupied by the S-type passive waveguide increases, and the integrated optical amplifier will further add to the chip size, which is not conducive to the performance and integration of the chip, and also increases the total required current [13].

At present, for parallel DFB laser arrays, the main problem is that it is difficult to achieve low coupling loss and volume reduction simultaneously, which requires a higher output optical power of the laser or a coupling mode with lower loss.

2.1.2. Main Research Progress and Status

A parallel DFB laser array was manufactured for the first time in 1984 [14] and has made great improvements in various aspects of performance over the years. Currently, in laboratory preparation and commercial application, DFB laser arrays are still mainly connected in parallel. This method can integrate 16 or more DFB lasers, covering C-band, L-band, or O-band, and is suitable for coarse wavelength-division multiplexing (CWDM) systems and dense wavelength-division multiplexing (DWDM) systems.

For DWDM systems, the parallel DFB laser array can achieve a channel spacing of 0.8 nm [15], which is difficult to achieve by other structures such as series and series–parallel combinations. However, at the same time, the coupling loss of a parallel DFB laser array increases when the number of integrated lasers increases. Additionally, the problem of large volume after the integration of the MMI and AWG has not been solved. One possible solution to this problem may be to manufacture smaller combination devices.

2.2. Series DFB Laser Array

2.2.1. Technical Principle, Advantages and Disadvantages

To reduce the coupling difficulty and power loss of the laser array, a series DFB laser array is introduced, and its structure is shown in Figure 3. The technological process combines several DFB lasers in series such that all the DFB lasers are on the same waveguide. Thus, the laser array need not be laterally coupled, which avoids power losses during the coupling process, effectively reduces the chip size, and improves integration. However, in a tandem DFB laser array, the single-mode stability and wavelength accuracy of a single laser are easily affected by reflections from other lasers, resulting in poor beam quality and wavelength shift. When the wavelength spacing is less than 100 GHz (0.8 nm), grating crosstalk causes mode hopping. Using linear-chirped gratings to perform multiple π-phase shifts in a full cavity is an effective method for reducing crosstalk and suppressing side modes.

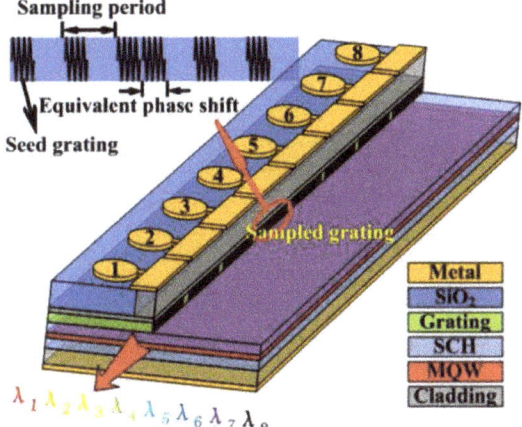

Figure 3. Series DFB laser array [16] © Elsevier. Copyright 2015 optics communications.

In a series DFB laser array, the wavelength spacing between the lasers is generally large, which requires a large temperature or current range to be tuned to cover all the desired wavelengths. Although, an excessive temperature or current tuning range many lead to device failure. Furthermore, decreases in the output power affects the reliability of the device. To reduce the influence of reflection, the phases of Bragg gratings in different lasers can be designed, and phase-shift gratings can be introduced between lasers to realise single longitudinal mode lasing.

In terms of coupling stability, if the lasers are coupled in series, the coupling stability is mainly affected by the two neighboring lasers. Consider a loop configuration and the coupling term $\sum Y_k$ in Equation (1) given by

$$\sum Y_k = Y_{j+1} + Y_{j-1} \tag{8}$$

From the characteristic equation [12]

$$\eta > \eta_c(n) = \frac{\alpha P}{2(1+2P)\sin^2\left(\frac{\pi n}{N}\right)} \tag{9}$$

The largest of all the $\eta_c(n)$ corresponds to the wave number $n = 1$. Thus, it is required that

$$\eta > \eta_s \equiv \frac{\alpha P}{2(1+2P)\sin^2\left(\frac{\pi n}{N}\right)} \tag{10}$$

By Equations (8) and (11), the critical coupling strengths η_s and η_p have been verified to be identical if $N = 3$, since series coupling and parallel coupling are the same for three coupled lasers in a ring configuration. If $N > 3$, η_s becomes larger than η_p. Thus, series coupling requires a larger value of the coupling strength q to have stable in-phase solutions.

At present, the number of DFB lasers that can be integrated into a series array is far less than that of a parallel array, and generally only 3–4 DFB lasers can be integrated with a channel spacing generally above 2 nm. For the series DFB laser array, it is easy to achieve a small volume, but methods to integrate more lasers, and ensure the single longitudinal mode and beam quality have not yet been developed.

2.2.2. Main Research Progress and Status

In a series DFB laser array, the number of lasers increases linearly as the required tuning range increases, which leads to an increasing length of the chip structure. By means of the shared grating sections of laser sections, that is, laser 1 passes through grating 1 and grating 2, laser 2 passes through grating 2 and grating 3, and laser N shines through grating N and grating $N + 1$, so that the total cavity length is significantly shortened, up to 40% in the array composed of 5 lasers [17].

In a series DFB laser array, the single-mode stability and wavelength reliability are poor due to the reflection of Bragg gratings in other lasers, especially when the wavelength spacing is very small. Lateral resonant modes can be suppressed by designing π phase shifts in the middle of each segment, and by inserting two mirrors on both ends of the laser to provide compensated reflection for both channels [18,19]. On this basis, a three-stage design was adopted to suppress the side mode by optimizing the current distribution and improving the priority of the dominant mode [20].

In 2021, Xiangfei et al. [21,22] proposed a tunable series-connected distributed feedback multi-wavelength laser array (DFB-MLA) based on reconstruction equivalent chirp (REC) technology. Its wavelength spacing was 2.4 nm and the SMSR was greater than 40 dB. The structure of the heat sink block was optimized for fast, continuous wavelength tuning.

2.3. Series–Parallel Combined DFB Laser Array

2.3.1. Technical Principle, Advantages, and Disadvantages

Regardless of a series or parallel onefold scheme of the DFB laser array, an increase in the number of lasers should consequently increase the difficulty of ensuring a single-mode output and low power loss. Therefore, when the number of integrated lasers is large, the use of a series or parallel structure alone makes the chip structure long and narrow, leading to difficulties when packaging. Therefore, a series–parallel combination is required.

At present, the series–parallel combination scheme is relatively new, and difficult to implement because it experiences the disadvantages of both types of arrays simultaneously. Nevertheless, it is still expected to become a popular topic in DFB laser array research.

2.3.2. Main Research Progress and Status

The series–parallel combination of a DFB laser array is not frequently implemented, but it may be the future development direction. In 2020, Xiangfei et al. [23,24] fabricated a 4×4 16-channel series–parallel DFB laser array that achieved 48 channels with a spacing of 100 GHz near 1550 nm and a temperature adjustment range below 20 °C. Compared with the series-only configuration, this matrix configuration reduces the potential interference from adjacent lasers, which reduces the overall power loss compared with the parallel-only configuration.

3. Implementation Method

The wavelength of the DFB laser is related to the refractive index of the active region and grating spacing, as shown in Equation (11). When fabricating a DFB laser array in which each DFB laser has a different emission wavelength, the fabrication methods can be divided into two categories: changing the grating structure and index of refraction modulation, and is represented as follows:

$$\lambda = \frac{2n_{eff} \Lambda}{m} \tag{11}$$

where λ is the lasing wavelength, n_{eff} is the effective index, Λ is the grating period, and m is the order of the grating.

3.1. Changing the Grating Structure

In the fabrication process of DFB laser arrays, various advanced fabrication methods are used to ensure the gratings of each laser have different periods. DFB laser arrays with different emission wavelengths can be fabricated based on one-time epitaxy of the same active region. Such technologies include electron-beam lithography, nano imprints, and REC technology.

3.1.1. Fabrication Methods
Electron Beam Lithography

1. Technical Principle

Electron beam lithography (EBL) is a lithography technique with the current highest known resolution, which can reach less than 10 nm. Currently, direct-write electron-beam lithography is primarily used. A focused electron beam bombards the photoresist to form the required pattern and scans, point by point, by moving and switching the electron beam to obtain the required grating structure.

2. Technical Advantages and Disadvantages

Technical advantages: EBL has high precision, its direct point-by-point scanning does not require a photolithography mask, it can produce highly complex patterns, diffraction effects are negligible since the electron beam has a short wavelength, it supports dry etching rather than wet etching, it is easier to fabricate high-quality nanostructures, and can be used in high-precision processing of micro-nano electronic and optoelectronic devices.

Technical disadvantages: Owing to the high maintenance cost of EBL equipment, the need to write grating lines one by one leads to an extremely slow etching rate. Therefore, it is expensive and time consuming, and it is difficult to apply to the large-scale manufacturing of lasers and arrays. At the same time, EBL also has blanking or deflection errors and splicing errors, and only 35% of the lasers can be controlled within a range of ±0.2 nm [25]. Processes associated with EBL can generate errors of up to 3 nm [26]. Therefore, it is difficult to ensure the yield of this method when fabricating DFB laser arrays with small channel spacing.

3. Main Research Progress and Status

EBL is a new lithography technology developed from the scanning electron microscope. Since the 1970s, it has been widely used in semiconductor integrated circuit manufacturing. Due to its slow processing speed, it is often used to manufacture application-specific integrated circuit (ASIC) with quick turn-around times and repeatable optical frames that require extremely high precision [27]. It has also been applied to laser grating manufacturing.

In the fabrication of the DFB laser array, a large number of studies have used electron beam lithography to fabricate grating structures [15,28–35], and the results are shown in Table 1. However, throughput capability, the most important problem of electron beam lithography in industrial applications, has not yet been solved. Hence, this technology is still mainly used in the laboratory and in the industry to build very small structures.

Table 1. Recent achievements in the fabrication of DFB laser arrays using EBL.

Year	Institution	Number of Lasers	Channel Spacing	Connection Method	Band	SMSR	Ref
2001	NIT Photonics Laboratories	16	3 nm	Parallel	1.55 μm		[36]
2011	Hitachi	9	3.7 nm	Parallel	1.3 μm	>42 dB	[28]
2017	University of Glasgow	8	0.8 nm	Parallel	1.55 μm	>36 dB	[31]
2017	University of Kassel	4	10 nm	Parallel	1.55 μm	>40 dB	[30]
2019	Russian Academy of Science	4	2.3 nm	Parallel	1.55 μm	>25 dB	[35]
2019	Huazhong University of Science and Technology	4	2.2 nm	Parallel	1.3 μm	>25 dB	[34]
2019	Tsinghua University	10	3.4 nm	Parallel	1.3 μm	>40 dB	[32]
2020	Tsinghua University	4	5.7 nm	Parallel	1.3 μm	>45 dB	[33]
2021	University of Glasgow	8	0.8 nm	Parallel	1.55 μm	>50 dB	[15]

Nano-Imprint Lithography

1. Technical Principle

Nanoimprint lithography (NIL) is a pattern transfer technology; under pressure or heat or UV assisted action, the pattern on the template is transferred to the resist, the resist on the substrate to produce thickness difference, and then by etching the pattern transferred to the substrate. NIL is a promising, high-resolution, low-cost, nanoscale replication patterning technology.

The NIL pattern transfer technology process is illustrated in Figure 4, and is as follows: a film is created by spin-coating a resist onto a substrate, and the film is heated to vitrification. Subsequently, specific pressure is applied to imprint the pattern from the resist-coated substrate to the mold. Afterward, the mold is exposed to heat, UV light, or other elements to solidify the pattern. Finally, the mold and the pattern are separated and the pattern resist is left on the substrate, completing the photonic lithography process. DFB lasers with different grating periods can be prepared by imprinting templates of different shapes using this technique.

2. Technical Advantages and Disadvantages

Technical advantages: NIL technology combines the advantages of EBL and holographic lithography and has the additional advantages of ultra-high resolution, easy mass production, low cost, fast speed, and high consistency. The imprint template is generally manufactured using high-precision EBL and can be used multiple times, thus significantly reducing costs. NIL technology changes the resist by imprinting instead of holographic lithography; therefore, its resolution is not affected by light diffraction, scattering, reflection, etc. Lithography pattern defects are suitable for large-scale manufacturing processes.

Technical disadvantages: during small-scale manufacturing, the cost of NIL technology is high owing to the high production cost of the imprint template. During the production process, the imprint template and the photoresist can easily lead to pattern defects, thereby decreasing precision. The mechanical brittleness of the InP substrate and the temperature difference between the substrate and the imprint template in hot imprinting results in nanoimprinting requiring a very refined process to ensure a good yield.

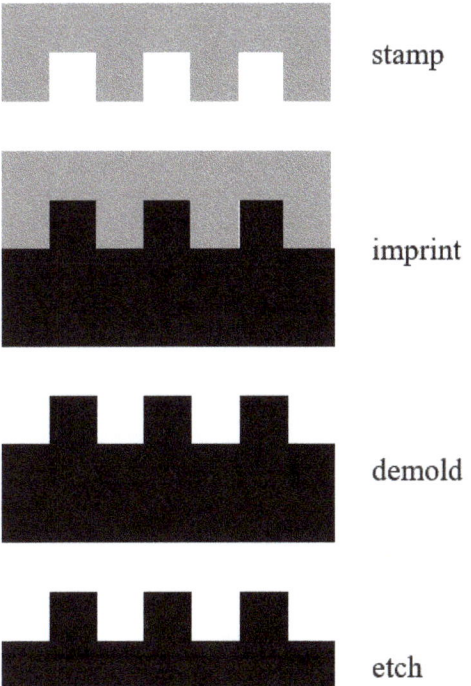

Figure 4. Schematic diagram of the principle of NIL technology.

3. Main Research Progress and Status

NIL technology mainly includes thermal and UV embossing.

Thermal NIL was first proposed by Chou in 1995 [37] and has been widely spread in the manufacturing process of optoelectronic devices such as gratings, microrings, and photonic crystals. At present, this technology is widely used in the fabrication of semiconductor laser gratings, which is a promising DFB laser array grating manufacturing technology. Thermal embossing is simple; however, the heating and cooling durations are long and time consuming, the thermoplastic polymer is prone to thermal expansion and contraction during the heating and cooling process, and the dimensional stability is poor, which easily leads to distortion of the copied graphics. This technique can be improved by employing laser-assisted direct imprinting [38].

There are two main challenges in laser grating thermal NIL. (1) The imprint process is very delicate because of the mechanical brittleness of the InP substrate and the thermal mismatch between the substrate and Si imprint. (2) Subsequent processing requires specific imprint compressive thicknesses after plate making, the crystallographic direction of the substrate, and arrangement of the grating for plate making [39].

Therefore, for thermal NIL, the most important physical properties of the mask are the thermal expansion coefficient and Poisson's ratio. In addition, properties such as hardness, durability, and surface roughness should also be considered. Material selection includes Si, SiO_2, SiC, Si_3N_4, metal, and sapphire [40].

For InP substrates, current studies have shown that the maximum pressure in hot stamping technology does not exceed 3 bar [39].

UV imprinting is also known as room-temperature imprinting. In 1996, UV nanoimprint lithography was designed by introducing a low viscosity UV-curable polymer layer to improve the fluidity of the imprint material [41]. The process involves cross-linking the polymer by UV light on a transparent die after embossing on a UV-cured layer, which significantly reduces the pressure during nano embossing without the need for high temperatures. Low temperature eliminates deformation errors and graphic distortions caused by differences in the thermal expansion coefficient, and low pressure makes the process more likely to succeed on more brittle substrates such as InP and GaAs. Therefore, the technology is more suitable for PICs and manufacturing DFB laser arrays.

UV soft-printing technology is used to fabricate gratings, as shown in Figure 5 [42]. Compared with nano-imprinting directly using hard master printing, the pattern quality is better, especially in the case of surface unevenness that often occurs during multiple epitaxy processes. The specific implementation method is as follows. First, a soft imprint is prepared by imprinting a hard master composed of Si on a flexible intermediate polymer imprint (IPS Obducat AB) using thermal NIL. Next, the epitaxial wafer is carefully cleaned and coated with a layer of UV-curable resist (stuc-220, Obducat AB). Home-made mechanical tools are used to precisely align the soft-printed grating pattern with the grating segments on the wafer. The pattern is then transferred onto the epitaxial structure using an intermediate polymer imprinting (IPS) process by performing simultaneous thermal and UV imprinting processes. After cleaning the residual resist using an oxidative reactive ion etching (RIE) process, inductively coupled plasma (ICP) RIE is used to prepare the grating, and O_2 plasma is used to remove the mask.

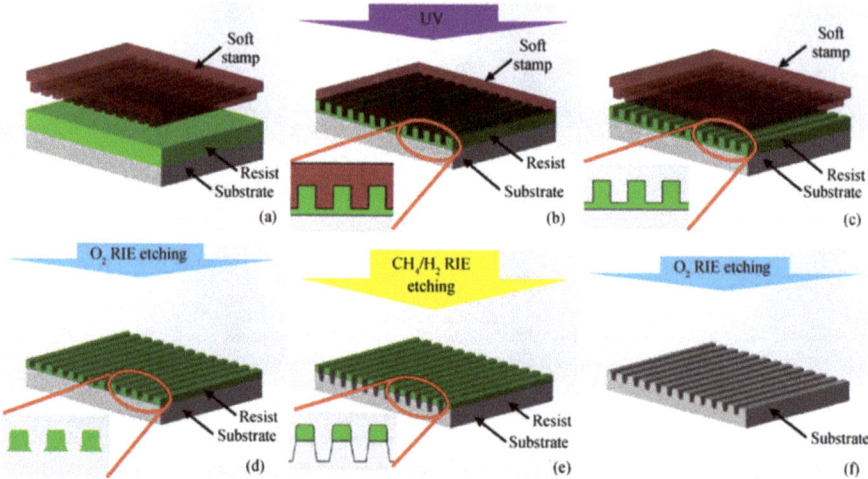

Figure 5. Preparation of gratings by UV soft printing (**a**–**c**) transfer the patterns onto the epitaxial structure (**d**) clean residual resist (**e**) fabricate grating (**f**) remove the mask [42]. © IOP Publishing. Copyright 2014 Journal of semiconductors.

However, as there is no heating or cooling process, polymer bubbles cannot be eliminated, which greatly affects the graphics. An improvement in this process uses a step-flash or thermal UV imprint technology that combines UV imprinting with stepping technology or thermal imprints, respectively. In this process, the corrosion inhibitor is the key problem. The selected resist must maintain low viscosity at room temperature, be able to cure under UV irradiation with minimal shrinkage to maximize pattern accuracy and be released from the mask after curing to adhere to the substrate. A modified poly-dimethyl-siloxane (PDMS) soft stamp has been used to reduce pattern deformation and residual layer thickness, and the residual layer thickness has been reduced by 50% [43].

NIL technology is also widely used in the fabrication of DFB lasers and array gratings, and there have been many recent innovations. Compared with Si substrates used in large-scale integrated circuits, compound semiconductor substrates such as GaAs and InP often have large thickness fluctuations, resulting in the uneven thickness of the residual layer in the imprinting area when NIL technology is applied, ultimately resulting in the graph changes. To solve this problem, reverse NIL technology has been developed and is based on step-and-repeat imprinting named step and flash imprint lithography (SFIL), which can inhibit the uneven residual layer caused by substrate fluctuation [44].

In the NIL process, the duty cycle of gratings may change due to the deformation of the resist and the soft mode. However, when the grating angle is less than 3° and the duty cycle is between 0.4 and 0.6, the effect of error can be ignored, which can be easily achieved with SFIL technology [45].

In NIL, the imprint template is usually prepared using the EBL method, which provides better stability. In addition, during the pattern-transfer process using the NIL method, the resist is in direct contact with the stamp with little deformation. Therefore, NIL can smooth raster edges. This has important implications for the fabrication of high-quality lasers, which can reduce the spectral width of higher-order Fourier components owing to imperfect gratings.

3.1.2. Special Grating Design Reconstruction

Reconstruction Equivalent Chirp

1. Technical Principle

The REC technique is based on the reflection response. By designing a sampling grating based on the sampling function, the chirp structure of the grating period that changes with position is obtained, and the function of the complex grating is realized effectively. The REC technique achieves equivalence with complex grating structures by fabricating specially designed sampling gratings that are superposed on a uniform gratings pattern. The REC technology utilizes sampled Bragg gratings (SBGs) to design and fabricate lasers and laser arrays with complex grating structures. By changing the sampling mode, an equivalent chirp (continuous variation) or equivalent displacement (discrete variation) of the grating neutron grating (Fourier component) can be generated. The Bragg wavelengths of the sub-ratings can also vary with the sampling period. Thus, various sub-grating with complex structures can be obtained by designing the sampling patterns in advance. The sampling period is usually a few microns and can be formed by ordinary photolithography [46].

According to Fourier analysis, the index modulation of a grating section with sampling period P and a basically uniform (seed) grating period Λ_0 can be expressed as [47]:

$$\Delta n(z) = \frac{1}{2}\Delta n_s \sum_m F_m \exp\left(j\frac{2\pi z}{\Lambda_0} + j\frac{2m\pi z}{P} + \varphi\right) + c.c \quad (12)$$

where Δn_s is the index modulation of the uniform seed grating, F_m is the Fourier coefficient of the mth order sub-grating z is the position along the laser cavity, Λ_0 is the seed grating period, P is the sampling period, φ is the initial phase of the seed grating, m is the order of Fourier component, and usually the $+1st$ or $-1st$ order sub-grating is used. The $\pm 1st$ order period can be expressed as,

$$\frac{1}{\Lambda_m} = \frac{1}{\Lambda_0} + \frac{m}{P} \quad (13)$$

Therefore, if the sampling period P is carefully designed for a suitable wavelength grid and each laser section with a specific wavelength is tuned by changing the chip temperature, a wide wavelength range can be covered and the wavelength tuning can be realized simultaneously. According to Equation (11), Λ can change by changing P to achieve different output wavelengths in different lasers.

2. Technical Advantages and Disadvantages

Technical advantages: the REC technology can achieve complex grating structures through micron-scale sampling patterns. In the grating fabrication process, only two steps are required: holographic exposure of the uniform base (seed) grating and lithography of the sampling pattern. All other processes are the same as those used for conventional DFB lasers. Therefore, the manufacturing costs are low. The most important advantage of this technique is that it provides precise fabrication of periodic structures for precise control of wavelengths, and its accuracy is 100-fold higher than that of general theoretical methods [48].

Technical disadvantages: owing to different sampling periods, the phase of the cavity surface of the laser will change, which destroys the uniformity of the wavelength spacing of the laser array. To obtain a uniform wavelength separation, minimal reflection is required on both laser faces; however, this results in a high threshold current and low slope efficiency. Without a reflection-enhancing coating, the wavelength deviation of the laser array can reach approximately 1 nm [47]. In addition, the REC technique reduces the SMSR of side-mode sampling.

3. Main Research Progress and Status

REC technology was initially proposed in fiber Bragg gratings [49] and has been applied to optical filters, fiber lasers, OCDMA en/decoders, and other important optical devices. Xiangfei et al. conducted extensive and in-depth research on the fabrication of DFB laser arrays using REC technology. REC technology in 2009 to fabricate a multi-wavelength DFB laser array, which attracted wide attention due to its low cost and simple method [50]. In 2012, they then made an 8-channel parallel DFB, with an error tolerance of approximately 520-fold higher than the previous array [48].

For the DFB laser array, because it integrates multiple DFB lasers, the yield drops sharply compared with a single DFB laser. The yield is a key issue in the fabrication of DFB laser arrays. The 60-channel DFB laser array produced by REC technology can control the wavelength error of 83% of the laser within ±0.20 nm and 93.5% of the laser within ±0.30 nm [51], greatly improving the yield. REC technology can not only be used in the manufacturing of parallel DFB laser arrays but also in the manufacturing of series DFB laser arrays [22] and series–parallel combined DFB laser arrays [52].

3.2. Index of Refraction Modulation

In the fabrication of DFB laser arrays, in addition to realizing different emission wavelengths by preparing gratings with different periods in different lasers, wavelength changes can also be achieved by changing the refractive index in different lasers. Such technologies include ridge width variation technology and selective area growth.

3.2.1. Ridge Width Variation Technology

Technical Principle

The systemic equivalent refractive index can be adjusted by adjusting the width of the ridge waveguide of the laser, according to Equation (12).

Technical Advantages and Disadvantages

Technical advantages: the process of changing the width of the laser waveguide is simple, and is suitable for manufacturing DFB laser arrays with narrow channel spacing.

Technical disadvantages: the scope of application is limited, the threshold value of the device is too wide and may cause the device to generate multiple transverse modes, the waveguide is too narrow, the series resistance is too high, the thermal effect is obvious, and it is difficult for the device to achieve consistent characteristics.

Main Research Progress and Status

Li et al. [53] proposed varying ridge widths to adjust the lasing wavelength of DFB laser arrays in 1996 and demonstrated that even if the ridge width was sufficient to support multiple lateral modes, single-mode lasers could be maintained. Therefore, the tuning range of the array could be increased without affecting the single-mode performance.

The method can be used to fabricate DFB laser arrays with very small channel spacing, which can reach 0.8 nm [54]. However, because the effective refractive index of the laser varies nonlinearly with the width of the ridge waveguide, and the width of the ridge waveguide cannot be too small (the machining accuracy is difficult to achieve) or too large (A high-order transverse mode is generated), the wavelength tuning range that can be achieved by this scheme is very limited.

3.2.2. Selective Area Growth
Technical Principle

Selective area growth (SAG) is performed as follows: a series of mask stripes are fabricated on a substrate, and then epitaxial layers are grown in areas without mask stripes, as shown in Figure 6. In the SAG process, SiO_2 mask stripes are formed on the substrate before the material is grown; thus, in the subsequent Metal–Organic Chemical Vapor Deposition (MOCVD) or Metal–Organic Vapor Phase Epitaxial Growth (MOPVE) process, the material is only grown where there are no mask stripes. In other words, SAG technology can be used to control the structural parameters of each laser active region, including the thickness and material composition, to change the refractive index of the laser active region, thereby changing the emission wavelength. The SAG technology can simultaneously control the multiple quantum well (MQWs) bandgap energy of the arrayed waveguide by changing the mask pattern's dummy stripe width and the width of the outer mask [55]. By using an asymmetric mask, an almost linear shift in emission wavelength can be obtained. For MQWs, because the emission wavelength is sensitive to the thickness of the well, extensive wavelength tuning can be obtained in addition to refractive index changes.

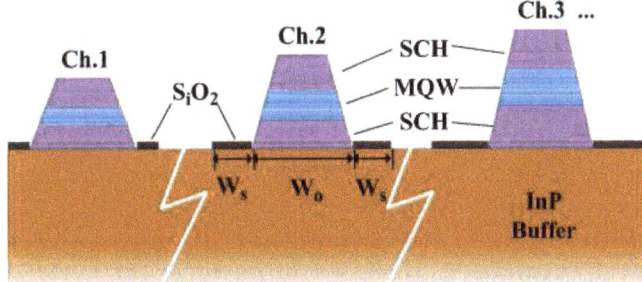

Figure 6. Selective area growth schematic (Zhang C et al. 2013 [11]). © Springer link. Copyright 2013 Science China Technological Sciences.

The thickness enhancement factor of the selectively grown layer is affected by the geometry of the mask pattern, including mask width and pitch. By designing the geometry of the mask, precise control of the thickness of the active region and the waveguide layer can be achieved, and thus control of the effective index of refraction can be achieved to achieve different wavelengths in the laser array [56].

The SAG mechanism is generally considered to be the surface diffusion component and the vapor phase diffusion component. Surface diffusion refers to the molecular diffusion of reactants from the mask surface to the exposed surface, while vapor phase diffusion refers to the vertical and lateral diffusion of reactant molecules in the gas phase and adhesion to the exposed surface.

In general SAG modeling, surface diffusion is usually ignored because it only occurs within a few micrometers of the edge of the mask.

Vapor phase diffusion can be described by the gas-phase diffusion model, and its diffusion equation is [57]:

$$\frac{\partial^2 C}{\partial x^2} + \frac{\partial^2 C}{\partial y^2} = 0 \tag{14}$$

where $C(y, z)$ is the vapor phase concentration.

Assuming no growth of material on the mask, and that the gas concentration just above the crystal surface in the exposed region between the masks remains constant at equilibrium [57]:

$$\begin{cases} z = d, \ C = C_\infty \\ y = \frac{y_0}{2} + w, z = \frac{r_s}{\pi}, \ C = C_c \\ z = 0, \ C = 0 \end{cases} \tag{15}$$

where d is the height of the stagnant layer, y_0 is the width of the mask, and w is the distance between the masks. r_s is the radius of the hemicylindrical structure of the grown SAG structure at the window. The value of r_s is equivalent to the window width, w, and is assumed to be much smaller than y_0 to simplify the following formula.

Under this boundary condition, the solution of equation (14) is [57]:

$$C = A \ln \left[\sin^2\left(\frac{\pi}{y_0}y\right) + \sinh\left(\frac{\pi}{y_0}z\right) \right]^{\frac{1}{2}} + B \tag{16}$$

where A and B are [57]:

$$A = \frac{C_\infty - C_c}{\ln\left[\sinh\left(\frac{\pi d}{y_0}\right)\right] - \ln\left[\sinh\left(\frac{r_s}{y_0}\right)\right]} \cong \frac{C_\infty - C_c}{\ln\left[\frac{y_0}{r_s}\sinh\left(\frac{\pi d}{y_0}\right)\right]} \tag{17}$$

$$B = C_\infty + A \ln\left[\sinh\left(\frac{\pi d}{y_0}\right)\right] \tag{18}$$

The growth rate v_g can be expressed as [57]:

$$v_g = -\frac{v D_g}{w} \frac{\pi A}{r_s} = \frac{v D_g \pi (C_\infty - C_c)}{w r_s \left\{\ln\left[\sinh\left(\frac{\pi d}{y_0}\right)\right] - \ln\left[\sinh\left(\frac{r_s}{y_0}\right)\right]\right\}} \tag{19}$$

where D_g is the diffusion coefficient of molecules in the vapor phase.

As a limiting case, $r_s \ll d$ and $y_0 \ll d$,

$$v_g \cong \frac{v D_g (C_\infty - C_c)}{w r_s d} y_0 \tag{20}$$

During the SAG growth process, the grown material generates a growth rate enhancement curve in the unmasked area, by formula (20), the growth rate is proportional to the width of the dielectric mask and inversely proportional to the square of the mask spacing. In addition, group III elements will have a certain compositional shift during growth, while group V elements will not; therefore, the SAG technique will lead to a certain degree of inhomogeneity [58].

The traditional MQW SAG technology forms a dielectric mask on the buffer substrate, and the selective region growth layer includes the lower separate confinement structure (SCH), MQW, and upper SCH layers. The difference is that in the new SAG technology, the buffer, lower SCH, and MQW layers are first grown on the substrate. Subsequently, mask strip pairs are formed on the MQW layer. In the following SAG process, only one upper SCH layer is grown, which effectively reduces the difficulty of the process by reducing the number of selectively grown layers. At the same time, SAG technology can be combined with EBL and REC technologies to achieve improved performance.

Technical Advantages and Disadvantages

Technical advantages: SAG technology is used to fabricate DFB semiconductor laser arrays because of its simplicity, low cost, and suitability for mass production. While forming a laser array, SAG can also be used for the integration of the laser array with other optical components, such as electro-absorption modulators (EAMs) [59].

Technical disadvantages: The SAG process requires good control of the epitaxial growth process and advanced technology. Additionally, it has low repeatability and it is difficult to guarantee the single longitudinal modulus rate, requiring other technologies to control it.

Main Research Progress and Status

SAG is widely used in the epitaxial growth of various semiconductor devices and was first applied in the preparation of the multi-wavelength DFB laser array in 1994 [60]. This method can prepare a DFB laser array with a wide wavelength span of up to 155 nm [61]. SAG technology is very flexible and can be used in combination with a variety of technologies. For example, gratings can be made using EBL based on SAG [62], and the width of ridge waveguides can be changed while SAG technology is used [63] or by combining SAG technology, and EBL technology, and reverse mesa ridge waveguide LD processing technology [64]. SAG technology can simultaneously integrate MMI, SOA, MOD, and other devices on the chip [65]. SAG technology can also be combined with bundled integrated waveguide (BIG) technology to obtain uniformly spaced multi-wavelength emission and low-loss passive waveguide materials in one MOCVD step, greatly simplifying the integrated fabrication of the array [66].

In the SAG process, the material layers affected by SAG include two SCH and MQW layers, which may affect the uniformity of the laser array [8]. Therefore, an improved SAG method has been proposed [67]. First, the buffer, low SCH, and MQW layers are grown on the substrate, and then the size is formed on the MQW layer, gradually changing mask–strip pairs. In the next step, only the upper SCH layer is grown, and only the thickness of the upper SCH layer is changed through the SAG mask to obtain different Bragg wavelengths. Materials including the SCH and MQW layers are very sensitive to different growth conditions; therefore, the wavelength spacing can be precisely controlled.

As the thickness of the material can be controlled with precision, SAG technology is especially suitable for fabricating laser arrays with small channel spacing, which can achieve good wavelength spacing uniformity. However, DFB lasers with a high single-longitudinal-mode rate are not guaranteed. Therefore, REC technology has been adopted to introduce an equivalent phase shift to ensure the single-mode laser output of the DFB laser [68].

4. Conclusions

In summary, the DFB laser array is among the most widely used and mature solutions in WDW. Compared with other lasers, its mode stability, high wavelength stability, and narrow linewidth make it stand out. The research on DFB laser arrays is also the most extensive. The multi-wavelength DFB laser array is a key component of the wavelength-division multiplexing system, and its manufacturing difficulties affect its mass production and use. This paper summarizes the structure and fabrication methods for multi-wavelength DFB laser arrays. At present, various DFB laser array structures and fabrication methods can still be improved. Finding a low-cost, high-yield, easy fabrication method, that can adopt a reasonable structure, are key factors to promote the further development of DFB laser arrays.

For DFB laser arrays, the main problems to be solved in the future are the wavelength accuracy and the wavelength spacing of different lasers. In particular for DWDM applications, the wavelength spacing of future DFB laser arrays may still be further reduced, which will put forward higher requirements for technical accuracy. On the one hand, we need to continue to improve the existing technologies such as EBL, NIL, and other processes; on the other hand, in the design of gratings, active regions, waveguides, etc., such as REC and SAG, it is also necessary to further reduce errors. In addition, the DFB laser arrays currently used in optical communications are mainly InP based, which is difficult to match with standard integrated circuit processes. In the future, emerging multi-wavelength WDM sources based on Si photonics are also very expected.

Author Contributions: Conceptualization, S.N. and Y.S.; methodology, C.Q. and Y.L.; validation, P.J., Y.C. and Y.W. (Ye Wang); formal analysis, Y.S. and L.Z.; investigation, Y.S., C.Q., Y.W. (Yubing Wang) and Y.L.; resources, S.N., Y.W. and Y.S.; writing—original draft preparation, S.N.; writing—review and editing, L.Z. and Y.S.; supervision, Y.C. and L.L.; project administration, Y.N.; funding acquisition, L.Q. and L.W. All authors have read and agreed to the published version of the manuscript.

Funding: This work is supported by the National Science and Technology Major Project of China (2021YFF0700500); National Natural Science Foundation of China (NSFC) (61904179, 62090051, 62090052, 62090054, 11874353, 61935009, 61934003, 62004194); Science and Technology Development Project of Jilin Province (20200401069GX, 20200401062GX, 20200501006GX, 20200501007GX, 20200501008GX); Key R&D Program of Changchun [21ZGG13, 21ZGN23]; Innovation and entrepreneurship Talent Project of Jilin Province [2021Y008]; Special Scientific Research Project of Academician Innovation Platform in Hainan Province (YSPTZX202034), and "Lingyan" Research Program of Zhejiang Province (2022C01108).

Institutional Review Board Statement: Not applicable.

Informed Consent Statement: Not applicable.

Data Availability Statement: Not applicable.

Conflicts of Interest: The authors declare no conflict of interest.

References

1. Nesset, D. NG-PON2 Technology and Standards. *J. Lightwave Technol.* **2015**, *33*, 1136–1143. [CrossRef]
2. Coldren, L.A.; Fish, G.A.; Akulova, Y.; Barton, J.S.; Johansson, L.; Coldren, C.W. Tunable semiconductor lasers: A tutorial. *J. Lightwave Technol.* **2004**, *22*, 193–202. [CrossRef]
3. Chan, C.K.; Sherman, K.L.; Zirngibl, M. A fast 100-channel wavelength-tunable transmitter for optical packet switching. *IEEE Photonics Technol. Lett.* **2001**, *13*, 729–731. [CrossRef]
4. Wang, Y.Y.; Lin, K.L.; Fang, T.; Chen, X.F. A tunable SFP optical module based on DFB laser array integrated with a SOA. *Optoelectron. Devices Integr. VII* **2018**, *10814*, 153–158. [CrossRef]
5. Kudo, K.; Morimoto, T.; Yashiki, K.; Sasaki, T.; Yokoyama, Y.; Hamamoto, K.; Yamaguchi, M. Wavelength-selectable microarray light sources of multiple ranges simultaneously fabricated on single wafer. *Electron. Lett.* **2000**, *36*, 745–747. [CrossRef]
6. Lee, S.L.; Pukhrambam, P.D. Wavelength division multiplexing laser arrays for applications in optical networking and sensing: Overview and perspectives. *Jpn. J. Appl. Phys.* **2018**, *57*, 08PA03. [CrossRef]
7. Luo, Y.Q.; Roberts, H.; Grobe, K.; Valvo, M.; Nesset, D.; Asaka, K.; Rohde, H.; Smith, J.; Wey, J.S.; Effenberger, F. Physical Layer Aspects of NG-PON2 Standards-Part 2: System Design and Technology Feasibility. *J. Opt. Commun. Netw.* **2016**, *8*, 43–52. [CrossRef]
8. Liang, S.; Lu, D.; Zhao, L.J.; Zhu, H.L.; Wang, B.J.; Zhou, D.B.; Wang, W. Fabrication of InP-based monolithically integrated laser transmitters. *Sci. China Inf. Sci.* **2018**, *61*, 080405. [CrossRef]
9. Pezeshki, B.; Vail, E.; Kubicky, J.; Yoffe, G.; Heanue, J.; Epp, P.; Rishton, S.; Ton, D.; Faraji, B.; Emanuel, M.; et al. 20-mW widely tunable laser module using DFB array and MEMS selection. *IEEE Photonics Technol. Lett.* **2002**, *14*, 1457–1459. [CrossRef]
10. Pan, P.; An, J.M.; Zhang, J.S.; Wang, Y.; Wang, H.J.; Wang, L.L.; Yin, X.J.; Wu, Y.D.; Li, J.G.; Han, Q.; et al. Flat-top AWG based on InP deep ridge waveguide. *Opt. Commun.* **2015**, *355*, 376–381. [CrossRef]
11. Zhu, H.L.; Ma, L.; Liang, S.; Zhang, C.; Wang, B.J.; Zhao, L.J.; Wang, W. InP based DFB laser array integrated with MMI coupler. *Sci. China Technol. Sci.* **2013**, *56*, 573–578. [CrossRef]
12. Li, R.D.; Erneux, T. Stability Conditions for Coupled Lasers—Series Coupling Versus Parallel Coupling. *Opt. Commun.* **1993**, *99*, 196–200. [CrossRef]
13. Macomber, S.H.; Mott, J.S.; Schwartz, B.D.; Setzko, R.S. Curved-grating, surface-emitting DFB lasers and arrays. *In-Plane Semicond. Lasers Ultrav. Midinfrared* **1997**, *3001*, 42–54. [CrossRef]

14. Okuda, H.; Hirayama, Y.; Furuyama, H.; Uematsu, Y. Simultaneous Cw Operation of 5-Wavelength Integrated Gainasp-Inp Dfb Laser Array with 50-a Lasing Wavelength Separation. *Jpn. J. Appl. Phys. Part 2 Lett.* **1984**, *23*, L904–L906. [CrossRef]
15. Hou, L.P.; Tang, S.; Marsh, J.H. Monolithic DWDM source with precise channel spacing. *J. Semicond.* **2021**, *42*, 042301. [CrossRef]
16. Li, L.Y.; Tang, S.; Lu, J.; Shi, Y.C.; Cao, B.L.; Chen, X.F. Study of cascaded tunable DFB semiconductor laser with wide tuning range and high single mode yield based on equivalent phase shift technique. *Opt. Commun.* **2015**, *352*, 70–76. [CrossRef]
17. Zhao, Y.; Shi, Y.C.; Li, J.; Liu, S.P.; Xiao, R.L.; Li, L.Y.; Lu, J.; Chen, X.F. A Cascaded Tunable DFB Semiconductor Laser With Compact Structure. *IEEE J. Quantum Electron.* **2018**, *54*, 2200111. [CrossRef]
18. Sun, Z.X.; Xiao, R.L.; Zhao, Y.; Lv, G.; Su, Z.R.; Shi, Y.C.; Chen, X.F. Design of Four-Channel Wavelength-Selectable In-Series DFB Laser Array With 100-GHz Spacing. *J. Lightwave Technol.* **2020**, *38*, 2299–2307. [CrossRef]
19. Sun, Z.X.; Xiao, R.L.; Zhao, Y.; Dai, P.; Lv, G.; Su, Z.R.; Shi, Y.C.; Chen, X.F. Design of Wavelength-selectable In-series DFB Laser Array Based on Chirped Bragg Grating. In Proceedings of the 2019 Asia Communications and Photonics Conference (ACP), Chengdu, China, 2–5 November 2019.
20. Sun, Z.X.; Xiao, R.L.; Su, Z.R.; Liu, K.; Hu, Z.Y.; Dai, P.; Lu, J.; Zheng, J.L.; Zhang, Y.S.; Shi, Y.C.; et al. High Single-Mode Stability Tunable In-Series Laser Array With High Wavelength-spacing Uniformity. *J. Lightwave Technol.* **2020**, *38*, 6038–6046. [CrossRef]
21. Sun, Z.X.; Xiao, R.L.; Su, Z.R.; Liu, K.; Lv, G.; Xu, K.; Fang, T.; Shi, Y.C.; Chiu, Y.J.; Chen, X.F. Experimental Demonstration of Wavelength-tunable In-Series DFB Laser Array with 100-GHz Spacing. *IEEE J. Sel. Top. Quantum Electron.* **2022**, *28*, 1500308. [CrossRef]
22. Dai, P.; Sun, Z.X.; Chen, Z.; Lu, J.; Wang, F.; Tong, H.; Xiao, R.L.; Chen, X.F. Enhanced Tuning Performance of In-Series REC-DFB Laser Array. *IEEE Photonics Technol. Lett.* **2021**, *33*, 1337–1340. [CrossRef]
23. Su, Z.; Xiao, R.; Sun, Z.; Yang, Z.; Chen, X. 48 channels 100-GHz tunable laser by integrating 16 DFB lasers with high wavelength-spacing uniformity. *arXiv* **2020**, arXiv:2001.01178.
24. Liu, W.; Wang, Q.M.; Lin, K.L.; Fang, T.; Chen, X.F. Fast wavelength-switching DFB laser array with 16 channels based on the REC technology. *Semicond. Lasers Appl. X* **2020**, *11545*, 42–48. [CrossRef]
25. Lee, T.P.; Zah, C.E.; Bhat, R.; Young, W.C.; Pathak, B.; Favire, F.; Lin, P.S.D.; Andreadakis, N.C.; Caneau, C.; Rahjel, A.W.; et al. Multiwavelength DFB laser array transmitters for ONTC reconfigurable optical network testbed. *J. Lightwave Technol.* **1996**, *14*, 967–976. [CrossRef]
26. Zanola, M.; Strain, M.J.; Giuliani, G.; Sorel, M. Post-Growth Fabrication of Multiple Wavelength DFB Laser Arrays With Precise Wavelength Spacing. *IEEE Photonics Technol. Lett.* **2012**, *24*, 1063–1065. [CrossRef]
27. Okazaki, S. High resolution optical lithography or high throughput electron beam lithography: The technical struggle from the micro to the nano-fabrication evolution. *Microelectron. Eng.* **2015**, *133*, 23–35. [CrossRef]
28. Adachi, K.; Shinoda, K.; Kitatani, T.; Fukamachi, Y.; Matsuoka, Y.; Sugawara, T.; Tsuji, S. 25-Gb/s Multichannel 1.3-mu Surface-Emitting Lens-Integrated DFB Laser Arrays. *J. Lightwave Technol.* **2011**, *29*, 2899–2905. [CrossRef]
29. Tsuruoka, K.; Kobayashi, R.; Ohsawa, Y.; Tsukuda, T.; Kato, T.; Sasaki, T.; Nakamura, T. Four-channel 10-Gb/s operation of AlGaInAs-MQW-BH-DFB-LD array for 1.3-mu m CWDM systems. *IEEE J. Sel. Top. Quantum Electron.* **2005**, *11*, 1169–1173. [CrossRef]
30. Becker, A.; Sichkovskyi, V.; Bjelica, M.; Rippien, A.; Schnabel, F.; Kaiser, M.; Eyal, O.; Witzigmann, B.; Eisenstein, G.; Reithmaier, J.P. Widely tunable narrow-linewidth 1.5 mu m light source based on a monolithically integrated quantum dot laser array. *Appl. Phys. Lett.* **2017**, *110*, 181103. [CrossRef]
31. Tang, S.; Hou, L.P.; Chen, X.F.; Marsh, J.H. Multiple-wavelength distributed-feedback laser arrays with high coupling coefficients and precise channel spacing. *Opt. Lett.* **2017**, *42*, 1800–1803. [CrossRef]
32. Li, A.K.; Wang, J.; Sun, C.Z.; Wang, Y.Q.; Yang, S.H.; Xiong, B.; Luo, Y.; Hao, Z.B.; Han, Y.J.; Wang, L.; et al. 1.3 mu m 10-Wavelength Laterally Coupled Distributed Feedback Laser Array with High-Duty-Ratio Gratings. *Phys. Status Solidi A Appl. Mater. Sci.* **2019**, *216*, 1800490. [CrossRef]
33. Wang, Q.C.; Wang, J.; Sun, C.Z.; Xiong, B.; Luo, Y.; Hao, Z.B.; Han, Y.J.; Wang, L.; Li, H.T.; Yu, J.D. A Directly Modulated Laterally Coupled Distributed Feedback Laser Array Based on SiO2 Planarization Process. *Appl. Sci.* **2021**, *11*, 221. [CrossRef]
34. Zhao, G.Y.; Liu, G.H.; Liu, C.; Lu, Q.Y.; Guo, W.H. Monolithically Integrated Directly Modulated ADR-DFB Laser Array in the O-Band. *IEEE Photonics Technol. Lett.* **2019**, *31*, 1495–1498. [CrossRef]
35. Dudelev, V.V.; Mikhailov, D.A.; Andreev, A.D.; Kognovitskaya, E.A.; Sokolovskii, G.S.J.Q.E. Tunable single-frequency radiation source based on an array of DFB lasers for the spectral range of 1.55 μm. *Quantum Electron.* **2019**, *49*, 1158–1162. [CrossRef]
36. Oohashi, H.; Shibata, Y.; Ishii, H.; Kawaguchi, Y.; Kondo, Y.; Yoshikuni, Y.; Tohmori, Y. 46.9-nm wavelength-selectable arrayed DFB lasers with integrated MMI coupler and SOA. In Proceedings of the 2001 International Conference on Indium Phosphide and Related Materials, Nara, Japan, 14–18 May 2001; pp. 575–578. [CrossRef]
37. Chou, S.Y.; Krauss, P.R.; Renstrom, P.J. Imprint lithography with 25-nanometer resolution. *Science* **1996**, *272*, 85–87. [CrossRef]
38. Chou, S.Y.; Keimel, C.; Gu, J. Ultrafast and direct imprint of nanostructures in silicon. *Nature* **2002**, *417*, 835–837. [CrossRef]
39. Smistrup, K.; Norregaard, J.; Mironov, A.; Bro, T.H.; Bilenberg, B.; Nielsen, T.; Eriksen, J.; Thilsted, A.H.; Hansen, O.; Kristensen, A.; et al. Nanoimprinted DWDM laser arrays on indium phosphide substrates. *Microelectron. Eng.* **2014**, *123*, 149–153. [CrossRef]

40. Lugli, P.; Harrer, S.; Strobel, S.; Brunetti, F.; Scarpa, G.; Tornow, M.; Abstreiter, G. Advances in Nanoimprint Lithography. In Proceedings of the 2007 7th IEEE Conference on Nanotechnology, Hong Kong, China, 2–5 August 2007; Volumes 1–3, pp. 1179–1184.
41. Haisma, J.; Verheijen, M.; van den Heuvel, K.; van den Berg, J. Mold-assisted nanolithography: A process for reliable pattern replication. *J. Vac. Sci. Technol. B* **1996**, *14*, 4124–4128. [CrossRef]
42. Zhao, J.Y.; Chen, X.; Zhou, N.; Huang, X.D.; Liu, W. Fabrication of four-channel DFB laser array using nanoimprint technology for 1.3 mu m CWDM systems. *J. Semicond.* **2014**, *35*, 114008. [CrossRef]
43. Viheriala, J.; Tommila, J.; Leinonen, T.; Dumitrescu, M.; Toikkanen, L.; Niemi, T.; Pessa, M. Applications of UV-nanoimprint soft stamps in fabrication of single-frequency diode lasers. *Microelectron. Eng.* **2009**, *86*, 321–324. [CrossRef]
44. Yanagisawa, M.; Tsuji, Y.; Yoshinaga, H.; Kono, N.; Hiratsuka, K. Evaluation of nanoimprint lithography as a fabrication process of phase-shifted diffraction gratings of distributed feedback laser diodes. *J. Vac. Sci. Technol. B* **2009**, *27*, 2776–2780. [CrossRef]
45. Wang, H.; Liu, W.; Zhang, Y.W.; Qiu, F.; Zhou, N.; Wang, D.L.; Xu, Z.M.; Zhao, Y.L.; Yu, Y.L. DFB LDs at DWDM wavelengths fabricated by a novel nanoimprint process for mass production and tolerance simulation. *Microelectron. Eng.* **2012**, *93*, 43–49. [CrossRef]
46. Jin, R.Q.; Chen, X.F. Precision photonic integration for future large-scale photonic integrated circuits. *J. Semicond.* **2019**, *40*, 050301. [CrossRef]
47. Shi, Y.C.; Li, S.M.; Li, L.Y.; Guo, R.J.; Zhang, T.T.; Rui, L.; Li, W.C.; Lu, L.L.; Song, T.; Zhou, Y.T.; et al. Study of the Multiwavelength DFB Semiconductor Laser Array Based on the Reconstruction-Equivalent-Chirp Technique. *J. Lightwave Technol.* **2013**, *31*, 3243–3250. [CrossRef]
48. Shi, Y.C.; Chen, X.F.; Zhou, Y.T.; Li, S.M.; Lu, L.L.; Liu, R.; Feng, Y.J. Experimental demonstration of eight-wavelength distributed feedback semiconductor laser array using equivalent phase shift. *Opt. Lett.* **2012**, *37*, 3315–3317. [CrossRef]
49. Dai, Y.T.; Chen, X.F.; Xia, L.; Zhang, Y.J.; Xie, S.Z. Sampled Bragg grating with desired response in one channel by use of a reconstruction algorithm and equivalent chirp. *Opt. Lett.* **2004**, *29*, 1333–1335. [CrossRef]
50. Li, J.S.; Wang, H.; Chen, X.F.; Yin, Z.W.; Shi, Y.C.; Lu, Y.Q.; Dai, Y.T.; Zhu, H.L. Experimental demonstration of distributed feedback semiconductor lasers based on reconstruction-equivalent-chirp technology. *Opt. Express* **2009**, *17*, 5240–5245. [CrossRef]
51. Shi, Y.C.; Li, S.M.; Chen, X.F.; Li, L.Y.; Li, J.S.; Zhang, T.T.; Zheng, J.L.; Zhang, Y.S.; Tang, S.; Hou, L.P.; et al. High channel count and high precision channel spacing multi-wavelength laser array for future PICs. *Sci. Rep.* **2014**, *4*, 7377. [CrossRef]
52. Chen, M.; Liu, S.P.; Shi, Y.C.; Dai, P.; Zhao, Y.; Chen, X.F. Study on DFB semiconductor laser based on sampled moire grating integrated with grating reflector. In Proceedings of the 2019 18th International Conference on Optical Communications and Networks (ICOCN), Huangshan, China, 5–8 August 2019.
53. Sarangan, A.M.; Huang, W.P.; Makino, T.; Li, G.P. Dynamic single-transverse-mode properties of varying ridge width DFB laser arrays. *IEEE Photonics Technol. Lett.* **1996**, *8*, 1305–1307. [CrossRef]
54. Ma, L.; Zhu, H.L.; Liang, S.; Wang, B.J.; Zhang, C.; Zhao, L.J.; Bian, J.; Chen, M.H. A 1.55-mu m laser array monolithically integrated with an MMI combiner. *J. Semicond.* **2013**, *34*, 044007. [CrossRef]
55. Hatakeyama, H.; Yokoyama, Y.; Naniwae, K.; Kudo, K.; Sasaki, T. Wavelength-selectable microarray light sources for wide-band DWDM. *Act. Passiv. Opt. Compon. WDM Commun. II* **2002**, *4870*, 153–160. [CrossRef]
56. Zhang, C.; Zhu, H.L.; Liang, S.; Cui, X.; Wang, H.T.; Zhao, L.J.; Wang, W. Ten-channel InP-based large-scale photonic integrated transmitter fabricated by SAG technology. *Opt. Laser Technol.* **2014**, *64*, 17–22. [CrossRef]
57. Ujihara, T.; Yoshida, Y.; Lee, W.S.; Takeda, Y. Pattern size effect on source supply process for sub-micrometer scale selective area growth by organometallic vapor phase epitaxy. *J. Cryst. Growth* **2006**, *289*, 89–95. [CrossRef]
58. Greenspan, J.E. Alloy composition dependence in selective area epitaxy on InP substrates. *J. Cryst. Growth* **2002**, *236*, 273–280. [CrossRef]
59. Zhang, C.; Liang, S.; Zhu, H.L.; Ma, L.; Wang, B.J.; Ji, C.; Wang, W. Multi-channel DFB laser arrays fabricated by SAG technology. *Opt. Commun.* **2013**, *300*, 230–235. [CrossRef]
60. Aoki, M.; Taniwatari, T.; Suzuki, M.; Tsutsui, T. Detuning Adjustable Multiwavelength Mqw-Dfb Laser Array Grown by Effective-Index Quantum Energy Control Selective-Area Movpe. *IEEE Photonics Technol. Lett.* **1994**, *6*, 789–791. [CrossRef]
61. Soares, F.; Baier, M.F.; Zhang, Z.; Gaertner, T.; Franke, D.; Decobert, J.; Achouche, M.; Schmidt, D.; Moehrle, M.; Grote, N.; et al. 155nm-Span Multi-Wavelength DFB Laser Array Fabricated by Selective Area Growth. In Proceedings of the 2016 Compound Semiconductor Week (CSW) Includes 28th International Conference on Indium Phosphide & Related Materials (IPRM) & 43rd International Symposium on Compound Semiconductors (ISCS), Toyama, Japan, 26–30 June 2016.
62. Darja, J.; Chan, M.J.; Sugiyama, M.; Nakano, Y. Four channel DFB laser array with integrated combiner for 1.55 mu m CWDM systems by MOVPE selective area growth. *IEICE Electron. Express* **2006**, *3*, 522–528. [CrossRef]
63. Cheng, Y.B.; Wang, Q.J.; Pan, J.Q. 1.55 mu m high speed low chirp electroabsorption modulated laser arrays based on SAG scheme. *Opt. Express* **2014**, *22*, 31286–31292. [CrossRef]
64. Kwon, O.K.; Leem, Y.A.; Han, Y.T.; Lee, C.W.; Kim, K.S.; Oh, S.H. A 10 × 10 Gb/s DFB laser diode array fabricated using a SAG technique. *Opt. Express* **2014**, *22*, 9073–9080. [CrossRef]
65. Kudo, K.; Yashiki, K.; Sasaki, T.; Yokoyama, Y.; Hamamoto, K.; Morimoto, T.; Yamaguchi, M. 1.55-mu m wavelength-selectable microarray DFB-LD's with monolithically integrated MMI combiner, SOA, and EA-Modulator. *IEEE Photonics Technol. Lett.* **2000**, *12*, 242–244. [CrossRef]

66. Han, L.S.; Liang, S.; Wang, H.T.; Xu, J.J.; Qiao, L.J.; Zhu, H.L.; Wang, W. Fabrication of Low-Cost Multiwavelength Laser Arrays for OLTs in WDM-PONs by Combining the SAG and BIG Techniques. *IEEE Photonics J.* **2015**, *7*, 1502807. [CrossRef]
67. Zhang, C.; Liang, S.; Zhu, H.L.; Han, L.S.; Wang, W. Multichannel DFB Laser Arrays Fabricated by Upper SCH Layer SAG Technique. *IEEE J. Quantum Electron.* **2014**, *50*, 92–97. [CrossRef]
68. Xu, J.J.; Liang, S.; Qiao, L.J.; Han, L.S.; Sun, S.W.; Zhu, H.L.; Wang, W. Laser Arrays with 25-GHz Channel Spacing Fabricated by Combining SAG and REC Techniques. *IEEE Photonics Technol. Lett.* **2016**, *28*, 2249–2252. [CrossRef]

Review

Research on Narrow Linewidth External Cavity Semiconductor Lasers

Keke Ding, Yuhang Ma, Long Wei, Xuan Li, Junce Shi, Zaijin Li *, Yi Qu, Lin Li, Zhongliang Qiao, Guojun Liu and Lina Zeng

Key Laboratory of Laser Technology and Optoelectronic Functional Materials of Hainan Province, Academician Team Innovation Center of Hainan Province, College of Physics and Electronic Engineering, Hainan Normal University, Haikou 571158, China; dingkeke1205@163.com (K.D.); mayuhang327@163.com (Y.M.); weilong0096@163.com (L.W.); lixuan13025794412@163.com (X.L.); jcs1075106215@163.com (J.S.); quyihainan@126.com (Y.Q.); licust@126.com (L.L.); qzhl060910@hainnu.edu.cn (Z.Q.); gjliu626@126.com (G.L.); zenglinahainan@126.com (L.Z.)
* Correspondence: lizaijin@126.com

Citation: Ding, K.; Ma, Y.; Wei, L.; Li, X.; Shi, J.; Li, Z.; Qu, Y.; Li, L.; Qiao, Z.; Liu, G.; et al. Research on Narrow Linewidth External Cavity Semiconductor Lasers. *Crystals* **2022**, *12*, 956. https://doi.org/10.3390/cryst12070956

Academic Editor: M. Ajmal Khan

Received: 6 June 2022
Accepted: 5 July 2022
Published: 8 July 2022

Publisher's Note: MDPI stays neutral with regard to jurisdictional claims in published maps and institutional affiliations.

Copyright: © 2022 by the authors. Licensee MDPI, Basel, Switzerland. This article is an open access article distributed under the terms and conditions of the Creative Commons Attribution (CC BY) license (https://creativecommons.org/licenses/by/4.0/).

Abstract: Narrow linewidth external cavity semiconductor lasers (NLECSLs) have many important applications, such as spectroscopy, metrology, biomedicine, holography, space laser communication, laser lidar and coherent detection, etc. Due to their high coherence, low phase-frequency noise, high monochromaticity and wide wavelength tuning potential, NLECSLs have attracted much attention for their merits. In this paper, three main device structures for achieving NLECSLs are reviewed and compared in detail, such as free space bulk diffraction grating external cavity structure, waveguide external cavity structure and confocal Fabry–Perot cavity structure of NLECSLs. The Littrow structure and Littman structure of NLECSLs are introduced from the free space bulk diffraction grating external cavity structure of NLECSLs. The fiber Bragg grating external cavity structure and silicon based waveguide external cavity structure of NLECSLs are introduced from the waveguide external cavity structure of NLECSLs. The results show that the confocal Fabry–Perot cavity structure of NLECSLs is a potential way to realize a lower than tens Hz narrow linewidth laser output.

Keywords: narrow linewidth; external cavity; FSBDG; FBG; silicon-based waveguide; confocal F-P cavity

1. Introduction

Semiconductor lasers have been applied in many fields, such as high-resolution spectroscopy and broadband communication network systems. Semiconductor lasers need to have the characteristics of a narrow linewidth, high-frequency modulation and wide tunable range at the same time. In other fields, it is required that lasers have the characteristics of narrower output linewidth, larger coherence length, and narrower spatial coherence [1]. By using the external cavity technology, semiconductor lasers can produce stabilized output with a single longitudinal mode and narrow linewidth, and they can also be tuned in the range of tens of nanometers to hundreds of nanometers [2]. At the same time, some other properties of semiconductor lasers are also improved, including a lower threshold, higher output power and larger side mode suppression ratio (SMSR) [3,4]. These properties meet the requirements of coherent optical communication, coherent detection and other applications of NLECSLs. In 1964, J.W. Crowe et al. [5] first proposed the external cavity theory of the semiconductor laser. In 1975, Heckscher H et al. [6] reported the compact and relatively inexpensive external cavity structure of the laser with the III-V compound semiconductor.

An NLECSL includes a semiconductor laser active section and an external cavity. The active section, which typically contains a III–V semiconductor quantum wells structure, is used to provide the optical gain for the whole cavity, and thereby determines the lasing wavelength range. The external cavity is used to select the lasing wavelength, while reducing the linewidth. The natural cleaving surface at both ends of the active section chip is the resonant cavity, which is called the internal cavity or the intrinsic cavity [7]. The cavity composed of the external feedback element and the chip cleaving surface is called the external cavity. Through the external cavity, part of the output light is fed back to the active region for multiple gain, thereby narrowing the linewidth and reducing the phase noise and intensity noise of the lasers [8]. There are many kinds of external feedback components, such as free space bulk diffraction grating, fiber Bragg grating (FBG), waveguide and Fabry–Perot (F-P) cavity and the combination of these components. NLECSLs have many advantages, such as good monochromaticity, high stability, long coherence length, and so on. Therefore, NLECSLs are widely used in the fields of photoelectric detection, coherent communication, precision measurement, optical frequency standards, absorption spectrum measurement and the study of the interaction between lights and matters [9]. In this paper, free space bulk diffraction grating (FSBDG) external cavity structure, waveguide external cavity structure and confocal F-P cavity structure, the three main device structures for achieving NLECELs, are expanded upon. Among them, the confocal F-P cavity can further narrow the linewidth. Lewoczko-Adamczyk W et al. [10] proposed the mode of optical self-locking with the external single-chip confocal F-P cavity; when the output power exceeds 50 mW, the corresponding Lorentz linewidth is only 15.7 Hz, which is the highest level in the world at present.

2. FSBDG Structure of NLECSLs

The FSBDG is one of the most widely used external mirrors. It has good performance, especially in its wide tuning range, high spectral resolution, and flexible and precise tunability. Great progress in the ECSL with FSBDG mirror have been achieved, and the results of research and development in the field are continuously transferred to industrialization. Wavelength selectivity and tunability of FSBDG ECSL can be realized by adjusting the incident angle to the FSBDG plane. Different applications put forward different tuning requirements; some require a large tuning range with hopping allowed, whereas others require fine continuous tuning without hopping. This paper focuses mainly on the latter. The continuous tuning range and precision are dependent on the design of optics and related mechanics. Two configurations are well developed and widely used, including the Littrow structure and Littman structure, which are introduced as follows.

2.1. Littrow Structure of NLECSLs

The Littrow structure of NLECSLs is shown in Figure 1. The output light of the Littrow ECSL is collimated by the lens group to obtain the horizontal parallel light, which is incident to the FSBDG external cavity for optical feedback. After FSBDG splitting, the first-order diffraction is fed back to the active region of the laser, and the light field in the active region interacts with each other, resulting in the gain difference between the longitudinal modes and the gain is larger. The longitudinal mode excitation satisfying the laser excitation condition is excited, and the mode with the small gain is lost. By changing the wavelength of the FSBDG external cavity feedback light, the laser output with different wavelengths can be obtained, so as to realize wavelength tuning [11].

In 2016, Shin D K et al. [12] used the SAF gain chip and FSBDG Littrow structure of ECSL and realized the maximum injection current was 195 mA, the maximum output power was 83 mW, the Lorentz linewidth was 4.2 kHz, and the Gaussian linewidth was 22 kHz.

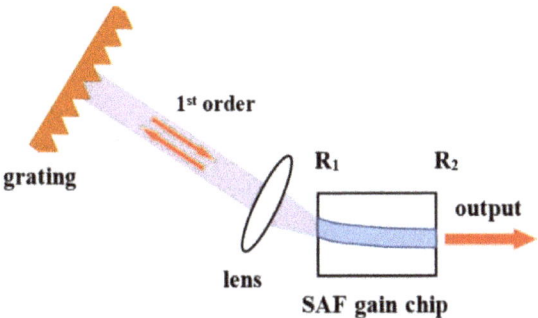

Figure 1. Littrow structure of NLECSLs.

In 2018, Xu B et al. [13] used a commercially available high-power green LD as a gain device and the influence of FSBDG parameters on the performance of external cavity laser was studied. For the Littrow structure of ECSL with the first-order diffraction beam as the feedback and the zero-order diffraction beam as the coupling output, the tuning bandwidth was 11.0 nm and the output power was close to 400 mW.

In 2019, Wang Yan, et al. [14] built an ECSL with Littrow structure using a 1200 groove/mm FSBDG with 91% first-order diffraction efficiency as an external cavity. The maximum SMSR could reach 65 dB and the tunable range could reach 209.9 nm.

In 2021, Lucia Duca et al. [15] reported an ECSL based on an improved Littrow structure, by placing a piezoelectric transducer behind the 780 nm diode laser; the wavelength adjustment by rotating the FSBDG was separated from the fine adjustment of the external cavity. The free spectral range was 3.6 GHz, the SMSR reached 48 dB, and the Lorentz linewidth was 540 kHz. Table 1 shows the performances of the Littrow structure of NLECSLs. Littrows structure of NLECSLs are widely used in many applications, due to its advantages of simple structure and convenient operation. However, the direction of output beam will rotate in tuning. This shortcoming must be overcome for many applications.

Table 1. The performances of Littrow structure of NLECSLs.

Central Wavelength	Tuning Range	SMSR	Line Width	Output Power	Publication Time
1080 nm	100 nm	-	4.2 kHz	83 mW	2016 [12]
525 nm	11.0 nm	-	0.08–0.18 nm	400 mW	2018 [13]
1550 nm	209.9 nm	65 dB	-	48.9 mW	2019 [14]
780 nm	3.6 GHz	48 dB	540 kHz	-	2021 [15]

Note: "-" denotes that the data are not available.

2.2. Littman Structure of NLECSLs

In the Littman structure, as shown in Figure 2, the FSBDG position is fixed; the wavelength of the semiconductor laser can only be changed by adjusting the position of the plane mirror. The laser output beam direction is constant, the linewidth becomes narrow, but the output laser power is less than Littrow structure of ECSL. In the Littman structure, the collimated laser grazes onto the FSBDG, and the incident angle of the beam is large. Compared with Littrow structure, when the laser irradiated the grating, more diffracted lasers are generated, anyone of them laser output linewidth is narrower [16].

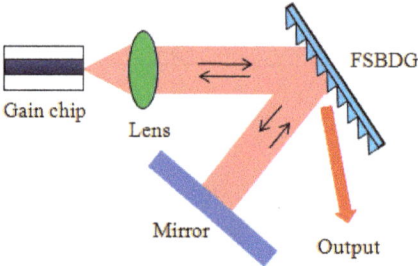

Figure 2. Littman structure of NLECSLs.

In 2018, N. Torcheboeuf et al. [17] reported a 222 nm tuning range, using a compact external cavity GaSb-based diode laser and micro-electro-mechanical system (MEMS) mirror. In the tuning range, the power range was 8–24 mW, the SMSR was 50 dB, and the mode hopping was controllable 18 GHz.

In 2020, Hoppe M et al. [18] optimized the ECSL of 1550 nm bent waveguide based on GaSb with the MEMS with the concept of ECSL cavity, and realized the tuning range of 106 nm, covering the wavelength range from near infrared to MIR.

In 2021, Morten Hoppe et al. [19] used a laser diode with a central wavelength of 2.02 μm. The collimated laser beam passed the MEMS mirror at approximately the 45° angle. It was reflected onto the reflection grating. The arrangement of the optical components was chosen to achieve optimal illumination of the grating. In the gain chip with curved waveguide, both facets are accessible, where the laser beam is couplet out via the rear facet of the laser diode, resulting in a higher efficiency of the resonator, with an SMSR of 2.02 μm and a central wavelength of 53 dB. Table 2 shows the performance of the Littman structure of NLECSLs. The Littman structure of NLECSLs provides an output beam with a stable direction. Tuning of the Littman structure of NLECSLs is realized by the rotating mirror. Since it does not change the incident angle to the grating, the direction of the output beam is stable.

Table 2. The performance of Littman structure of NLECSLs.

Central Wavelength	Tuning Range	SMSR	Line Width	Output Power	Publication Time
2221 nm	222 nm	50 dB	18 GHz	8~24 mW	2018 [17]
1550 nm	106 nm	55 dB	-	30 mW	2020 [18]
2.02 μm	110 nm	53 dB	-	7.1 mW	2021 [19]

Note: "-" denotes that the data are not available.

3. Waveguide Structure of NLECSLs

There are two types of waveguide structure of NLECSLs, including FBG and silicon-based waveguide. When the fiber grating is used as the feedback element of the external cavity laser, the linewidth performance is excellent and the tuning range is wider, but the volume is larger, the refractive index is smaller, the size is larger, and the absorption loss of material is larger [20]. Using the external low loss waveguide as the optical feedback element can effectively reduce the linewidth of the semiconductor laser and obtain low noise spectral characteristics. Due to its small size, low energy consumption, low loss and the ability to integrate with other optical components, NLECSLs based on silicon-based waveguides have become a competitive and attractive candidate laser in many coherent applications [21].

3.1. FBG Structure of NLECSLs

FBG structures of NLECSLs in optical fiber transmission systems have become a research hotspot. The FBG structure of NLECSL has its AR facet facing the FBG, the FBG end coupled directly to the gain chip and the other end of the FBG acts as the end reflector of the external cavity. With AR coating on the gain chip, the lasing wavelength may be selected by choosing the appropriate FBG, as shown in Figure 3.

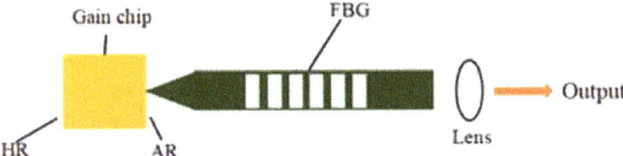

Figure 3. FBG structure of NLECSLs.

In 2011, Loh W et al. [22] reported a 1550 nm InGaAlAs/InP quantum well, high power, low noise encapsulated ECSL demonstration. The laser consisted of a dual-channel curved channel plate coupled with an optical waveguide amplifier and a 2.5 GHz narrow bandwidth FBG passive cavity using a lens fiber. Under the bias current of 4A, ECSL generates 370 mW of fiber-coupled output power, and its Gaussian linewidth and Lorentz linewidth are 35 kHz and 1 kHz, respectively.

In 2016, Lynch S G et al. [23] demonstrated a new integrated platform with FBG. The high thermal conductivity of silicon substrate contributes to the heat dissipation and thermalization of the device. The geometric shape of the device is precisely designed with a small inclined plane, which connects the end of integrated platform to eliminate unnecessary optical feedback, and its layout can minimize the angular coupling loss between waveguides. The laser works in a single mode at 1532.83 nm, with an output power of 9 mW and a linewidth of 14 kHz.

In 2017, Li Zhang et al. [24] combined a semiconductor gain chip and FBG with enhanced thermal sensitivity, and demonstrated a mode-free external cavity laser design. The compact ECSL had a narrow linewidth of 35 kHz, SMSR greater than 50 dB, and the mode-free tuning range was 62.5 GHz.

In 2019, Huang D et al. [25] demonstrated an ultra-low loss silicon based waveguide (0.16 dB/cm) with a linewidth of 1 kHz and an output power of more than 37 mW, and a long FBG fully integrated extended distributed Bragg reflector laser with a narrow bandwidth. The combination of narrow linewidth and high power enables it to be used in coherent communication, radio frequency photonics and optical sensing.

In 2021, Antoine Congar et al. [26] realized a 400 nm FBG InGaN-based laser diode. A narrow band FBG was fabricated under near ultraviolet light. The device has a SMSR of 44 dB and an inherent linewidth of 16 kHz.

In 2022, Suqs et al. [27] reported a laser based on the FBG ECSL module near the wavelength of 1550 nm, using the combination of narrow linewidth technology and frequency stable transfer technology to narrow the laser intrinsic Lorentz linewidth to 15 kHz. Table 3 shows the performances of the FBG structure of NLECSLs. The FBG structure of NLECSL is easily obtain narrow linewidth, high SMSR and high wavelength thermal stability. It is easy to design and screen the gain chip and FBG separately; the performances of FBG structure of NLECSL can be optimized and it is very convenient to be used in fiber systems.

Table 3. Performance of FBG structure of NLECSLs.

Central Wavelength	Tuning Range	Line Width	Output Power	Publication Time
1550 nm	-	1 kHz	370 mW	2011 [22]
1532.83 nm	20 pm	14 kHz	9 mW	2016 [23]
1550.4 nm	62.5 GHz	35 kHz	-	2017 [24]
1565 nm	2.9 GHz	1 kHz	37 mW	2019 [25]
400 nm	-	16 kHz	-	2021 [26]
1572 nm	22 GHz	15 kHz	25 mW	2022 [27]

Note: "-" denotes that the data are not available.

3.2. Silicon-Based Waveguide Structure of NLECSLs

With the maturity of the silicon optical chip design and process platform, the external cavity feedback elements based on silicon, Si_3N_4 and other materials endlessly emerge, and with the help of various microcavity structures, the linewidth of ECSL can be further compressed. It has the characteristics of high reliability and low power consumption of the monolithic integrated structure, as well as the narrow linewidth and wide tuning characteristics of the external cavity structure, and has gradually become a hotspot in the research field of NLECSLs [28]. Vissers E et al. [29] studied the hybrid integrated mode-locked laser diode with silicon nitride expansion cavity in 2021, coupled the silicon nitride external cavity with the InP active chip, and obtained the line width of 31 Hz. In 2018, Guan H et al. [30] studied III-V/Si hybrid external cavity lasers. The Si_3N_4 edge-coupled silicon chip is mixed into the spot size converter in the silicon chip. The maximum output power of the laser is 11 mW, the measured minimum linewidth is 37 kHz (maximum < 80 Hz), and the SMSR is 55 dB.

In 2019, Guo Y et al. [31] demonstrated a III-V/silicon nitride hybrid external cavity laser. The tuning range of ECSL is 45 nm, the SMSR is 60 dB, and the linewidth is about 100 kHz.

In 2020, Kharas D et al. [32] showed a high-power on-chip 1550 nm laser, which was integrated into a silicon nitride waveguide and distributed Bragg reflector grating photonic integrated circuit by a bending channel and a two-way InGaAsP/InP plate coupled with an optical waveguide amplifier. The driving current of the single-mode emission optical power of 312 kW was 2.5 A, and the linewidth of 192 kHz was integrated.

In 2020, Guo Y Y et al. [33] demonstrated hybrid lasers by using InP reflective semiconductor optical amplifier chips coupled with Si_3N_4 tunable reflector chips. The laser wavelength tuning range was 160 nm, and the linewidth was 30 kHz.

In 2020, Sia Jx et al. [34] adopted the hybrid integration of the III-V optical amplifier and extended, low-loss wavelength-tunable silicon cursor cavity, and first reported the III-V/silicon hybrid wavelength-tunable laser in the rich wavelength region of 1647–1690 nm. When the continuous wave operates at room temperature, the output power can reach 31.1 mW, the maximum SMSR is 46.01 dB, and the line width is 0.7 kHz.

In 2021, Guo Y Y et al. [35] reported a widely tunable III-V/Si_3N_4 hybrid integrated external cavity laser. Under 500 mA injection current, the maximum output power was 34 mW. In the tuning range of 58.5 nm, the SMSR exceeds 70 dB. The laser linewidth is 2.5 kHz. The same structure was used in the optical fiber communication conference and exhibition next year, but its performance was higher than last year, reaching a record of about 170 nm tuning range. The linewidth of the laser decreased slightly less than 2.8 kHz [36].

In 2021, Zhao R L et al. [37] reported a wavelength tunable hybrid integrated external cavity laser for C-band. Two parallel reflective semiconductor optical amplifier gain channels are composed of Y branches in the Si_3N_4 photonic circuit to increase the optical gain. The SMSR is about 67 dB and the pump current is 75 mA. The linewidth of the unpackaged laser is 6.6 kHz, and the on-chip output power is 23.5 mW.

In 2021, Mckinzie K A et al. [38] demonstrated the hybrid integration of an InP-based laser and amplifier array PIC and high-quality factor silicon nitride microring resonator. Laser emission based on the gain of the interference combination amplifier array in the external cavity was formed by the feedback from the silicon nitride micro resonator chip; the linewidth was reduced to 3 kHz, and the average output power was 37.9 mW. Table 4 shows the performances of silicon-based waveguide structures of NLECSLs. Silicon-based waveguide structures of NLECSLs have excellent characteristics, such as compact structure, low cost, mass production, integrated packaging, small size, etc., and have a wide tuning range, while achieving a narrow linewidth. At present, the technical difficulty of silicon-based waveguide structures of NLECSL involves how to improve the coupling efficiency and reduce the reflectivity at the coupling. In addition, the heat accumulation and dissipation during the thermo-optic effect tuning process takes a certain amount of time, which affects the high-speed tuning. How to further improve the modulation speed is a big challenge.

Table 4. Performance of silicon-based waveguide structure of NLECSLs.

Central Wavelength	Tuning Range	SMSR	Line Width	Output Power	Publication Time
1565 nm	1560–1570 nm	-	31 Hz	300 mW	2021 [29]
1550 nm	60 nm	55 dB	37 kHz	11 mW	2018 [30]
1540 nm	45 nm	60 dB	100 kHz	0.78 mW	2019 [31]
1550 nm	-	55 dB	192 kHz	312 mW	2020 [32]
1559 nm	160 nm	55 dB	30 kHz	-	2020 [33]
1670 nm	1647–1690 nm	46.01 dB	0.7 kHz	31.1 mW	2020 [34]
1550 nm	58.5 nm	70 dB	2.5 kHz	34 mW	2021 [35]
1546 nm	44 nm	67 dB	6.6 kHz	23.5 mW	2021 [36]
1542 nm	1513–1564 nm	42 dB	3 kHz	37.9 mW	2021 [37]
1550 nm	170 nm	64 dB	2.8 kHz	24.8 mW	2022 [38]

Note: "-" denotes that the data are not available.

4. Confocal F-P Cavity Structure of NLECSLs

To further narrow the linewidth on the basis of the structure of the ECSL, it is necessary to use the mode selection element with a narrow bandwidth. The interference filter or F-P cavity and narrow-band filter are the structures of the external cavity optical feedback technology that are commonly used to narrow the linewidth [39]. Compared with the optical cavities used in the traditional fiber narrow linewidth laser, solid narrow linewidth laser and chip external cavity narrow linewidth laser, the high quality factor F-P cavity has an extremely low thermal effect, nonlinear effect and ultra-high temperature stability [40].

Confocal F-P cavity structures of NLECSLs are designed as a monolithic confocal F-P cavity and the focused laser beam is coupled with the tilted monolithic confocal F-P cavity. The tilt angle of the cavity, with regard to the optical axis of the laser system, prevents the non-resonant feedback from the cavity being re-injected into the emitter; this ensures resonant-only optical feedback in the laser diode, as shown in Figure 4.

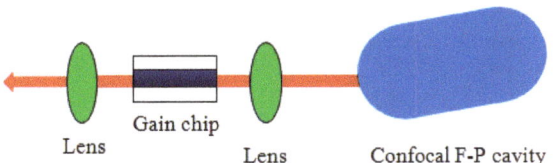

Figure 4. Confocal F-P Cavity structure of NLECSLs.

A 657 nm ECSL system with stable output frequency was proposed in 2011 [41]. Through a narrowband high transmission interference filter, the instantaneous linewidth of the laser emitted by this new diode laser system was 7 kHz and the linewidth was 432 kHz. In the same year, Yang et al. [42] proposed a wide-cavity ECSL with a linewidth of kilohertz using optical feedback from a single folded F-P cavity. The linewidth of the ECSL was successfully reduced to 6.8 kHz.

In 2012, Yang Z et al. [43] proposed a NLECSL with high-precision dual-mirror non-confocal cavity optical feedback. Through Lorentz fitting, the full width half maximum linewidth of the laser was reduced to 100 Hz, and the instantaneous linewidth was reduced to 30 Hz.

In 2014, Luo Z et al. [44] proposed an extended cavity diode laser with MHz linewidth. The optical feedback technology of the folded Fabry–Perot cavity was used to replace the mirror in the traditional ECSL configuration. The effective laser reduced the linewidth and stable frequency, and the linewidth of the laser was reduced from about 20 GHz to 15 MHz.

In 2015, Lewoczko-Adamczyk W et al. [10] proposed a compact, ultra-narrow linewidth semiconductor laser based on a 780 nm distributed feedback diode laser, which was self-locked to an external single-chip confocal F-P cavity mode. When the output power exceeds 50 mW, the Lorentz linewidth corresponding to the resonant optical feedback laser is 15.7 Hz.

In 2015, Pyrlik C et al. [45] proposed a DFB based on 1.5 mm length and 780 nm with a single confocal Fabry–Perot cavity. Both surfaces of DFB are coated with anti-reflection coating. The tilt of the external resonator cavity relative to the optical axis of the laser system is 15°, which can prevent the non-resonant feedback of the cavity from being reinjected into the transmitter. The line width of 31 Hz is obtained in the paper.

In 2017, Christopher H et al. [46] focused the light emission of the DFB semiconductor laser chip into a confocal resonant feedback cavity. Therefore, the resonant feedback is re-injected into the DFB diode laser chip. The light emitted from the other side of the DFB laser chip is collimated through an optical isolator and coupled to the single-mode fiber. The Lorentz linewidth of about 630 Hz is obtained by the self-delayed heterodyne device. The corresponding FWHM level technical linewidth is about 30 kHz.

In 2018, the ultra-narrow bandwidth dual filter was used as the ECSL of the laser longitudinal mode selection element developed by the Institute of Optoelectronics, Shanxi University. For the angle of the rotating narrow band filter, the laser wavelength coarse tuning range was 14 nm. The linewidth of the narrow-band filter ECSL is measured by the fiber delay beat method. The linewidth is about 187 kHz [47].

In 2018, Yu Li et al. [48] developed a new on-chip semiconductor laser by introducing the cursor effect and self-injection locking effect between the F-P diode laser on the silicon chip and the external micro resonator. The narrow linewidth of the laser is 8 kHz, and the wide switching range is 17 nm.

In 2020, Zhang L et al. [49] used a narrow-band interference filter for spectral selection, and used a cat-eye reflector for optical feedback to design an ECSL. The ECSL works near 698.45 nm. The tuning range of the current control is more than 40 GHz, and the tuning range of piezoelectric control is 3 GHz. The ECSL line width measured by heterodyne beat frequency is about 180 kHz.

In 2021, YongXiang Zheng et al. [50] demonstrated a method of laser frequency stabilization in a wide tuning range by installing piezoelectric ceramic actuators into the Fabry–Perot cavity to stabilize the ultraviolet laser. In order to suppress the piezoelectric drift, the piezoelectric actuator adopts a two-layer symmetrical structure to achieve a tuning range of 14.7 GHz. It can be extended to the wavelength from ultraviolet to infrared. The wavelength of ECSL is 369.5 nm and the linewidth is 20 MHz.

In 2021, Jakup Ratkoceri et al. [51] observed the stable locking region of the injection-locked FP laser by using the delay self-zero difference technique and the RF spectrum of the external cavity FP laser. The center wavelength is 1546.244 nm, and the 3 dB Lorentz linewidth is 100 MHz. Table 5 shows the performance of the confocal F-P cavity structure of

NLECSLs. The confocal F-P cavity structure of NLECSLs shows wide band frequency noise suppression characteristics with a narrow linewidth; the confocal cavity length and the cavity mirror's curvature radius must be matched to avoid breaking the mode degeneracy, which means higher requirements for accuracy when using higher finesse cavities.

Table 5. Performance of confocal F-P cavity structure of NLECSLs.

Central Wavelength	Tuning Range	Line Width	Output Power	Publication time
657 nm	0.5 GHz	432 kHz	-	2011 [41]
689 nm	3.97 GHz	6.8 kHz	20 mW	2011 [42]
689 nm	4 MHz	100 Hz	-	2012 [43]
635 nm	5–20 GHz	15 MHz	5 mW	2014 [44]
780 nm	-	15.7 Hz	50 mW	2015 [10]
780 nm	-	31 Hz	38 mW	2015 [45]
1064.49 nm	-	630 Hz	4 mW	2017 [46]
852 nm	1.5 GHz	187 kHz	56 mW	2018 [47]
1555 nm	17 nm	8 kHz	-	2018 [48]
698.45 nm	40 GHz	180 kHz	36 mW	2020 [49]
369.5 nm	14.7 Hz	20 MHz	-	2021 [50]
1547 nm	20 nm	100 MHz	-	2021 [51]

Note: "-" denotes that the data are not available.

5. Conclusions

In this paper, the three main device structures of NLECSLs are expanded upon. By comparing a large number of data, we conclude that the confocal F-P cavity structure of NLECSLs is the best structure to achieve a narrow linewidth, and could obtain the narrowest linewidth, which is more precise and more suitable for applications that require a high accuracy of the linewidth. NLECSLs are developing towards high power and narrower linewidth. Through the continuous development of new optical feedback elements and optical resonator design, the ultra-narrow linewidth laser below 20 Hz has been realized. Combined with its characteristics of small volume, light weight, high conversion efficiency and wide spectral range, it will be widely used in the fields of ultra-high precision lidar, inter satellite communication, coherent optical communication, laser spectroscopy, atomic clock pumping, atmospheric absorption measurement and optical fiber communication. How to realize the wide tuning range, narrow linewidth laser output is a main research direction for the future development of NLECSLs. In addition, a narrow linewidth laser is critical for its application as a pump source for generating an extremely narrow linewidth Brillouin output [52]. Currently, different approaches to narrow linewidth lasers have distinct characteristics. In the future, new technologies will lead to further compression of the laser linewidth, improvement of frequency stability, expansion of wavelength, and increase in power, which will pave the way for human beings to explore the unknown world.

Author Contributions: Conceptualization, K.D. and L.W.; methodology, Y.M., X.L. and J.S.; writing—original draft preparation, K.D. and Z.L.; writing—review and editing, L.Z. and L.L.; visualization, Y.Q., Z.Q. and G.L.; supervision, Z.Q.; funding acquisition, G.L. All authors have read and agreed to the published version of the manuscript.

Funding: This work was supported in part by specific research fund for Innovation Platform for Academicians of Hainan Province under Grant YSPTZX202034 and Grant YSPTZX202127; in part by the Major Science and Technology Program of Hainan Province of China under Grant ZDKJ 2019005; in part by Scientific Research Projects of Higher Education Institutions in Hainan Province under Grant hnky2020-24, Grant Hnjg2021ZD-22, Grant hnky2020ZD-12; in part by the Hainan Provincial

Natural Science Foundation of China under Grant 622RC671, Grant 120MS031, Grant 2019RC190, Grant 2019RC192; in part by the National Natural Science Foundation of China under Grant 61774024, Grant 61864002, Grant 11764012, Grant 62174046,Grant 62064004 and Grant 61964007; in part by the Key Research and Development Projects in Hainan Province under Grant ZDYF2020020, Grant ZDYF2020036, and Grant ZDYF2020217; in part by the Open Fund for Innovation and Entrepreneurship of college students under Grant 202111658021X, Grant 202111658022X, Grant 202111658023X, Grant 202111658013.

Institutional Review Board Statement: Not applicable.

Informed Consent Statement: Not applicable.

Data Availability Statement: Not applicable.

Acknowledgments: The authors thank Dongxin Xu, Hao Chen, and Yanbo Liang for helping with this article.

Conflicts of Interest: The authors declare no conflict of interest.

References

1. Schremer, A.T.; Tang, C.L. External-cavity semiconductor laser with 1000 GHz continuous piezoelectric tuning range. *IEEE Photon. Technol. Lett.* **1990**, *2*, 3–5. [CrossRef]
2. Kobayashi, K.; Mito, I. Single frequency and tunable laser diodes. *J. Light. Technol.* **1988**, *6*, 1623–1633. [CrossRef]
3. Aoyama, K.; Yoshioka, R.; Yokota, N.; Yasaka, H.; Kobayashi, W. Narrow-linewidth laser diode with compact optical-feedback system. In Proceedings of the Microwave Photonics (MWP) and the 2014 9th Asia-Pacific Microwave Photonics Conference (APMP) 2014 International Topical Meeting, Hokkaido, Japan, 20–23 October 2014; pp. 79–81.
4. O'Carroll, J.; Phelan, R.; Kelly, B.; Byrne, D.; Smyth, F.; Cardiff, B.; Anandarajah, P.M.; Barry, L.P. Narrow linewidth discrete mode laser diodes at 1550 nm. In *Proceedings Volume 8432, Semiconductor Lasers and Laser Dynamics V*; SPIE: Bellingham, WA, USA, 2012; pp. 166–174.
5. Crowe, J.W.; Craig, R.M., Jr. Small-signal amplification in GaAs lasers. *Appl. Phys. Lett.* **1964**, *4*, 57–58. [CrossRef]
6. Heckscher, H.; Rossi, J.A. Flashlight-size external cavity semiconductor laser with narrow-linewidth tunable output. *Appl. Opt.* **1975**, *14*, 94–96. [CrossRef]
7. Duraev, V.P.; Marmalyuk, A.A.; Petrovskiy, A.V. Tunable laser diodes for the 1250–1650 nm spectral range. *Spectrochim. Acta Part A Mol. Biomol. Spectrosc.* **2007**, *66*, 846–848. [CrossRef]
8. Saliba, S.D.; Scholten, R.E. Linewidths below 100 kHz with external cavity diode lasers. *Appl. Opt.* **2009**, *48*, 6961–6966. [CrossRef] [PubMed]
9. Bennetts, S.; McDonald, G.D.; Hardman, K.S.; Debs, J.E.; Kuhn, C.C.; Close, J.D.; Robins, N.P. External cavity diode lasers with 5 kHz line width and 200 nm tuning range at 1.55 µm and methods for line width measurement. *Opt. Express* **2014**, *22*, 10642–10654. [CrossRef] [PubMed]
10. Lewoczko-Adamczyk, W.; Pyrlik, C.; Häger, J.; Schwertfeger, S.; Wicht, A.; Peters, A.; Erbert, G.; Tränkle, G. Ultra-narrow linewidth DFB-laser with optical feedback from a monolithic confocal Fabry-Perot cavity. *Opt. Express* **2015**, *23*, 9705–9709. [CrossRef] [PubMed]
11. Akhavan, F.; Saini, S.; Hu, Y.; Kershaw, E.; Wilson, M.; Krainak, M.; Leavitt, R.; Heim, P.J.S.; Dagenais, M. High power external cavity semiconductor laser with wavelength tuning over C, L, and S-bands using single-angled-facet gain chip. In Proceedings of the Summaries of Papers Presented at the Lasers and Electro-Optics. CLEO'02. Technical Diges, Long Beach, CA, USA, 24 May 2002; pp. 761–763.
12. Shin, D.K.; Henson, B.M.; Khakimov, R.I.; Ross, J.A.; Dedman, C.J.; Hodgman, S.S.; Baldwin, K.G.; Truscott, A.G. Widely tunable, narrow linewidth external-cavity gain chip laser for spectroscopy between 1.0–1.1 µm. *Opt. Express* **2016**, *24*, 27403–27414. [CrossRef]
13. Xu, B.; Lv, X.; Ding, D.; Lv, W.; Zhang, Y.; Zhang, J. High-power broadly tunable grating-coupled external cavity laser in green region. *Rev. Sci. Instrum.* **2018**, *89*, 125106. [CrossRef]
14. Wang, Y.; Wu, H.; Chen, C.; Zhou, Y.; Wang, Y.; Liang, L.; Tian, Z.; Qin, L.; Wang, L. An Ultra-High-SMSR External-Cavity Diode Laser with a Wide Tunable Range around 1550 nm. *Appl. Sci.* **2019**, *9*, 4390. [CrossRef]
15. Duca, L.; Perego, E.; Berto, F.; Sias, C. Design of a Littrow-type diode laser with independent control of cavity length and grating rotation. *Opt. Lett.* **2021**, *46*, 2840–2843. [CrossRef] [PubMed]
16. Hoppe, M.; Jiménez, Á.; Rohling, H.; Schmidtmann, S.; Grahmann, J.; Tatenguem, H.; Milde, T.; Schanze, T.; Sacher, J.R. Construction and Characterization of External Cavity Diode Lasers Based on a Mi-croelectromechanical System Device. *IEEE J. Sel. Top. Quant. Electron.* **2019**, *25*, 1–9. [CrossRef]
17. Torcheboeuf, N.; Droz, S.; Šimonytè, I.; Miasojedovas, A.; Trinkunas, A.; Vizbaras, K.; Vizbaras, A.; Boiko, D.L. MEMS Tunable Littman-Metcalf Diode Laser at 2.2 µm for Rapid Broadband Spec-troscopy in Aqueous Solutions. In Proceedings of the 2018 IEEE International Semiconductor Laser Conference (ISLC), Santa Fe, NM, USA, 16–19 September 2018; pp. 1–2.

18. Hoppe, M.; Rohling, H.; Schmidtmann, S.; Honsberg, M.; Tatenguem, H.; Grahmann, J.; Milde, T.; Schanze, T.; Sacher, J.R. Wide and fast mode-hop free MEMS tunable ECDL concept and realization in the NIR and MIR spectral regime. In *Proceedings Volume 11293, MOEMS and Miniaturized Systems XIX*; SPIE: Bellingham, WA, USA, 2020; p. 112930C.
19. Hoppe, M.; Schmidtmann, S.; Aβmann, C.; Honsberg, M.; Tatenguem, H.; Milde, T.; Schanze, T.; Sacher, J.; Gu-Stoppel, S.; Senger, F. Innovative ECDL design based on a resonant MEMS scanner for ultra-fast tuning in the MIR range. In *Proceedings Volume 11697, MOEMS and Miniaturized Systems XIX*; SPIE: Bellingham, WA, USA, 2021; p. 1169709.
20. Huang, Y.; Li, Y.; Zhu, H.; Tong, G.; Zhang, H.; Zhang, W.; Zhou, S.; Sun, R.; Zhang, Y.; Li, L.; et al. Theoretical investigation into spectral characteristics of a semiconductor laser with dual-FBG external cavity. *Opt. Commun.* **2011**, *284*, 2960–2965. [CrossRef]
21. Satyan, N.; Rakuljic, G.; Vilenchik., Y.; Yariv, A. A hybrid silicon/III-V semiconductor laser with sub-kHz quantum linewidth. In Proceedings of the 2015 IEEE Summer Topicals Meeting Series (SUM), Nassau, Bahamas, 13–15 July 2015; pp. 154–155.
22. Loh, W.; O'Donnell, F.J.; Plant, J.J.; Brattain, M.A.; Missaggia, L.J.; Juodawlkis, P.W. Packaged, high-power, narrow-linewidth slab-coupled optical waveguide external cavity laser (SCOWECL). *IEEE Photon. Technol. Lett.* **2011**, *23*, 974–976. [CrossRef]
23. Lynch, S.G.; Holmes, C.; Berry, S.A.; Gates, J.C.; Jantzen, A.; Ferreiro, T.I.; Smith, P.G. External cavity diode laser based upon an FBG in an integrated optical fiber platform. *Opt. Express* **2016**, *24*, 8391–8398. [CrossRef]
24. Zhang, L.; Wei, F.; Sun, G.; Chen, D.J.; Cai, H.W.; Qu, R.H. Thermal Tunable Narrow Linewidth External Cavity Laser with Thermal Enhanced FBG. *IEEE Photon Technol. Lett.* **2017**, *29*, 385–388. [CrossRef]
25. Huang, D.; Tran, M.A.; Guo, J.; Peters, J.; Komljenovic, T.; Malik, A.; Morton, P.A.; Bowers, J.E. High-power sub-kHz linewidth lasers fully integrated on silicon. *Optica* **2019**, *6*, 745–752. [CrossRef]
26. Congar, A.; Gay, M.; Perin, G.; Mammez, D.; Simon, J.C.; Besnard, P.; Rouvillain, J.; Georges, T.; Lablonde, L.; Robin, T.; et al. Narrow linewidth near-UV InGaN laser diode based on external cavity fiber Bragg grating. *Opt. Lett.* **2021**, *46*, 1077–1080. [CrossRef]
27. Su, Q.; Wei, F.; Sun, G.; Li, S.; Wu, R.; Pi, H.; Chen, D.; Yang, F.; Ying, K.; Qu, R.; et al. Frequency-Stabilized External Cavity Diode Laser at 1572 nm Based on Frequency Stability Transfer. *IEEE Photon Technol. Lett.* **2022**, *34*, 203–206. [CrossRef]
28. Kita, T.; Tang, R.; Yamada, H. Compact silicon photonic wavelength-tunable laser diode with ultra-wide wavelength tuning range. *Appl. Phys. Lett.* **2015**, *106*, 111104. [CrossRef]
29. Vissers, E.; Poelman, S.; de Beeck, C.O.; Van Gasse, K.; Kuyken, B. Hybrid integrated mode-locked laser diodes with a silicon nitride extended cavity. *Opt. Express* **2021**, *29*, 15013–15022. [CrossRef] [PubMed]
30. Guan, H.; Novack, A.; Galfsky, T.; Ma, Y.; Fathololoumi, S.; Horth, A.; Huynh, T.N.; Roman, J.; Shi, R.; Caverley, M.; et al. Widely-tunable, narrow-linewidth III-V/silicon hybrid external-cavity laser for coherent communication. *Opt. Express* **2018**, *26*, 7920–7933. [CrossRef] [PubMed]
31. Guo, Y.; Zhou, L.; Zhou, G.; Zhao, R.; Lu, L.; Chen, J. Edge-coupled III-V/Si 3 N 4 hybrid external cavity laser. In Proceedings of the 2019 18th International Conference on Optical Communications and Networks (ICOCN), Huangshan, China, 5–8 August 2019; pp. 1–3.
32. Kharas, D.; Plant, J.; Bramhavar, S.; Loh, W.; Swint, R.; Sorace-Agaskar, C.; Heidelberger, C.; Juodawlkis, P. High Power (>300 mW) 1550 nm On-Chip Laser Realized Using Passively Aligned Hybrid Integration. In Proceedings of the 2020 Conference on Lasers and Electro-Optics (CLEO), San Jose, CA, USA, 10–15 May 2020.
33. Guo, Y.; Zhou, L.; Zhou, G.; Zhao, R.; Lu, L.; Chen, J. Hybrid external cavity laser with a 160-nm tuning range. In Proceedings of the 2020 Conference on Lasers and Electro-Optics (CLEO), San Jose, CA, USA, 10–15 May 2020; pp. 1–2.
34. Sia JX, B.; Li, X.; Wang, W.; Qiao, X.; Guo, X.; Zhou, J.; Littlejohns, C.G.; Liu, C.; Reed, G.T.; Wang, H. Sub-kHz linewidth, hybrid III-V/silicon wavelength-tunable laser diode operating at the ap-plication-rich 1647–1690 nm. *Opt. Express* **2020**, *28*, 25215–25224.
35. Guo, Y.; Zhao, R.; Zhou, G.; Lu, L.; Stroganov, A.; Nisar, M.S.; Chen, J.; Zhou, L. Thermally Tuned High-Performance III-V/Si 3 N 4 External Cavity Laser. *IEEE Photon. J.* **2021**, *13*, 1–13.
36. Zhao, R.; Guo, Y.; Lu, L.; Nisar, M.S.; Chen, J.; Zhou, L. Hybrid dual-gain tunable integrated InP-Si 3 N 4 external cavity laser. *Opt. Express* **2021**, *29*, 10958–10966. [CrossRef]
37. McKinzie, K.A.; Wang, C.; Al Noman, A.; Mathine, D.L.; Han, K.; Leaird, D.E.; Hoefler, G.E.; Lal, V.; Kish, F.; Qi, M.; et al. InP high power monolithically integrated widely tunable laser and SOA array for hybrid integration. *Opt. Express* **2021**, *29*, 3490–3502. [CrossRef]
38. Guo, Y.; Li, X.; Xu, W.; Liu, C.; Jin, M.; Lu, L.; Xie, J.; Stroganov, A.; Chen, J.; Zhou, L. A hybrid-integrated external cavity laser with ultra-wide wavelength tuning range and high side-mode suppression. In Proceedings of the 2022 Optical Fiber Communications Conference and Exhibition (OFC), San Diego, CA, USA, 6–10 March 2022; pp. 1–3.
39. Iwata, Y.; Cheon, D.; Miyabe, M.; Hasegawa, S. Development of an interference-filter-type external-cavity diode laser for resonance ionization spectroscopy of strontium. *Rev. Sci. Instrum.* **2019**, *90*, 123002. [CrossRef]
40. Zhang, X.M.; Wang, N.; Gao, L.; Feng, M.; Chen, B.; Tsang, Y.H.; Liu, A.Q. Narrow-linewidth external-cavity tunable lasers. In Proceedings of the 10th International Conference on Optical Communications and Networks (ICOCN 2011), Guangzhou, China, 5–7 November 2011; pp. 1–2.
41. Wang, Z.; Lv, X.; Chen, J. A 657-nm narrow bandwidth interference filter-stabilized diode laser. *Chin. Opt. Lett.* **2011**, *9*, 041402. [CrossRef]
42. Zhao, Y.; Peng, Y.; Yang, T.; Li, Y.; Wang, Q.; Meng, F.; Cao, J.; Fang, Z.; Li, T.; Zang, E. External cavity diode laser with kilohertz linewidth by a monolithic folded Fabry–Perot cavity optical feedback. *Opt. Lett.* **2010**, *36*, 34–36. [CrossRef]

43. Zhao, Y.; Li, Y.; Wang, Q.; Meng, F.; Lin, Y.; Wang, S.; Lin, B.; Cao, S.; Cao, J.; Fang, Z.; et al. 100-Hz Linewidth Diode Laser with External Optical Feedback. *IEEE Photon- Technol. Lett.* **2012**, *24*, 1795–1798. [CrossRef]
44. Luo, Z.; Long, X.; Tan, Z. External folded cavity optical feedback diode laser with megahertz relative linewidth. In Proceedings of the International Symposium on Optoelectronic Technology and Application 2014: Development and Application of High Power Lasers, Beijing, China, 13–15 May 2014; Volume 9294, pp. 201–208.
45. Pyrlik, C.; Lewoczko-Adamczyk, W.; Schwertfeger, S.; Häger, J.; Wicht, A.; Peters, A.; Erbert, G.; Tränkle, G. Ultra-narrow linewidth, micro-integrated semiconductor external cavity diode laser module for quantum optical sensors in space. In Proceedings of the 2015 Conference on Lasers and Electro-Optics (CLEO), San Jose, CA, USA, 10–15 May 2015.
46. Christopher, H.; Arar, B.; Bawamia, A.; Kürbis, C.; Lewoczko-Adamczyk, W.; Schiemangk, M.; Smol, R.; Wicht, A.; Peters, A.; Tränkle, G. Narrow linewidth micro-integrated high power diode laser module for deployment in space. In Proceedings of the 2017 IEEE International Conference on Space Optical Systems and Applications (ICSOS), Naha, Okinawa, Japan, 14–16 November 2017; pp. 150–153.
47. Yul, J.; Jun, H.; Min, W.J. Optimization of 852-nm External-Cavity Diode Laser with Narrow-bandwidth Filter. *Acta Sin. Quant. Opt.* **2018**, *24*, 98–106.
48. Li, Y.; Zhang, Y.; Chen, H.; Yang, S.; Chen, M. Tunable self-injected Fabry–Perot laser diode coupled to an external high-Q Si_3N_4/SiO_2 microring resonator. *J. Lightwave Technol.* **2018**, *36*, 3269–3274. [CrossRef]
49. Zhang, L.; Liu, T.; Chen, L.; Xu, G.; Jiang, C.; Liu, J.; Zhang, S. Development of an interference filter-stabilized external-cavity diode laser for space applications. *Photonics* **2020**, *7*, 12. [CrossRef]
50. Zheng, Y.X.; Cui, J.M.; Ai, M.Z.; Qian, Z.H.; Cao, H.; Huang, Y.F.; Jia, X.J.; Li, C.F.; Guo, G.C. Large-tuning-range frequency stabilization of an ultraviolet laser by an open-loop piezo-electric ceramic controlled Fabry–Pérot cavity. *Opt. Express* **2021**, *29*, 24674–24683. [CrossRef]
51. Ratkoceri, J.; Batagelj, B. Determining the Stable Injection Locking of a Fabry-Pérot Laser by Observing the RF Spectral Components Generated by a Low-Reflectivity External Cavity. *Photonics* **2021**, *8*, 487. [CrossRef]
52. Cui, C.; Wang, Y.; Lu, Z.; Yuan, H.; Wang, Y.; Chen, Y.; Wang, Q.; Bai, Z.; Mildren, R.P. Demonstration of 2.5 J, 10 Hz, nanosecond laser beam combination system based on noncollinear Brillouin amplification. *Opt. Express* **2018**, *26*, 32717–32727. [CrossRef]

Review

Processes of the Reliability and Degradation Mechanism of High-Power Semiconductor Lasers

Yue Song [1,2], Zhiyong Lv [3], Jiaming Bai [4], Shen Niu [1,2], Zibo Wu [5], Li Qin [1,2], Yongyi Chen [1,2,6,*], Lei Liang [1,2], Yuxin Lei [1,2], Peng Jia [1,2], Xiaonan Shan [1,2,*] and Lijun Wang [1,2,7,8]

1. State Key Laboratory of Luminescence and Applications, Changchun Institute of Optics, Fine Mechanics and Physics, Chinese Academy of Sciences, Changchun 130033, China; songyue@ciomp.ac.cn (Y.S.); 13623558919@163.com (S.N.); qinl@ciomp.ac.cn (L.Q.); liangl@ciomp.ac.cn (L.L.); leiyuxin@ciomp.ac.cn (Y.L.); jiapeng@ciomp.ac.cn (P.J.); wanglj@ciomp.ac.cn (L.W.)
2. Daheng College, University of Chinese Academy of Sciences, Beijing 100049, China
3. School of Physics and Microelectronics, Zhengzhou University, Zhengzhou 450001, China; lvzhiyong@stu.zzu.edu.cn
4. School of Physics, Jilin University, Changchun 130015, China; baijm1119@mails.jlu.edu.cn
5. School of Opto-Electronics Information Science and Engineering, Changchun College of Electronic Technology, Changchun 130061, China; wzb23923651592021@163.com
6. Jlight Semiconductor Technology Co., Ltd., Changchun 130102, China
7. Peng Cheng Laboratory, No.2, Xingke 1st Street, Nanshan, Shenzhen 518000, China
8. Academician Team Innovation center of Hainan Province, Key Laboratory of Laser Technology and Optoelectronic Functional Materials of Hainan Province, School of Physics and Electronic Engineering of Hainan Normal University, Haikou 570206, China
* Correspondence: chenyy@ciomp.ac.cn (Y.C.); shanxn@ciomp.ac.cn (X.S.); Tel.: +86-180-4304-7205 (Y.C.)

Abstract: High-power semiconductor lasers have attracted widespread attention because of their small size, easy modulation, and high conversion efficiency. They play an important role in national economic construction and national defense construction, including free-space communication; industrial processing; and the medical, aerospace, and military fields, as well as other fields. The reliability of high-power semiconductor lasers is the key point of the application system. Higher reliability is sought in the military defense and aerospace fields in particular. Reliability testing and failure analysis help to improve the performance of high-power semiconductor lasers. This article provides a basis for understanding the reliability issues of semiconductor lasers across the whole supply chain. Firstly, it explains the failure modes and causes of failure in high-power semiconductor lasers; this article also summarizes the principles and application status of accelerated aging experiments and lifetime evaluation; it also introduces common techniques used for high-power semiconductor laser failure analysis, such as the electron beam-induced current (EBIC) technique and the optical beam-induced current (OBIC) technique, etc. Finally, methods used to improve the reliability of high-power semiconductor lasers are proposed in terms of the preparation process, reliability screening, and method application.

Keywords: high-power semiconductor laser; failure mechanisms; accelerated aging test; failure analysis techniques

1. Introduction

High-power semiconductor lasers have the advantages of small size, light weight, high electro-optical conversion efficiency, and easy monolithic integration, and are widely used in free-space communication; industrial processing; and the medical, aerospace, and military fields, as well as in other fields. High power characteristics and the long-term stability of the laser's wavelength and bandwidth are important prerequisites for the broad application of semiconductor lasers. Generally speaking, levels above 100 mW for narrow-stripe, single-mode devices and levels above 1 W for all other single- and multi-emitter lasers can be considered to be high power [1]. The reliability of high-power semiconductor lasers is limited by the optical power density at the output facet, heat dissipation, and the current density in the semiconductor. Understanding the reliability and failure mechanisms of high-power semiconductor lasers is essential for the development of high-performance and highly reliable application systems.

Research on the degradation mechanisms of semiconductor lasers dates back to the 1960s. In 1966, internal self-damage in gallium arsenide lasers was investigated by D.P. Cooper et al. from Services Electronics Research Lab. Baldock, Herts, UK. They suggested that the damage occurred as a result of the interaction between a critical high flux density and some randomly distributed structural properties such as defects or diffusion irregularities [2]. In 1967, H. Kressel and H. Mierop from RCA Laboratories postulated that the catastrophic damage in a GaAs injection laser resulted from the effects generated by stimulated Brillouin emissions [3].

In the mid-1970s, the Nippon Telegraph and Telephone Corporation (NTT) and the Nippon Electric Corporation (NEC) in Japan formed two research groups for collaboration in the study of the reliability of semiconductor lasers for optical communications [4]. However, for the next two decades, reliability research studies were limited by the technology, and the degradation failure phenomenon was not fully understood at the atomic level [5,6]. In the 21st century, IBM Zurich Research Laboratory (Switzerland) [7–9], Sumitomo Electric Industries, Ltd. (Japan) [10], Agilent (USA) [11], and the American Aerospace Corporation [12] have conducted systematic studies on the reliability and degradation of semiconductor lasers. The degradation mechanisms are gradually being addressed.

Analysis of the failure mechanism of high-power semiconductor lasers is an important basis for studying their reliability. In this paper, the failure mechanisms of high-power semiconductor lasers are introduced in detail, including three failure modes and the causes of performance degradation, such as internal degradation, mirror facet degradation, electrode degradation, packaging-related degradation, and the influence of environmental factors. The principles of accelerated aging experiments in high-power semiconductor lasers and their applications are also summarized. Failure analysis techniques such as the electron beam-induced current (EBIC) technique, the optical beam-induced current (OBIC) technique, the thermally induced voltage alteration (TIVA) technique, electroluminescence (EL), microphotoluminescence mapping (μ-PL), emission microscopy (EMMI), cathodoluminescence (CL), electron channeling contrast imaging (ECCI), transmission electron microscopy (TEM), and Raman are introduced in detail. The advantages and limitations of each technique are compared. Finally, methods by which to improve the reliability of high-power semiconductor lasers are proposed in terms of the preparation process, reliability screening, and method application. It is hoped that this study can provide a reference for research on the failure mechanisms of high-power semiconductor lasers, as well as improvements to their reliability.

2. Failure Mechanism

The performance of semiconductor lasers decreases with an increasing operating time. The most obvious manifestation of this is the decrease in output power and electro-optical conversion efficiency of the semiconductor laser at a constant drive current. In addition, catastrophic damage to the laser leads to a sudden drop in optical power [13]. Therefore, an increase in the drive current is required to prevent the degradation of the laser during constant power operation. The change in output power during laser degradation is mainly caused by the decrease in the lifetime of the injected carriers and the increase in internal optical losses. The mathematical model of the optical power output of a semiconductor laser is as follows [1]:

$$P_{out} = \eta_e (I - I_{th}) \frac{\hbar \omega}{q} \quad (1)$$

$$\eta_e = \eta_i \frac{\alpha_m}{\alpha_m + \alpha_i} \quad (2)$$

$$\alpha_m = \frac{1}{2L} \ln \frac{1}{R_1 R_2} \qquad (3)$$

where P_{out} is the optical output power, η_e is the external differential quantum efficiency, I is the drive current, I_{th} is the threshold current, q is the electron charge, and $\hbar\omega$ is the energy quantum. η_i is the internal differential quantum efficiency; α_i is the internal loss, mainly caused by the free-carrier absorption of the waveguide material and the scattering loss due to the roughness of the optical waveguide layer; and α_m is the mirror loss. L is the cavity length, R_1 and R_2 are the front cavity mirror reflectivity and the rear cavity mirror reflectivity, respectively. Then, the root cause parameters directly affecting the decrease in output power are the increase in the threshold current and the decrease in the slope efficiency or external differential quantum efficiency.

2.1. Failure Mode

In most operating systems, semiconductor lasers usually operate at a constant output power and lasing wavelength. Therefore, when degradation occurs, variations in the drive current show different patterns, as shown in Figure 1.

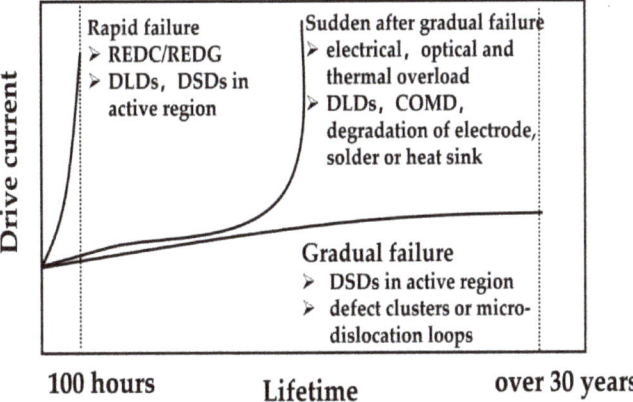

Figure 1. Failure mode of the laser diode at constant output power.

The failure modes of high-power semiconductor lasers are similar to those of ordinary semiconductor lasers. According to the relationship between drive current and lifetime, the failure modes of high-power semiconductor lasers are manifested in the following three forms: rapid failure, gradual failure, and sudden failure after gradual failure [14].

2.1.1. Rapid Failure

Rapid failure is usually observed within the first hundred hours of operation of a semiconductor laser. It manifests itself as a rapid drop in the output power and a rapid increase in the threshold current. Rapid failure is associated with defects present within the laser. These defects are introduced during the epitaxial growth of the wafer or the fabrication of the chip. The defects (such as the precipitation of impurity atoms and dislocations) are accelerated by non-radiative recombination processes; there is also vibration energy generated by recombination-enhanced dislocation climb or glide (REDC or REDG) [15,16]. As defects propagate in the diode structure, dislocation networks form and grow in the crystal structure, and these defects can destroy the active region of the semiconductor laser and make the laser ineffective. The types and formation mechanism of defects in degraded high-power semiconductor lasers are shown in Table 1 below.

Table 1. Defect types in degraded high-power semiconductor lasers.

Lasing Wavelength	Semiconductor Lasers	Defect Types	Mechanism
440–450 nm	InGaN/GaN QW lasers	<11-20> a-type dislocations [17]	Climb mechanism involving point defect
808 nm	AlGaAs/GaAs QW-SCH lasers	<110> DLDs Dislocations dipoles [18]	Gliding mechanism
980 nm	InGaAs/AlGaAs QW lasers	<100> DLDs Dislocation dipoles or climbed dipoles [18] <1-10> DLDs Edge dislocation dipoles	Climb mechanism involving point defect Gliding mechanism not involving point defect
1300 nm	InGaAsP/InP QW lasers	<100> DLDs Dislocation dipoles [19]	Climb mechanism involving point defect
1550 nm	InGaAs/InP QW lasers	V-shaped defects DLDs Misfit dislocations [20]	Climb mechanism

To eliminate rapid failure, it is necessary to apply low-defect or defect-free substrates to eliminate dark-line defects (DLDs) or dark-spot defects (DSDs) formed by the REDC or REDG mechanism. In addition, the selection of lattice-matched epitaxial films to reduce internal stresses in the laser structure and the selection of a support solder to reduce external stress during bonding are useful methods. Last but not least, avoiding defect formation during the epitaxial growth and fabrication processes is very important for reducing the probability of rapid failure.

2.1.2. Gradual Failure

This failure mode manifests itself in several forms, including a gradual and slow long-term decrease in the optical output power, and an increase in the threshold current with an increasing operation or aging time. To reach a certain output power, the drive current of the semiconductor laser needs to be increased. During this degradation process, the slope efficiency of the laser remains constant, and, for a constant drive current, the output power of the laser experiencing this failure mode can be seen to decrease over time. In general, the optical power of the laser must be recovered at the cost of a higher drive current [21]. The gradual failure mode often determines the maximum operating lifetime of semiconductor lasers. The mechanism of gradual failure is related to the enhancement of internal stress and the increase in non-radiative recombination centers in the active region. This failure is always associated with the formation of DSDs, caused by the growth of dislocation networks, and internal stresses, which are caused by defect clusters or micro-dislocation loops in the active region of high-power semiconductor lasers [4].

To avoid the gradual failure of semiconductor lasers, it is important to reduce the formation of non-radiative deep-level defect centers during epitaxial growth and device fabrication [22]. For example, a diffusion barrier layer is created during the preparation of metal electrodes to eliminate the inter-diffusion of metals and semiconductor materials.

Elimination of internal and external stresses around the active region, such as the selection of an appropriate strained quantum well (QW) structure, is an effective way to avoid long-term degradation. When the thickness of the QW layer is designed below a certain level, no misfit dislocation is formed due to strain relaxation. A heat sink material with a coefficient of thermal expansion that matches that of the semiconductor laser chip should be selected.

2.1.3. Sudden Failure after Gradual Failure

Sudden failure after gradual failure is manifested by a semiconductor laser that has been operating for a long period under normal operating conditions, but suddenly has no light intensity output. This is the most serious failure mode in high-power semiconductor lasers. Sudden failure after gradual failure in most devices is associated with DLDs [23]. This degradation mode is closely related to electrical, optical, and thermal overloads and also limits the maximum power of high-power semiconductor lasers. In addition, catastrophic optical mirror damage (COMD) and electrode, solder, or heat sink degradation may also lead to sudden failure after gradual failure [24].

Effective cavity surface passivation methods, vacuum cleavage, and coating techniques or the fabrication of non-absorbing mirror structures, such as the use of QW intermixing technologies, are applied to prevent the COMD phenomenon. Appropriate composite heat sinks with high thermal conductivity and less thermal expansion stress can effectively improve the reliability of semiconductor lasers due to their good matching of thermal expansion efficiency to the chip.

2.2. Reasons for Degradation

As with many other types of electronic devices, the reliability and lifetime of high-power semiconductor lasers depend heavily on their operating conditions, the suitability of the epitaxial materials, and device structure. The high optical power density at the output facet, heat accumulation, and the large current density in the semiconductor have a significant influence on the reliability of high-power semiconductor lasers. Depending on the type and location of the degradation in semiconductor lasers, degradation can be classified as internal, mirror facet, electrode, degradation related to packaging, and degradation caused by environmental factors.

2.2.1. Internal Degradation

As the operating time increases, local gain saturation or the spatial hole burning effect lead to the formation of filamentation, self-focusing, or thermal lensing in the chip. Optically induced heating enhances defects, which are introduced by the substrate, epitaxial growth, or fabrication processes migrating to the active region. The non-radiative recombination rate of the carrier is accelerated [12]. The energy generated by the non-radiative recombination process is converted into an Auger recombination or lattice vibrations, leading to recombination-enhanced defect reactions (REDRs). The propagation of defects through the REDC and REDG mechanisms leads to the formation of dislocation networks known as DLDs [4]. The REDR process is further enhanced by the strong optically induced heating phenomenon. The slow propagation of internal defects within the epitaxial material leads to an increase in non-radiative recombination centers in the active region, accompanied by a decrease in the electro-optical conversion efficiency and the output power. As a result, degradation of the semiconductor laser occurs. Internal degradation also includes DLDs and DSDs, which often lead to the rapid failure of semiconductor lasers. Figure 2 shows a TEM image of a typical DLD observed in a rapidly degraded GaAlAs/GaAs double-heterojunction laser [25]. DLD can also cause catastrophic optical bulk damage (COBD), leading to the rapid failure of the laser [26]. COBD refers to the catastrophic damage occurring inside the laser, which is different from that on the mirror facet.

2.2.2. Mirror Facet Degradation

Mirror facet degradation is a serious problem for high-power semiconductor lasers. It is a combination of COMD and chemical corrosion. COMD is considered one of the major limiting factors for achieving ultra-high optical power and is the main bottleneck limiting its reliability [27].

Figure 3 below illustrates the process of COMD generation, which is directly caused by high cavity surface temperatures. The relevant heating mechanism is the non-radiative recombination process induced by impurity oxides formed at the cleaved facet surface and the strong optical absorption process of the emitted laser light [28].

Figure 2. TEM image of dislocation dipole associated with <1 0 0> direction DLD in a rapidly degraded GaAlAs/GaAs double-heterojunction laser. Reprinted with permission from ref. [25] © Elsevier. Copyright 1999 Microelectronics Reliability.

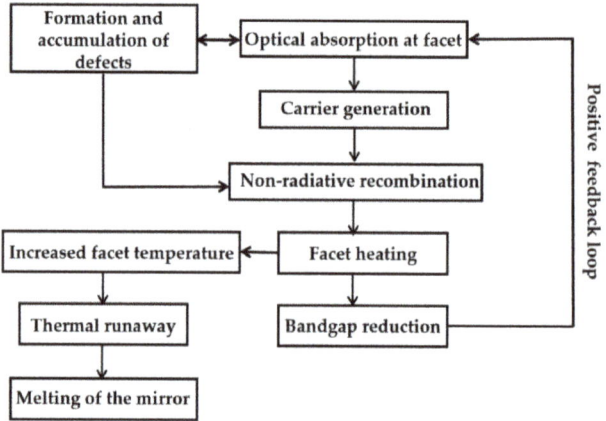

Figure 3. COMD generation mechanism.

In addition, the type of package may lead to an increase in the cavity surface temperature: packages of semiconductor lasers usually have an "overhang" of about 10 µm to protect the mirror facet's desorption from the solder material. As a result, the temperature of the mirror facet may be higher than that of the cavity [29]. The critical local temperature may be reached when the temperature is further increased by increasing the operating current or heat sink temperature, or when defects caused by aging are generated by long-term operation. At this point, intrinsic or extrinsic mechanisms leading to additional temperature increases from a positive feedback loop and thermal runaway start to occur.

When COMD occurs, DLDs are generated during the laser's operation; these DLDs are areas of non-radiative combination centers in the active region of the laser; they are generated locally, both on the cavity surface and inside the cavity, and can propagate along the cavity driven by the optical field, eventually leading to laser failure.

The photochemical action leads to slow oxidation of the mirror facet, which, in turn, leads to a gradual increase in local defects on the cavity surface and a change in their corresponding local reflection coefficients. An increase in optical absorption at the defects leads to an increase in local non-radiative recombination and a corresponding increase in temperature. Semiconductor lasers containing aluminum in the active region are more susceptible to catastrophic damage because aluminum adsorbs water and oxygen, making the laser cavity surface more susceptible to oxidation, which leads to the formation of local defects, and the device suffers catastrophic damage with increasing operation time.

2.2.3. Electrode Degradation

The main causes of electrode degradation are the degradation of ohmic contact and the degradation of thermal resistance. Ohmic contact degradation is mainly due to ion diffusion and electrical migration between the electrode material and the semiconductor material. In order to obtain a high output power, the electrodes of semiconductor lasers usually operate at very high current densities. At high current densities, electrical migration of metal ions in the electrodes occurs [30]. Metal ions undergo directional migration driven by high-density electron flow, and undergo significant mass transport, which, in turn, creates cavities voids or mounds (or whiskers) in the contact structures of the metal and semiconductor materials, producing electromigration failures. Electromigration failure often occurs together with processes such as chemical migration, stress migration, and thermal migration [31], leading to degradation failure of the electrode. In addition, impurity particles introduced during the welding process, as well as thermal stresses due to excessive temperature rising and falling, can also lead to degradation of the electrode.

2.2.4. Packaging-Related Degradation

The packaging stresses generated during the manufacturing process of a high-power semiconductor lasers can seriously affect their service life. Encapsulation stress is mainly caused by the mismatch of thermal expansion coefficients between the chip and the heat sink materials. Encapsulation stress changes the bandwidth of the semiconductor material and introduces a large number of dislocations in the active region, affecting the threshold current, wavelength, linewidth, and polarization of the laser [32]. Then, the service life of the laser is reduced, leading to transient failure. Solder (e.g., indium) creeps and climbs during operation, accumulating at the mirror facet and causing side leakage, leading to short circuits. Gaps created during the soldering process and oxidation of the solder can lead to laser degradation.

2.2.5. Influence of Environmental Factors

High-power semiconductor lasers are widely used in both terrestrial and space systems. For application scenarios in space systems, the exacerbating effect of radiation damage from the environment on laser degradation needs to be considered. Proton irradiation [33], electron irradiation [34], and γ-ray irradiation [35] can lead to crystal defects in semiconductor materials (e.g., Frenkel-type defects), which can affect the reliability of semiconductor lasers. The ambient temperature range for terrestrial systems is typically -40 to $+85\ ^\circ C$. The temperature conditions in space are much harsher than those in terrestrial applications, e.g., from $-120\ ^\circ C$ to $+120\ ^\circ C$. A drastic temperature variation range can also deteriorate the reliability.

The reliability of high-power semiconductor lasers can also be greatly affected by electrostatic discharge damage (ESD) generated during manufacturing, transportation, and use. ESD damage is typically caused by one of three events: direct static electricity from a person to a device, static electricity from a charged device, or field-induced discharge. Therefore, suitable models for simulating ESD-induced damage are called the Human Body Model (HBM), the Charged Device Model (CDM), and the Machine Model (MM), respectively. High-power semiconductor lasers are classified into edge-emitting lasers (EELs) and vertical cavity surface-emitting lasers (VCSELs) according to the emitting direction. VCSEL and VCSEL arrays are extremely sensitive to ESD [36,37]. If an ESD event occurs, catastrophic damage in the active area may occur leading to sudden failure of the semiconductor laser, and the laser may then degrade in its subsequent operation with the occurrence of other failure mechanisms, which can be easily confused with other causes of failure.

3. Accelerated Aging Test and Lifetime Test Method

3.1. Reliability

3.1.1. Reliability Overview

Reliability is the ability of a product to perform a "specified function" under "specified conditions" and for a "specified time". Reliability is usually reflected indirectly by the failure rate. According to Equation (1), failure of a semiconductor laser is generally defined as a failure when the drive current I remain constant and the output power P_{out} decreases to a certain percentage (e.g., 50% of the initial value, which can also be specified), or when the output power P_{out} remains constant and the drive current I rises to a certain degree (e.g., 120% of the initial value).

The failure rate of a semiconductor laser (instantaneous failure rate) $\lambda(t)$ is the probability of failure per unit of time after the laser has been in operation up to the moment t.

Figure 4 shows a typical failure rate function for a 1550 nm distributed feedback (DFB) laser [38].

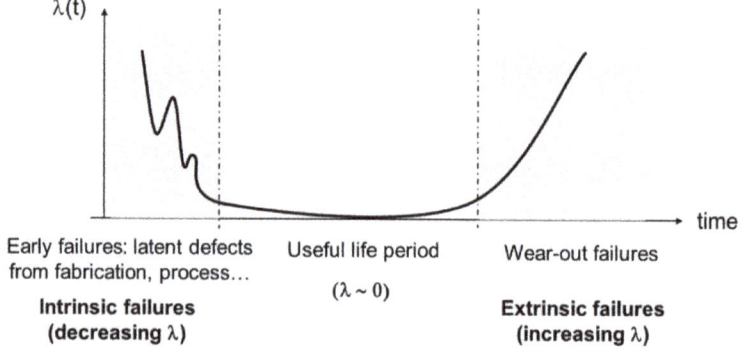

Figure 4. Variations of the time-dependent failure rate. Reprinted with permission from ref. [38] © Elsevier. Copyright 2021 Advanced Laser Diode Reliability.

The failure pattern of most optoelectronic devices is similar to this curve and is divided into three phases called the early failure period, the random failure period, and the wear period.

(1) Early failure: The device has a high failure rate and a very short operating lifetime, usually due to rapid degradation caused by the rapid growth of the internal defects within the device. These defects are mainly generated during the manufacturing process.
(2) Random failure: This stage has a low failure rate, is difficult or impossible to predict, and is associated with chance factors.
(3) Wear and tear failure: The device shows wear and ages in different operating environments, reaching its service life, and eventually failing.

3.1.2. Reliability Experiments

Reliability experiments are an important method by which to study the reliability of high-power semiconductor lasers. Reliability experiments are divided into four main categories: environmental experiments, lifetime experiments, special test experiments, and field application experiments. The lifetime of a high-power semiconductor laser is an important indicator of its reliability. Lifetime experiments are common methods used to evaluate the reliability and degradation analysis of semiconductor lasers.

The lifetime experiments are divided into long-term lifetime experiments and accelerated aging experiments. With the development of high-power semiconductor lasers, their lifetimes have been continuously improved, and most of them have reached a lifetime on the order of 10,000 h. If the lifetime experiment is completed according to an actual application environment, the test time can be long and the device cost can be high. In order to save time and costs, accelerated aging experiments are often used to quickly determine the reliability of the device.

3.2. Accelerated Aging Experiment

3.2.1. Theoretical Basis for the Accelerated Aging Test

The idea of the accelerated aging experiment is to create an experimental environment at high stress levels, so the necessary experimental data of the test device can be obtained. Then, one can apply statistics or other data processing methods to extrapolate the lifetime characteristics of the device operating at normal stress [39]. Therefore, the relationship between lifetime characteristics and stress levels needs to be established. The common accelerated stresses used in aging experiments of high-power semiconductor lasers are mainly temperature stresses and electrical stresses.

Arrhenius Model

An important model is the Arrhenius equation that expresses the temperature stress and reaction rate as follows [40]:

$$\frac{dM(T)}{dt} = A_0 exp\left(\frac{-E_a}{kT}\right) \quad (4)$$

$\frac{dM(T)}{dt}$ is the rate of process to failure, indicating the speed of the reaction; k is the Boltzmann constant; E_a is the activation energy that causes the failure or degradation process; T is the absolute temperature; and A_0 is a non-thermal constant factor. A cumulative failure distribution diagram at different constant temperatures is usually used to obtain the median life (mean time to failure) at each temperature. According to the Arrhenius curve of median life and temperature, the failure activation energy and the extrapolated working life of the devices can be obtained.

Inverse Power Law Model

When the life of a system is an inverse power function of the accelerating stress variable, the inverse power law is commonly used. The inverse power law model describes the relationship between the applied stresses (e.g., voltage, current, optical power, temperature cycling, or mechanical vibration, etc.) and the lifetime of a semiconductor laser [41].

$$L_s = Av^{-C} \quad (5)$$

L_S represents the life at a stress of v; A is a constant typical for laser type; C is an exponent characteristic of the laser device, which is a positive constant related to the activation energy; and v is the accelerating stress.

Different stresses produce different failure mechanisms, and the choice of stresses in the experiment is determined by the actual situation. The Arrhenius model and the inverse power law model can be linearized and written uniformly in the following form:

$$L_s = a + b \ln \phi(s) \quad (6)$$

L_S is the characteristic lifetime; ϕ is a function related to the s stress; and a and b are coefficients that can be calculated from the experimental data.

3.2.2. Classification of Accelerated Aging Test

The accelerated aging test can be divided into constant stress, step stress, and sequential stress accelerated aging experiments according to the variation law of the accelerated stress applied in accelerated aging experiments [42].

Constant stress experiments are performed by dividing the stress levels into different groups and then testing the devices in different groups until all the devices fail. The stress level does not change during the whole experiment.

Step stress experiments are also divided into different stress levels. All the test devices are placed at the same level. After a period of time, the stress level is switched to a higher level. The failed devices are removed, then the experimental conditions are switched to another higher stress level, and so on, until a certain percentage of devices fail.

The sequential stress experiment is similar to the step stress experiment, except that the stress level of the sequential stress experiment increases continuously with time, which can also be regarded as the limiting case of the step stress, in which the time interval of the stress transition is considered to be very small.

Among these three experimental methods, the experimental environment setting of the constant stress experiment is relatively simple, but the test is very time-consuming; the experimental operation of the step stress experiment is more complicated than that of a constant stress experiment, but it is more time saving. The experimental environment setting of the sequential stress experiment is complicated, and there are fewer related reports.

3.3. Example of Accelerated Aging Test

Research on the lifetime of high-power semiconductor lasers around the world has focused on several large laser manufacturing companies and research institutions, such as IBM Zurich Research Laboratory [43,44], NASA [45,46], COHERENT [47], JOLD in Germany [48], the 13th Research Institute of China Electronics Technology Group Corporation [49], the Research Institutes of Chinese Academy of Sciences [50], nlight Corporation [51,52], and the American Aerospace Corporation [53,54], etc.

In 1991, A. Moser and E.E. Latta of the IBM Research Division, Zurich Research Laboratory, determined the apparent Arrhenius parameters for the rate process of COMD failure. They compared the rate processes for various cleaved facets with and without a subsequent plasma oxidation step. A tentative model for facet heating that ultimately leads to COMD in AlGaAs/GaAs QW lasers was established [43].

In 1994, A. Oosenbrug and E.E. Latta of the IBM Research Division, Zurich Research Laboratory studied the high-power operational stability of 980 nm InGaAs/AlGaAs QW lasers. In total, 60 high-power semiconductor lasers were tested for their long-term lifetimes, with individual hours ranging from 6000 to 32,000 h, some at power levels above 200 mW (up to 300 mW continuous wave). For a group of devices operating at 200 mW continuous wave and at 50 °C, they found a log-normal distribution with a median life of >150 kh [44].

In 2005, Guoguang Lu et al. of Changchun Institute of Optics and Mechanics, Chinese Academy of Sciences, investigated the reliability of 808 nm high-power InGaAsP/InP lasers using constant stress accelerated aging tests [49]. They conducted constant current aging tests on six randomly selected lasers at 70 °C and 80 °C with an operating current of 1000 mA, and used the Arrhenius equation to derive a lifetime of 30,000 h at 25 °C.

In 2008, Hongde Wang et al. of the 13th Research Institute of China Electronics Technology Group Corporation conducted aging experiments on 808 nm AlGaInAs/AlGaAs/GaAs QW lasers [50]. They performed step stress experiments based on the inverse power law model, with electrical stress levels of 1, 1.3, 1.5, 1.7, and 2 A for a duration of 600 h. By comparing the failure rates of the devices with constant stress and step stress experiments, they found the same failure modes and similar lifetime estimates for both methods. This experiment verified that the step stress test is more time efficient than the constant stress test.

In 2011, nLight corporation performed reliability tests on 976 nm single-emitter laser diodes [51,52]. Seven sets of test conditions were performed: 18 A, 52 °C (drive current, junction temperature); 12 A, 62 °C; 16 A, 55 °C; 14 A, 58 °C; 15.7 A, 88 °C; 15 A, 73 °C; and 13.7 A, 42 °C. Over 15,000 h of accelerated life test reliability data were collected. The effects of temperature and power acceleration were evaluated by accelerated aging tests, and it was concluded that the mean time to failure was greater than 30 years at an output power of 10 W and a junction temperature of 353 K (80 °C), with a statistical confidence level of 90%.

In 2016, the American Aerospace Corporation reported their study of long-term accelerated aging tests on high-power, single-mode and multi-mode InGaAs/AlGaAs strained QW lasers [53,54]. They tested 64 single-mode 975 nm QW lasers with four experimental conditions: a drive current of 1.5 A and a junction temperature of 70 °C; a drive current of 1.5 A and a junction temperature of 120 °C; a drive current of 1.8 A and a junction temperature of 70 °C; a drive current of 2.1 A and a junction temperature of 70 °C. The test time for each test condition was 6500 h, and the test accumulation was 25,000 h (nearly 3 years). They tested 32 multi-mode lasers with lasing wavelengths of 920–960 nm at a drive current of 4 A and a junction temperature of 55 °C, for a total of over 35,000 h. This is the longest reported lifetime test for a single-mode or multi-mode laser in the 915–980 nm band. Their failure analysis of the devices showed that the main cause of failure was internal degradation.

In 2017, Zhiwen Wang et al. of the Institute of Semiconductors, Chinese Academy of Sciences, conducted a lifetime test of their self-developed 975 nm high-power semiconductor laser using a current step stress aging experiment [55]. They set the current stresses to 10, 12, and 14 A, and the aging times to 1200 h, 500 h, and 500 h. The test results were analyzed according to the inverse power law model, and the average lifetime of the device at a current of 8 A was calculated to be 28,999 h.

In 2019, constant temperature and constant current aging tests were conducted on 18 conduction-cooled packaged 60 W 808 nm high-power diode lasers by the Xi'an Institute of Optics and Precision Mechanics, Chinese Academy of Sciences, and Xi'an Focuslight Technologies Inc [56]. The average laser lifetime was 1022, 620, and 298 h at three different heat sink temperatures of 55, 65, and 80 °C with a constant current of 60 A. The lifetime at room temperature was calculated to be 5762 h according to the Arrhenius formula. The number of lasers in this experiment was relatively large, and the test temperature and output power were high. Because of the high costs and long duration of this experiment, such reports are quite rare.

4. Failure Analysis Techniques

The performance and reliability of semiconductor lasers can be significantly improved by precisely defining the location and cause of damage. Many techniques have been employed to characterize the failure modes and degradation mechanisms of high-power semiconductor lasers: the EBIC technique, the OBIC technique, the TIVA technique, EL, µ-PL, EMMI, CL, ECCI, TEM, and Raman, etc.

4.1. EBIC

EBIC is a technique commonly used to detect defect localization in semiconductor lasers for in-depth failure analysis [57,58]. It is a scanning electron microscopy (SEM)-based technique used to measure the current flowing through a semiconductor. When an electron beam is shone on a semiconductor chip, electron–hole pairs are created in a certain range within the semiconductor. The induced current of carriers can be collected by the internal electric field, which detects electrical defects with reduced carriers due to recombination. The intensity of the EBIC signal corresponds to the strength of the internal electric field around the p–n junction. Defects that are non-radiative recombination centers show a significantly lower EBIC signal [59,60].

Since the depth of the generated carriers depends on the accelerating voltage, a wide voltage range from 5 to 40 kV is essential for the quantitative study of the defect activity. Typically, the lateral resolution of EBIC varies from 20 to 500 nm, depending on the SEM conditions and material composition.

In 2018, Yong Kun Sin et al. from the California Aerospace Corporation El Segundo, employed the EBIC technique for the first time to determine the failure modes of 980 nm degraded single-mode InGaAs/AlGaAs strained QW lasers by observing DLDs [61]. Figure 5 shows the EBIC images of the high-power InGaAs/AlGaAs QW laser under different aging conditions: (a) 2.1 A/70 °C with a fail time of 10,205 h, (c) 1.8 A/70 °C with a fail time of 2560 h, and (**d**) 2.1 A/70 °C with a fail time of 1180 h. This clearly shows that the onset of the DLDs is confined within the 4 μm wide waveguide. The combination of the SEM and EBIC images indicates that the degradation process occurs in the active layer.

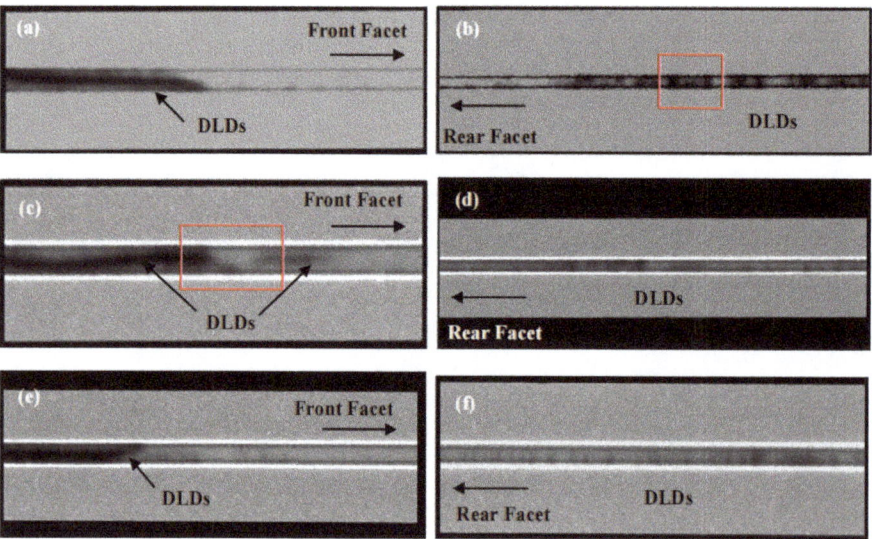

Figure 5. EBIC images were captured from high-power InGaAs/AlGaAs QW lasers with different aging conditions near the front facet. (**a**) 2.1 A/70 °C with a fail time of 10,205 h, (**c**) 1.8 A/70 °C with a fail time of 2560 h, and (**e**) 2.1 A/70 °C with a fail time of 1180 h and corresponding EBIC images near the rear facet (**b,d,f**). Reprinted with permission from ref. [61] © SPIE. Copyright 2018 Proceedings of SPIE.

In 2020, the rapid degradation behavior and failure mechanisms of InGaN/GaN green laser diodes (LDs) were investigated by the University of Shanghai for Science and Technology and the Chinese Academy of Sciences using the EBIC technique to assist in identifying the location of defects [62]. The SEM and EBIC images of the non-aged LD are shown in Figure 6a,d, and the SEM images of the two possible locations where the LD underwent severe degradation after the aging experiments are shown in Figure 6b,c. Compared with the non-aged LD, bubble-like defects can be observed in the images marked as P1 and P2, in which, combined with Figure 6e,f, the EBIC signal is slightly reduced. According to the EBIC principle, a higher carrier trap density exists in the active region of the laser near the bubble-like defects.

4.2. OBIC

OBIC is a non-destructive, highly sensitive, and high-resolution technique that is widely used to characterize defects present in semiconductor lasers such as stacking faults, dislocations, diffusion spikes, diffusion pipes, electrical over stress (EOS), and ESD damage. OBIC is a scanning optical microscopy imaging mode that locates regions of Fermi-level transitions. When the active region of a semiconductor laser is illuminated by a focused and scanned beam, the electron–hole pairs generated in the active region are separated by the built-in electric field in the p–n junction, and then collected by the electrodes to form a photocurrent that serves as the OBIC signal. Defects in the semiconductor material produce local variations in the Fermi level or the built-in potential that can enhance or weaken the recombination current, and, hence, the OBIC signal.

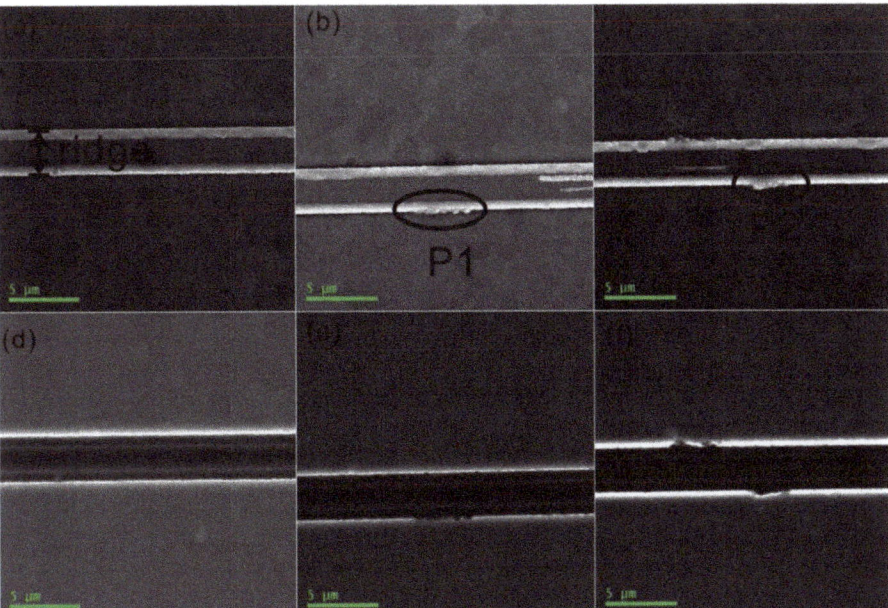

Figure 6. Typical secondary electron images of (**a**) non-aged LD and (**b**) and (**c**) rapidly degraded LD and the corresponding EBIC images of (**d**) non-aged LD and (**e**) and (**f**) rapidly degraded LD (the two black ovals mark two possible heavily degraded positions in the LD). Reprinted with permission from ref. [62] © Elsevier. Copyright 2020 Superlattices and Microstructures.

In 2007, Tatsuya Takeshita et al. of NTT, Japan, proposed a novel OBIC measurement technique to analyze the degradation position of a 2.5 Gbps directly modulated 1.55 μm uncooled DFB laser [63]. Incident beam sources with wavelengths longer than the band edge of the InGaAsP active layer were applied to improve the sensitivity of the OBIC. They confirmed that the degradation mechanism of the 1.55 μm InGaAsP/InP strained QW laser was mainly governed by diffusion defects on the waveguide rather than defects near the anti-reflective surface.

Figure 7a shows a digital OBIC scan image of the anti-reflective facet for the 1.55 μm InGaAsP/InP strained QW laser [63]. It underwent an aging time of 9000 h at 95 °C. The bright areas in the figure represent the active region. To characterize the degradation of the semiconductor laser, the peak OBIC intensity of the laser was normalized to 1.0 before aging. Figure 7b shows the normalized OBIC (nor-OBIC) signal intensity profile of the laser perpendicular to the p–n junction before and after aging. The peak of the nor-OBIC intensity dropped to 0.95 after aging, which indicates the degradation of the active layer after aging.

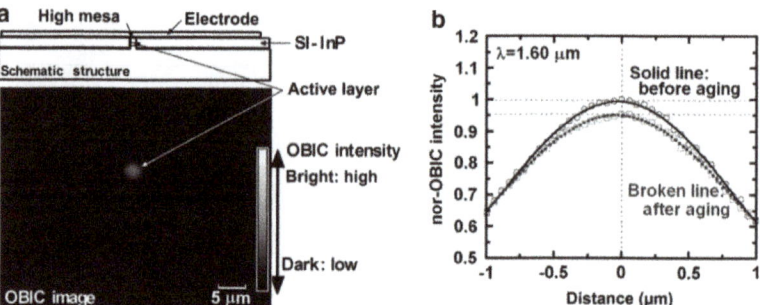

Figure 7. OBIC technique for monitoring the internal degradation of the laser. (**a**) OBIC image of the laser and (**b**) normalized OBIC signal intensity profile. Reprinted with permission from ref. [63] © Elsevier. Copyright 2007 Microelectronics Reliability.

In 2021, Che Lun Hsu et al. from National Yang Ming Chiao Tung University presented a modified OBIC microscope based on a tunable ultrafast laser to spectrally resolve the failure point of an electrostatic discharge-damaged VCSEL [64]. A spectrally resolved OBIC image of a normal VCSEL is shown in Figure 8 [64].

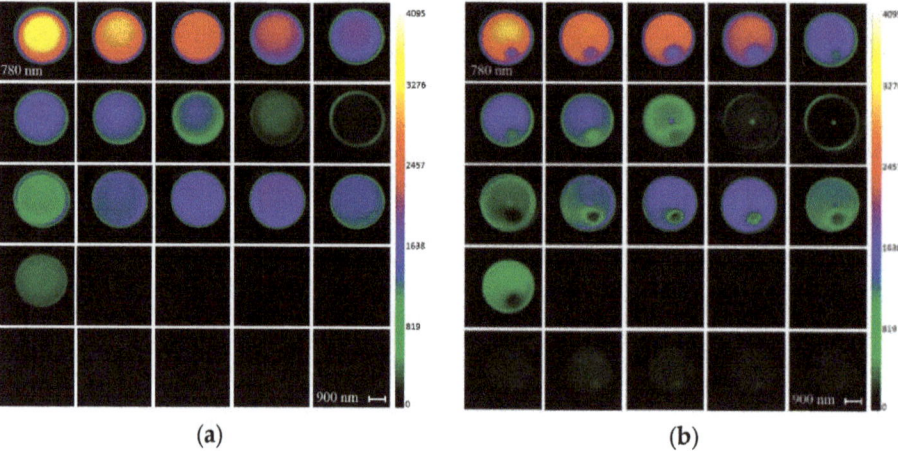

Figure 8. Spectrally resolved OBIC for ESD detection. (**a**) Normal and (**b**) Spectrally resolved OBIC image of the EDS-HBM VCSEL. Reprinted with permission from ref. [64] © The Optical Society. Copyright 2021 OSA Continuum.

The OBIC image was obtained by adjusting the incident wavelength at 5 nm intervals between 780 and 900 nm. The photocurrent signal was adjusted by a voltage preamplifier before being fed to the A/D converter for mapping. The obtained OBIC images are presented in a color-coded format. The active region of the OBIC confined by the oxide layer can be identified, resulting in a circular pattern. In a normal VCSEL (Figure 8a), the photocurrent distribution in the active region is very uniform. The maximum intensity of the photocurrent appears when the incident wavelength is 780 nm. When the incident wavelength is longer than 860 nm, there is no photocurrent response.

Figure 8b shows the spectrally resolved OBIC image of the ESD-HBM (human body model) VCSEL, in which it can be seen that there are two defect points in the active region: a large one at the edge and a small one near the center of the edge. Therefore, the OBIC technique can precisely define the damage location and failure cause of ESD-damaged VCSELs.

4.3. TIVA

Due to it being non-invasive and having a simple sample preparation, the TIVA technique is widely recognized as a fast and effective tool for locating defects in current biased devices, especially those with poor optical emission [65,66]. It is a type of thermal laser stimulation (TLS) technique. A focused 1.064 µm laser beam is used as an active probe to scan and locate defects in group III–V semiconductor lasers. The focused laser scan causes a localized change in thermal gradient at the scan location, which results in a change in local resistance. The change in resistance is captured when monitoring the voltage change across a fixed current source.

TIVA has been widely used to perform topside and backside inspections of failed VCSELs. Robert W. Herrick et al. of the Intel Corporation used the "backside TIVA" technique to inspect the DLDs in VCSELs [67]. Figure 9 shows TIVA images of mesa oxide in VCSELs before and after aging tests [67]. The reflected bright image (Figure 9a) shows a mesa with a diameter of about 42 µm and an oxide pore size of 12 µm. Figure 9b shows the TIVA image of the back side of the device under normal operation conditions. It shows some evidence of damage in the upper left corner, but it has not yet spread. The TIVA image of the degraded device (Figure 9c) shows the DLD propagating from the edge of the mesa in the lower right corner and then causing a failure as it reaches the emitting region in its center.

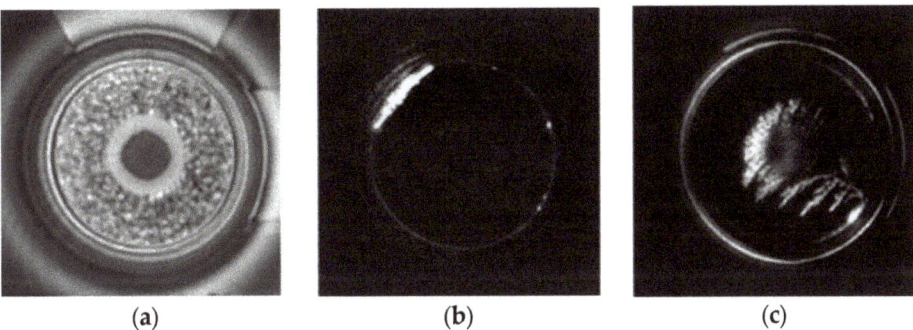

Figure 9. TIVA for detecting DLDs in VCSEL. (a) Reflected light image, (b) TIVA image of a normal VCSEL, and (c) TIVA image of a VCSEL damaged by aging test. Reprinted with permission from ref. [67] © Springer Nature. Copyright 2013 Springer eBook.

4.4. EL

EL is always employed to detect hidden cracks, black blocks, dislocations, and stacking faults in semiconductor lasers. The use of EL is usually sufficient to classify the location of the damage in EELs. When a fixed field is applied to the active region of a laser chip, the part without defects emits photons due to electron and hole recombination, while the defective part without electron and hole recombination shows a dark region. EL imaging is particularly suited to study the development of DLDs caused by luminescence-killing dislocation networks in laser cavities. EL line scans across a defect structure provide useful quantitative information.

In 2018, Yongkun Sin et al. from the Aerospace Corporation, El Segundo, found a new failure mode in high-power multi-mode InGaAs/AlGaAs strained QW lasers using EL techniques for short-term step stress tests and long-term accelerated aging tests [68]. Figure 10 shows the top surface EL images of two multi-mode InGaAs/AlGaAs strained QW lasers [68]. Figure 10a shows the DLD starting from the front mirror, indicating the degradation of the facets (COMD failure). Additionally, the DLDs of the bulk failure in Figure 10b start from the inside of the cavity, indicating catastrophic optical bulk damage (COBD failure).

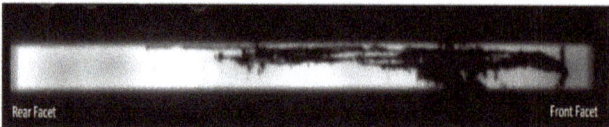

Figure 10. EL images of degraded multi-mode lasers show a facet failure and a bulk failure. Reprinted with permission from ref. [68] © Springer Nature. Copyright 2018 MRS Advances.

4.5. μ-PL Mapping

The μ-PL mapping technique is a simple, non-destructive, and non-contact method used to detect the defect concentrations and crystalline quality of semiconductor materials. The semiconductor material is irradiated by a certain frequency of light. The electrons in the defect-free parts are excited to a higher electronically excited state, and then radiate a photon as the electrons return to a lower energy state and release energy (photons). PL intensity mapping and peak wavelength mapping in semiconductor wafers as grown and processed can provide information on material uniformity, dislocation density, the density of deep traps, the distribution of residual impurities, and anneal uniformity.

In 2006, Marwan Bou Sanayeh et al. of OSRAM Opto Semiconductor GmbH employed the μ-PL mapping technique to investigate COMD-induced defects in high-power AlGaInP broad-area lasers [69]. Figure 11 shows the μ-PL mapping of the COMD laser. From the magnification of the defects near the output facet, highly non-radiative defects start from the output facet and propagate parallel to the waveguide direction inside the cavity.

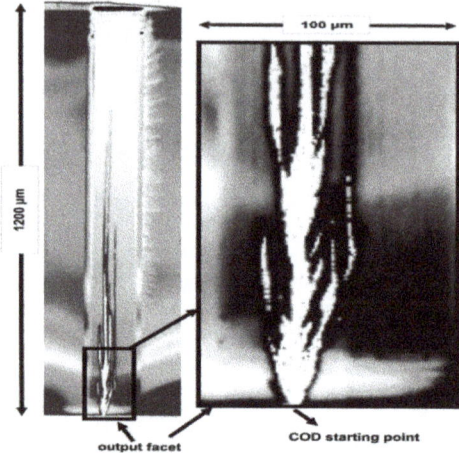

Figure 11. μ-PL mapping of a laser after COMD with a magnification of the output facet (white = low PL intensity and black = high PL intensity). Reprinted with permission from ref. [69] © AIP Publishing. Copyright 2006 Applied Physics Letters.

4.6. EMMI

The EMMI technique is a highly efficient and high-precision, non-destructive method used to detect leakage currents from device defects, ESD failures, junction leakage, contact spiking, and hot electrons, etc. By detecting the photons excited by the recombination of electron and hole, the failure position and mechanism can be deduced. Based on the precise positioning of the EMMI, focused ion beam (FIB) slitting enables direct observation and analysis of the localization of failure points.

In 2021, Roert Fabbro et al. from AMS, Unterpremsteatten, AG, Austria, and Graz University of Technology, Graz, applied the reverse-bias EMMI technique to detect defect localization in high-power VCSEL arrays. They performed failure detection on a high-power 2D-VCSEL array consisting of 932 emitters, as shown in Figure 12a [70].

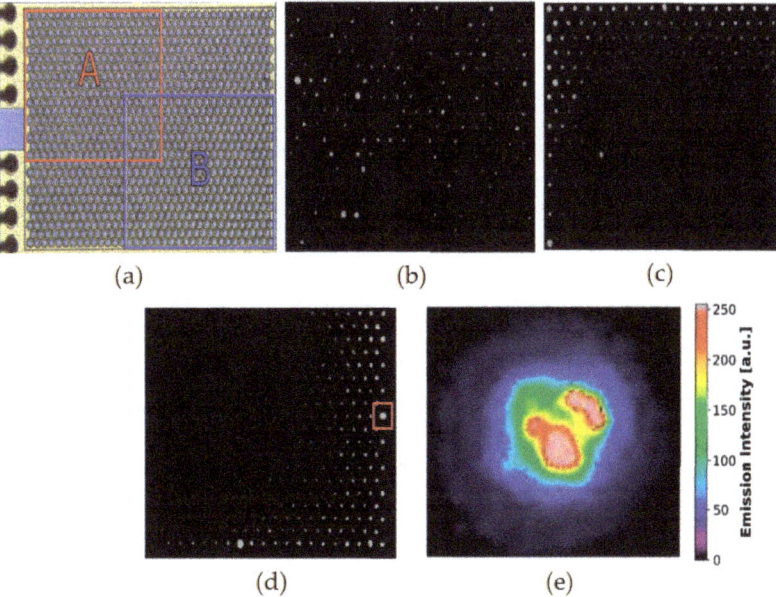

Figure 12. (a) High-power 2D-VCSEL array consisting of 932 emitters. (b) Reverse-bias EMMI image of region A of the VCSEL array before stress; (c) reverse-bias EMMI image of region A of the VCSEL array after stress; (d) reverse-bias EMMI image of region B in the VCSEL array after stress; and (e) colorized close-up of the marked emitter in (d). Reprinted with permission from ref. [70] © IOP Publishing. Copyright 2021 Measurement Science and Technology.

Figure 12b,c show the photon emission changes of individual emitters on the VCSEL array in a reverse-bias EMMI before and after stress [70]. They found that the most promising emitters showing degradation and defects were those showing a change from dark to bright (i.e., the change shown between Figure 12b,c) The emission intensity of each emitter depends on the intrinsic shunt resistance; defects such as cracks, ESD damage, and DLDs also act as leakage current pathways by damaging the internal crystal structure. These additional pathways increase the overall leakage current flowing through the emitters during PN junction reverse bias, and avalanche breakdown occurs, thus increasing the emission intensity of the damaged emitters during reverse bias. Figure 12d presents the reverse-bias EMMI image of region B in the VCSEL array after stress, where one of the emitters has high brightness (red square on the right side of the figure). A colorized close-up of the reverse-bias emission pattern of one of the emitters with a high brightness shows emissions across the whole emitter with two main defect spots (Figure 12e).

4.7. CL

The cathodoluminescence (CL) spectroscopy technique has a nanoscale spatial resolution (30–50 nm), which is normally applied to characterize the changes in composition, strains, and crystalline defects in semiconductor materials, such as dislocations, stacking faults, and impurities, etc. It can track elements and impurities and analyze the failure causes of semiconductor materials and defect localization within lasers. The CL technique is a powerful tool for observing dislocation defects in the active region of a semiconductor laser.

Due to the interaction of high-energy electrons (cathode rays) with a semiconductor material, electromagnetic radiation or a light ranging from visible to near-infrared is created by electron transitions in bandgaps or defect localizations. Using a specialized SEM, the electromagnetic radiation or light can be collected.

In 2013, V. Hortelano from the University of Valladolid investigated the defect signatures in degraded high-power laser diodes [18]. The degradation mechanism of an aged 980 nm single-ridge waveguide InGaAs/AlGaAs-based QW laser and broad-area emitter 808 nm AlGaAs/GaAs QW laser were studied using the CL technique. The characteristics are shown in Figure 13. Figure 13a,b show the CL images of DLDs in the 980 nm InGaAs/AlGaAs QW laser. The <100> DLDs at 45° to the cavity and the <1–10> DLDs perpendicular to the cavity can be observed in Figure 13 (a) and (b), respectively. The propagation of the DLD in Figure 13a is driven by the REDC mechanism. The propagation of the DLD in Figure 13b is driven by the REDG mechanism. Figure 13c shows the V-shaped defect in an aged 808 nm AlGaAs/GaAs QW laser.

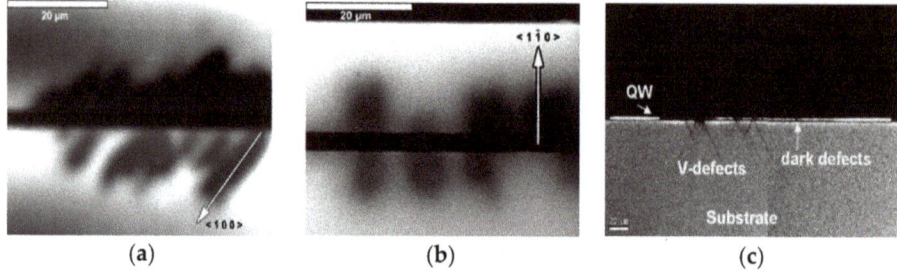

Figure 13. DLDs observed by the CL technique. (a) CL image of DLDs at 45° to the cavity, (b) CL image of DLDs perpendicular to the cavity, and (c) CL image of V-shaped defects. Reprinted with permission from ref. [18] © Elsevier. Copyright 2013 Microelectronics Reliability.

4.8. TEM

TEM is based on the interaction between a high-energy electron beam and a very thin sample. It is a high-resolution technique used to reveal the crystallographic phase at the nanometer scale; to characterize crystal defects such as dislocation, grain boundaries, voids, and stacking faults in a III–V super lattice; and to identify nanometer-sized defects on integrated circuits, including embedded particles and via residues.

In 2010, Shigetaka Tomiya et al. from Advanced Materials Laboratories, Sony Corporation, investigated the structural defects and degradation phenomena in high-power pure-blue InGaN-based laser diodes. They proposed that the 440–450 nm InGaN-based LDs were degraded by an increase in the capture cross-section of non-radiative recombination centers at the active regions due to the diffusion of point defects and/or impurities during laser operation [17].

They confirmed that the introduction of a current injection-free region at/near the laser facet had a significant effect on the suppression of COMD degradation. Additionally, the increase in non-radiative recombination centers was caused by the current injection and not by an optical effect. The multiple defects consisted of a combination of columnar defects, and dislocations were demonstrated by cross-sectional TEM images of the multiple quantum wells (MQWs).

4.9. ECCI

ECCI relies on electron diffraction inside crystalline materials. The information of the crystalline defect is carried by elastically backscattered electrons, which can be significantly modulated by the crystal orientation near the top surface. It is a helpful method used to examine the dislocation distribution and cross-sectional facet conditions of aged devices. It allows the detection and characterization of dislocation substructures near the surface of bulk specimens, which can be retested and re-examined, and it allow a large area of the specimen to be examined, thus allowing statistical information to be collected for large numbers of persistent slip bands and cracks.

The advantage of adopting the ECCI method compared to cross-sectional TEM image is that no sample preparation is required, and, moreover, one can continuously monitor the device aging process without damaging the device structure.

In 2021, Bei Shi et al. from the University of California Santa Barbara, USA, applied the ECCI technique to examine the dislocation distribution of aged 1550 nm InGaAs/InP QW laser diodes monolithically grown on silicon [20]. A high density of V-shaped defects originating from the multiple-QW region was detected before device aging. Extended DLDs were observed in the QW region, the InP buffer, and the strained-layer super lattices. They found that the DLDs introduced new misfit dislocations and promoted the climbing of threading dislocations, leading to device degradation.

4.10. Raman

Laser Raman microprobe spectroscopy is a non-destructive detection method that measures damaged portions of semiconductor materials and characterizes the compositional changes and lattice quality of the epitaxial layers [71]. It is an analytical technique in which scattered light is used to measure the vibrational energy pattern of the sample. The stresses and strains present in the sample can be obtained by comparing the changes in the peak positions of the Raman peaks; the polarization direction of Raman is used to determine the crystal symmetry and lattice orientation. By comparing the intensity of the Raman peaks, the total amount of substances in the material can also be known.

In addition to the above methods, there are many other techniques that can be applied to analyze the degradation mechanism of high-power semiconductor lasers, such as deep level transient spectroscopy (DLTS) [72,73], X-ray diffraction (XRD) [74,75], scanning electron microscope (SEM) [76], secondary ion mass spectroscopy (SIMS) [77,78], X-ray photoelectron spectroscopy (XPS) [79], and infrared thermal image detection technology (IRT) [80]. The principles of the different detection techniques are quite different, so there are big differences in the difficulty of the detection techniques, sample preparation methods, and matters needing attention.

Table 2 lists the detection category, selectivity, and technical advantages and limitations of different detection techniques. In practical applications, the advantages and strategies of the various techniques should be comprehensively considered to find the root cause of the device's failure.

Table 2. Comparison of various detection methods.

Method	Detection category	Selectivity	Advantages	Limitation
EBIC	Non-destructive (VCSEL, topside) Non-destructive (EEL, topside and front facet) Destructive (EEL, backside)	High	Detecting the DLDs, stacking faults and precipitates in the active region nearby	Defects in cladding or contact layer are not visible
OBIC	Non-destructive (VCSEL, topside) Non-destructive (EEL, topside and front facet) Destructive (EEL, backside)	High	Detecting degradation area and degree in the buried heterojunction and waveguide layers	For high-speed devices, the ability to observe the spectral response adequately is limited

Table 2. *Cont.*

Method	Detection category	Selectivity	Advantages	Limitation
TIVA	Non-destructive (VCSEL, topside) Non-destructive (EEL, topside)	High	Sensitive to leakage current and can detect the failure both on the surface and below	Signal might be blocked by metallic layers
EL	Non-destructive (VCSEL, topside) Non-destructive (EEL, topside and front facet) Destructive (EEL, backside)	Medium	Good at detecting the development of DLDs caused by luminescence-killing dislocation networks in the laser cavity	Not suited to detecting lasers with blanket metal over the epi-side, or mounted epi-side-down
μ-PL	Non-destructive (epitaxial wafer, topside) Non-destructive (VCSEL, topside) Non-destructive (EEL, topside) Destructive (EEL, backside)	High	Measures isolated defects and residual mechanical stress induced by the packaging process	Cannot provide information of luminescence features with dimensions below the classical diffraction limit
CL	Non-destructive (epitaxial wafer, topside) Non-destructive (VCSEL, topside) Non-destructive (EEL, topside) Destructive (EEL, backside)	High	Sensitive to the presence of non-recombination centers, high lateral and in-depth resolution	Result depends greatly on the device resistances
EMMI	Non-destructive (VCSEL, topside) Non-destructive (EEL, topside and front facet)	Medium	The luminescent defects can be located, low leakage current requirement	Poor resolution of non-luminous defects in metallic shaded areas
DLTS	Non-destructive (VCSEL, topside) Non-destructive (EEL, topside)	High	Sensitive to deep level defects, comprehensive characterization information	Cannot obtain deep energy level electron wave functions or defect internal information
Raman	Non-destructive (VCSEL, topside) Non-destructive (EEL, topside)	Medium	No need for sample processing, short time, high sensitivity	Weak signal strength
FIB-TEM	Destructive (VCSEL, topside) Destructive (EEL, topside)	High	Maximum magnification, high image quality, deep failure analysis; root cause identification	Significant sample preparation time, small sampling volumes, unable to detect the materials' instability in high-energy electron beams
XRD	Non-destructive (VCSEL, topside) Non-destructive (EEL, topside)	High	Small sample size and simple interpretation result	Susceptible to interference of mutual elements and superposition peaks
SEM	Non-destructive (epitaxial wafer, VCSEL, EEL, topside) Non-destructive (epitaxial wafer, cross-section) Non-destructive (EEL, front facet) Destructive (EEL, backside)	High	Detects failures such as delamination, oxidation contamination, and melting at the surface	Cannot detect root cause hidden below surface

Table 2. Cont.

Method	Detection category	Selectivity	Advantages	Limitation
ECCI	Non-destructive (epitaxial wafer, topside) Non-destructive (VCSEL, topside) Non-destructive (EEL, topside) Destructive (EEL, backside)	High	High precision, no need for sample preparation, allows the examination of large area of specimen	Resolution is limited by SEM system; the background signal is strong; and dislocation detection and signal analysis are prone to interference
SIMS	Destructive (epitaxial wafer, VCSEL, EEL, topside)	High	Can acquire the information of several atomic layers or even single elements	Difficulties in quantitative analysis
XPS	Non-destructive (VCSEL, topside) Non-destructive (EEL, topside)	High	Good repeatability of spectral line, fast speed and high sensitivity	Low energy resolution and low peak-to-back ratio
IRT	Non-destructive (VCSEL, topside) Non-destructive (EEL, front facet) Destructive (EEL, topside)	Medium	Strong signal and easy to measure	Large quantitative analysis errors and low sensitivity

5. Failure Improvement Measures

By analyzing the failure modes and causes of high-power semiconductor lasers, methods by which to improve the reliability of high-power semiconductor lasers are proposed in terms of the preparation process, reliability screening, and method application.

5.1. Preparation Process

To avoid internal degradation, it is very important to avoid defects caused by epitaxial growth and to minimize internal and external stresses in semiconductor lasers.

Substrates with low or no dislocation density are commercially available to avoid the propagation of threading dislocations into the active region. Designing an appropriate buffer layer or an etching stop layer can reduce the defects caused by the segregation process between atomic layers. Minimizing the oxygen content in the epitaxial growth atmosphere and selecting lattice-matched epitaxial films are useful methods by which to reduce the internal stresses and defects in laser diodes.

To reduce the external stresses in the bonding process, a low-melting-point Aurum–stannum (Au–Sn) alloy solder with strong oxidation, good thermal conductivity, and less susceptibility to metallurgical reactions can be used to replace the original Indium (In) solder. Appropriate composite heat sinks with high thermal conductivity and less thermal expansion stress can effectively improve the reliability of semiconductor lasers due to their good match with the thermal expansion efficiency of the chip. Appropriate designs of the high heat flux microchannel heat sink exchanger can be used for the cooling of semiconductor laser diode arrays. High-efficiency coupling and package technique are necessary to increase the lifetime of laser diodes.

The application of effective surface passivation; vacuum cleavage and coating techniques; or the fabrication of non-absorbing mirror structures, such as the use of QW intermixing techniques, can be used to prevent the COMD phenomenon.

5.2. Reliability Screening

High-power semiconductor lasers are generally required to operate for long periods with continuous electricity input and are subject to changes in environmental conditions (radiation, temperature, and humidity, etc.), and therefore require a high degree of reliability and stability.

To ensure the quality of the devices in the application system, semiconductor lasers must undergo rigorous inspection and aging screening prior to assembly. In the manufacturing process of lasers, early failures are usually accelerated by aging screening to screen out early failing devices and to ensure the reliability of semiconductor lasers.

Aging screening refers to the method of applying stress during the early operation of the device to accelerate the discovery of early failing devices and to screen out potential faults in time [81]. The applied stress can be thermal stress, electrical stress, mechanical stress, or a combination of different stresses. The applied stress should be controlled within an appropriate range. If it is too small, it may not achieve the screening effect, and if it is too large, it will introduce new failure causes. At present, the main methods used are high-temperature storage, high- and low-temperature shock, high-temperature power aging, mechanical vibration, centrifuge acceleration, leakage, and humidity and heat, etc., among which high-temperature power aging is the most common method [82].

The benefit of aging screening is that it can greatly reduce the risk of early failure, but it can also increase the overall production costs. For example, for a 144-pin package product, 24-h aging is used, and the cost is about 30% of the total cost [83].

Typical failure criteria of pump semiconductor lasers include, for example, a 5% reduction in ex-facet power during accelerated life tests at high temperatures and power or a 50% increase in the threshold current at the rated power [1].

5.2.1. High-Temperature Storage

High-temperature storage is a static aging mode. The devices are aged and screened after being placed in a high-temperature environment for certain time. The method mainly examines the effect of temperature on the device, which can accelerate chemical reactions on the chip's surface, the extension of defects inside the device, and the oxidation process of virtual solder joint, etc. [84]. The influences of high temperatures on the performance of the device include electrical parameter changes, heat dissipation difficulties, thermal aging, and chemical decomposition.

The devices are subjected to environmental tests of high-temperature storage between 120 °C to 300 °C; the storage time can be from dozens of hours to hundreds of hours. Generally speaking, the higher the temperature and the longer the storage time, the better the effect. The test depends on the reliability requirements of the device.

5.2.2. High- and Low-Temperature Shock

High- and low-temperature shock means that the device is stored in alternating high- and low-temperature environments without powering up the device. It is a useful way to check the match conditions of thermal expansion coefficients between different epitaxial materials. Potential quality-related problems (cracks, sealing performance, and pressure welding) can be found.

The rigor of the test depends on the selected high and low temperatures, the length of exposure time, and the number of cycles.

5.2.3. High-Temperature Power Aging

High-temperature power aging refers to a certain high-temperature stress and electrical stress on the device. It is an effective aging method used to simulate the operating conditions of the device in an actual circuit, coupled with high-temperature (+80 °C–180 °C) aging.

High-temperature power aging is suitable for filtering semiconductor lasers with surface contamination, wire welding, current leakage, cracks, oxide layer defects, and local hot spots, etc.

5.3. *Method Application*

When the semiconductor laser is in operation for a long time, electrical or voltage overload, electrical surges, and thermal overload often occur. These phenomena could be decreased, such as intentionally reducing the electrical stress, thermal stress, or mechanical stress in the laser.

High-power semiconductor lasers must have good thermal resistance, and the internal and external thermal resistance of the device should be reduced as much as possible. Scratches and mechanical damage should be prevented during the operation and use of the laser diode.

It is important to minimize the generation of static electricity, for example, by designing the physical circuit of the electrostatic discharge outside the laser; by using an anti-static carpet, workbench, and clothing; or by controlling the humidity of the environment. The welding and testing instruments must be properly grounded.

6. Conclusions

This paper introduces the failure mechanisms, accelerated aging experiments, failure detection techniques, and improvement methods of high-power semiconductor lasers. It can be seen that the reliability of high-power semiconductor lasers is influenced by many factors, and the reliability can be improved by using a comprehensive analysis of multiple techniques. Choosing the right experimental aging conditions not only saves time but also reveals the correct failure mechanism in semiconductor lasers and obtains the true reliability level of the devices. Choosing the proper failure detection techniques helps one to correctly analyze the causes of failure in high-power semiconductor lasers so as to improve their lifetime and reliability.

Author Contributions: Conceptualization, Y.S., S.N. and Z.L.; methodology, J.B. and Z.W.; validation, Y.S. and Y.C.; formal analysis, L.L. and P.J.; investigation, Y.S, Z.L., S.N., and Y.L.; resources, Y.S., Z.L. and J.B.; writing—original draft preparation, Y.S., S.N. and Z.L.; writing—review and editing, Y.C.; supervision, Y.C. and L.L.; project administration, L.Q., X.S. and L.L.; funding acquisition, L.Q., X.S. and L.W. All authors have read and agreed to the published version of the manuscript.

Funding: This work was funded by National Science and Technology Major Project of China (2021YFF0700500); National Natural Science Foundation of China (NSFC) (62090051, 62090052, 62090054, 11874353, 61935009, 61934003, 61904179, 62004194); Science and Technology Development Project of Jilin Province (20200401069GX, 20200401062GX, 20200501006GX, 20200501007GX, 20200501008GX); Key R&D Program of Changchun (21ZGG13, 21ZGN23); Innovation and entrepreneurship Talent Project of Jilin Province (2021Y008); Special Scientific Research Project of Academician Innovation Platform in Hainan Province (YSPTZX202034); Lingyan Research Program of Zhejiang Province (2022C01108); Science and Technology Innovation Program of Changchun Institute of Optics, Fine Mechanics and Physics, Chinese Academy of Sciences.

Data Availability Statement: Not applicable.

Acknowledgments: The authors thank Ligong Zhang and Yongqiang Ning for helping with the article, and Chaoshuai Zhang and Bokai Zhang for their constant advice and support.

Conflicts of Interest: The authors declare no conflict of interest.

References

1. Epperlein, P.W. *Semiconductor Laser Engineering, Reliability and Diagnostics: Basic Diode Laser Engineering Principles*; John Wiley & Sons Ltd.: Hoboken, NJ, USA, 2013; pp. 35–39, 72–74, 326–332.
2. Cooper, D.P.; Gooch, C.H.; Sherwell, R.J. Internal self-damage of gallium arsenide lasers. *IEEE J. Quantum Electron.* **1966**, *2*, 329–330. [CrossRef]
3. Kressel, H.; Mierop, H. Catastrophic Degradation in GaAs Injection Lasers. *J. Appl. Phys.* **1968**, *38*, 5419–5421. [CrossRef]
4. Orton, J.W. Reliability and Degradation of Semiconductor Lasers and LEDs. *Opt. Acta Int. J. Opt.* **1992**, *39*, 1799–1800. [CrossRef]
5. Ueda, O. On Degradation Studies of III–V Compound Semiconductor Optical Devices over Three Decades: Focusing on Gradual Degradation. *Jpn. J. Appl. Phys.* **2010**, *49*, 090001–090008. [CrossRef]
6. Ueda, O. Reliability and Degradation of III–V Optical Devices. In *Materials and Reliability Handbook for Semiconductor Optical and Electron Devices*; Springer: New York, NY, USA, 2013; pp. 87–120.
7. Epperlein, P.W.; Buchmann, P.; Jakubowicz, A. Lattice disorder, facet heating and catastrophic optical mirror damage of AlGaAs quantum well lasers. *Appl. Phys. Lett.* **1993**, *62*, 455–457. [CrossRef]

8. Jakubowicz, A.; Oosenbrug, A.; Forster, T. Laser operation-induced migration of beryllium at mirrors of GaAs/AlGaAs laser diodes. *Appl. Phys. Lett.* **1993**, *63*, 1185–1187. [CrossRef]
9. Jakubowicz, A.; Oosenbrug, A. SEM/EBIC characterization of degradation at mirrors of GaAs/AlGaAs laser diodes. *Microelectron. Eng.* **1994**, *24*, 189–194. [CrossRef]
10. Ichikawa, H.; Kumagai, A.; Kono, N.; Matsukawa, S.; Fukuda, C.; Iwai, K.; Ikoma, N. Dependence of facet stress on reliability of AlGaInAs edge-emitting lasers. *J. Appl. Phys.* **2010**, *107*, 083109. [CrossRef]
11. Krueger, J.; Sabharwal, R.; Mchugo, S.; Nguyen, K.; Tan, N.; Janda, N.; Mayonte, M.; Heidecker, M.; Eastley, D.; Keever, M.; et al. Studies of ESD-related failure patterns of Agilent oxide VCSELs. *Proc. SPIE* **2003**, *4994*, 162–172.
12. Sin, Y.; Ives, N.; Lalumondiere, S.; Presser, N.; Moss, S.C. Catastrophic optical bulk damage (COBD) in high power multi-mode InGaAs-AlGaAs strained quantum well lasers. *Proc. SPIE* **2011**, *7918*, 791803-1–791803-11.
13. Pitts, O.J.; Benyon, W.; Springthorpe, A.J. Modeling and process control of MOCVD growth of InAlGaAs MQW structures on InP. *J. Cryst. Growth* **2014**, *393*, 81–84. [CrossRef]
14. Chu, S.; Nakahara, S.; Twigg, M.E.; Koszi, L.; Flynn, E.; Chin, A.; Segner, B.; Johnston, W. Defect mechanisms in degradation of 1.3-µm wavelength channeled-substrate buried heterostructure lasers. *J. Appl. Phys.* **1988**, *63*, 611–623. [CrossRef]
15. Jianping, J. *Semiconductor Laser*, 1st ed.; Publishing House of Electronics Industry: Beijing, China, 2000; pp. 330–331.
16. Jiménez, J. Laser diode reliability: Crystal defects and degradation modes. *Comptes Rendus Phys.* **2003**, *4*, 663–673. [CrossRef]
17. Tomiya, S.; Goto, O.; Ikeda, M. Structural Defects and Degradation Phenomena in High-Power Pure-Blue InGaN-Based Laser Diodes. *Proc. IEEE* **2010**, *98*, 1208–1213. [CrossRef]
18. Hortelano, V.; Anaya, J.; Souto, J.; Jimenez, J.; Perinet, J.; Laruelle, F. Defect signatures in degraded high power laser diodes. *Microelectron. Reliab.* **2013**, *53*, 1501–1505. [CrossRef]
19. Ueda, O.; Umebu, I.; Yamakoshi, S.; Kotani, T. Nature of dark defects revealed in InGaAsP/InP double heterostructure light emitting diodes aged at room temperature. *J. Appl. Phys.* **1982**, *53*, 2991–2997. [CrossRef]
20. Shi, B.; Pinna, S.; Zhao, H.; Zhu, S.; Klamkin, J. Lasing Characteristics and Reliability of 1550 nm Laser Diodes Monolithically Grown on Silicon. *Phys. Status Solidi (A)* **2021**, *218*, 2000374. [CrossRef]
21. Ahrens, R.G.; Jaques, J.J.; Piccirilli, A.B.; Camarda, R.M.; Fields, A.B.; Lawrence, K.R.; Dutta, N.K.; Luvalle, M.J. Failure mode analysis of high-power laser diodes. *Proc. SPIE* **2002**, *4648*, 30–42.
22. Epperlein, P.W. *Semiconductor Laser Engineering, Reliability and Diagnostics: A Practical Approach to High Power and Single Mode Devices*; John Wiley & Sons Ltd.: Hoboken, NJ, USA, 2013; pp. 229–231.
23. Shi, J.W. Semiconductor laser degradation and its screening. *Semicond. Optoelectron.* **1997**, *1*, 14–19.
24. Fukuda, M. Historical overview and future of optoelectronics reliability for optical fiber communication systems. *Microelectron. Reliab.* **2000**, *40*, 27–35. [CrossRef]
25. Ueda, O. Reliability issues in III–V compound semiconductor devices: Optical devices and GaAs-based HBTs. *Microelectron. Reliab.* **1999**, *39*, 1839–1855. [CrossRef]
26. Liu, Q.K.; Kong, J.X.; Zhu, L.N.; Xiong, C.; Ma, X.Y. Failure Mode Analysis of High-power Laser Diodes by Electroluminescence. *Chin. J. Lumin.* **2018**, *39*, 180–187.
27. Fukuda, M.; Mura, G. Laser Diode Reliability. *Adv. Laser Diode Reliab.* **2021**, 1–49. [CrossRef]
28. Tomm, J.W.; Ziegler, M.; Hempel, M.; Elsaesser, T. Mechanisms and fast kinetics of the catastrophic optical damage (COD) in GaAs-based diode lasers. *Laser Photonics Rev.* **2011**, *5*, 422–441. [CrossRef]
29. Yoo, J.S.; Lee, H.H.; Zory, P. Temperature rise at mirror facet of CW semiconductor lasers. *Quantum Electron. IEEE J. Quantum Electron.* **1992**, *28*, 635–639. [CrossRef]
30. Chuang, T.H.; Lin, H.J.; Wang, H.C.; Chuang, C.H.; Tsai, C.H. Mechanism of Electromigration in Ag-Alloy Bonding Wires with Different Pd and Au Content. *J. Electron. Mater.* **2014**, *44*, 623–629. [CrossRef]
31. Korhonen, M.A.; Bo/Rgesen, P.; Tu, K.N.; Li, C.Y. Stress evolution due to electromigration in confined metal lines. *J. Appl. Phys.* **1993**, *73*, 3790–3799. [CrossRef]
32. Tomm, J.W.; Tien, T.Q.; Oudart, M.; Nagle, J. Aging properties of high-power diode laser arrays: Relaxation of packaging-induced strains and corresponding defect creation scenarios. In *Proceedings of the Cleo/Europe Conference on Lasers and Electro-optics Europe, Munich, Germany, 12–17 June 2005*; IEEE: Piscataway Township, NJ, USA, 2005; p. 112.
33. Okada, H.; Nakanishi, Y.; Wakahara, A.; Yoshida, A.; Ohshima, T. 380 keV proton irradiation effects on photoluminescence of Eu-doped GaN. *Nucl. Instrum. Methods Phys. Research. B Beam Interact. Mater. At.* **2008**, *266*, 853–856. [CrossRef]
34. O'Neill, J.; Ross, I.M.; Cullis, A.G.; Wang, T.; Parbrook, P.J. Electron-beam-induced segregation in InGaN/GaN multiple-quantum wells. *Appl. Phys. Lett.* **2003**, *83*, 1965–1967. [CrossRef]
35. Yamaguchi; Masafumi; Okuda; Takeshi. Minority-carrier injection-enhanced annealing of radiation damage to InGaP solar cells. *Appl. Phys. Lett.* **1997**, *70*, 2180–2182. [CrossRef]
36. Neitzert, H.C.; Piccirillo, A.; Gobbi, B. Sensitivity of proton implanted VCSELs to electrostatic discharge pulses. *IEEE J. Sel. Top. Quantum Electron.* **2002**, *7*, 231–241. [CrossRef]
37. Mchugo, S.A.; Krishnan, A.; Krueger, J.J.; Yong, L.; Tan, N.; Osentowski, T.; Xie, S.; Mayonte, M.S.; Herrick, R.W.; Deng, Q. Characterization of failure mechanisms for oxide VCSELs. *Proc. SPIE—Int. Soc. Opt. Eng.* **2003**, *4994*, 55–66.

38. Mendizabal, L.; Verdier, F.; Deshayes, Y.; Ousten, Y.; Danto, Y.; Béchou, L. Reliability of Laser Diodes for High-rate Optical Communications—A Monte Carlo-based Method to Predict Lifetime Distributions and Failure Rates in Operating Conditions. In *Advanced Laser Diode Reliability*; Elsevier: Amsterdam, The Netherlands, 2021; pp. 79–137.
39. Mao, S.S. Acceleration model for accelerated life test. *Qual. Reliab.* **2003**, *02*, 15–17.
40. Jennifer, L.K.; Joel, G.K. Erroneous Arrhenius: Modified Arrhenius Model Best Explains the Temperature Dependence of Ectotherm Fitness. *Am. Nat.* **2010**, *176*, 227–233.
41. Kececioglu, D.; Jacks, J.A. The Arrhenius, Eyring, inverse power law and combination models in accelerated life testing. *Reliab. Eng.* **1984**, *8*, 1–9. [CrossRef]
42. Hakamipour, N. Comparison between constant-stress and step-stress accelerated life tests under a cost constraint for progressive type I censoring. *Seq. Anal.* **2021**, *40*, 17–31. [CrossRef]
43. Moser, A.; Latta, E.E. Arrhenius parameters for the rate process leading to catastrophic damage of AlGaAs-GaAs laser facets. *J. Appl. Phys.* **1992**, *71*, 4848–4853. [CrossRef]
44. Oosenbrug, A.; Latta, E.E. High-power operational stability of 980 nm pump lasers for EDFA applications. *Lasers Electro-Opt. Soc. Meet.* **1994**, *2*, 37–38.
45. Amzajerdian, F.; Meadows, B.L.; Ba Ker, N.R.; Ba Ggott, R.S.; Singh, U.N.; Kavaya, M.J. Advancement of High Power Quasi-CW Laser Diode Arrays For Space-based Laser Instruments. In *Proceedings of the SPIE's Fourth International Asia-Pacific Environmental Remote Sensing Symposium, Honolulu, HI, USA, 8–12 November 2004*; SPIE: Washington, DC, USA, 2005; Volume 5659.
46. Stephen, M.; Krainak, M.; Dallas, J. Quasi-cw Laser Diode Bar Life Tests. In *Proceedings of the SPIE—Laser-Induced Damage in Optical Materials*; SPIE: Washington, DC, USA, 1997; Volume 3244.
47. Fouksman, M.; Lehconen, S.; Haapamaa, J.; Kennedy, K.; Li, J. High-performance high-reliability 880-nm diode laser bars and fiber-array packages. *Proc. SPIE* **2006**, *6104*, 61040A-1–61040A-6.
48. Bonati, G. Exploring failure probability of high-power laser diodes. *Photonics Spectra* **2003**, *37*, 56–58.
49. Lu, G.G.; To, G.T.; Yao, S.; Shan, X.N.; Sun, Y.F.; Liu, Y.; Wang, L.J. Reliability study of high-power semiconductor lasers. *Laser J.* **2005**, *4*, 14–15.
50. Wang, D.H.; Li, Y.J.; An, Z.F. Study on step-accelerated aging of high-power semiconductor lasers. *Micro-Nanoelectron.* **2008**, *9*, 508–511.
51. Bao, L. Reliability of high performance 9xx-nm single emitter laser diodes. *Proc. SPIE-Int. Soc. Opt. Photonics* **2010**, *7583*, 758302-1–758302-10.
52. Bao, L.; Leisher, P.; Wang, J.; Devito, M.; Xu, D.; Grimshaw, M.; Dong, W.; Guan, X.; Zhang, S.; Bai, C. High reliability and high performance of 9xx-nm single emitter laser diodes. *Proc. SPIE—Int. Soc. Opt. Eng.* **2011**, *7918*, 791806.
53. Sin, Y.; Lingley, Z.; Brodie, M.; Presser, N.; Moss, S.C. Catastrophic optical bulk degradation (COBD) in high-power single- and multi-mode InGaAs-AlGaAs strained quantum well lasers. *IEEE J. Sel. Top. Quantum Electron.* **2017**, *23*, 1500813. [CrossRef]
54. Sin, Y.; Presser, N.; Lingley, Z.; Brodie, M.; Moss, S.C. Reliability, failure modes, and degradation mechanisms in high power single- and multi-mode InGaAs-AlGaAs strained quantum well lasers. *Proc. SPIE—Int. Soc. Opt. Eng.* **2016**, *9733*, 973304.
55. Wang, W.C.; Ji, H.Q.; Qi, Q.; Wang, C.L.; Ni, Y.X.; Liu, S.P.; Ma, S. Reliability study and failure analysis of high-power semiconductor lasers. *J. Lumin.* **2017**, *38*, 165–169.
56. Nie, Z.Q.; Wang, M.P.; Sun, Y.B.; Li, S.N.; Wu, D. Thermally accelerated lifetime test of conduction-cooled high-power semiconductor laser single-bar devices in CW operation mode. *J. Emit. Light* **2019**, *40*, 1136–1145.
57. Vanzi, M.; Salmini, G.; Palo, R.D. New FIB/TEM evidence for a REDR mechanism in sudden failures of 980 nm SL SQW InGaAs/AlGaAs pump laser diodes. *Microelectron. Reliab.* **2000**, *40*, 1753–1757. [CrossRef]
58. Akamatsu, B. Electron beam-induced current in direct band-gap semiconductors. *J. Appl. Phys.* **1981**, *52*, 7245–7250. [CrossRef]
59. Henoc, P.; Benetton-Martins, R.; Akamatsu, B. Ebic and CL Study of Laser Degradation. *J. De Phys. IV* **1991**, *1*, C6-317–C6-322. [CrossRef]
60. Boudjani, A.; Sieber, B.; Cleton, F.; Rudra, A. Cl and EBIC analysis of a p + −InGaAs/n-InGaAs/n-InP/n + −InP heterostructure. *Mater. Sci. Eng. B* **1996**, *42*, 192–198. [CrossRef]
61. Sin, Y.; Ayvazian, T.; Brodie, M.; Lingley, Z. Catastrophic optical bulk degradation in high-power single- and multi-mode InGaAs-AlGaAs strained QW lasers: Part II. *Proc. SPIE* **2018**, *10514*, 10514-1–10514-14.
62. Xiu, H.; Xu, P.; Wen, P.; Zhang, Y.; Yang, J. Rapid degradation of InGaN/GaN green laser diodes. *Superlattices Microstruct.* **2020**, *142*, 106517. [CrossRef]
63. Takeshita, T.; Iga, R.; Yamamoto, M.; Sugo, M. Analysis of interior degradation of a laser waveguide using an OBIC monitor. *Microelectron. Reliab.* **2007**, *47*, 2135–2140. [CrossRef]
64. Hsu, C.L.; Das, S.; Wu, Y.S.; Kao, F.J. Spectrally resolved optical beam inducedcurrent imaging of ESD induced defects on VCSELs. *OSA Contin.* **2021**, *4*, 711–719. [CrossRef]
65. Cole, E.I., Jr.; Tangyunyong, P.; Benson, D.A.; Barton, D.L. TIVA and SEI developments for enhanced front and backside interconnection failure analysis. *Microelectron. Reliab.* **1999**, *39*, 991–996. [CrossRef]
66. Reissner, M. Fault localization at high voltage devices using thermally induced voltage alteration (TIVA). *Microelectron. Reliab.* **2007**, *47*, 1561–1564. [CrossRef]
67. Herrick, R.W. Reliability and degradation of vertical-cavity surface-emitting lasers. In *Materials and Reliability Handbook for Semiconductor Optical and Electron Devices*; Ueda, O., Pearton, S.J., Eds.; Springer: New York, NY, USA, 2013; pp. 195–197.

68. Yongkun, S.; Zachary, L.; Talin, A.; Miles, B.; Neil, I. Catastrophic Optical Bulk Damage—A New Failure Mode in High-Power InGaAs-AlGaAs Strained Quantum Well Lasers. *Mrs Adv.* **2018**, *3*, 3329–3345.
69. Sanayeh, M.B.; Jaeger, A.; Schmid, W.; Tautz, S.; Brick, P.; Streubel, K.; Bacher, G. Investigation of dark line defects induced by catastrophic optical damage in broad-area AlGaInP laser diodes. *Appl. Phys. Lett.* **2006**, *89*, 10111. [CrossRef]
70. Fabbro, R.; Haber, T.; Fasching, G.; Coppeta, R.; Michael, P.; Werner, G. Defect localization in high-power vertical cavity surface emitting laser arrays by means of reverse biased emission microscopy. *Meas. Sci. Technol.* **2021**, *32*, 095406. [CrossRef]
71. Peng, J.; Li, Q.; Xing, Z.; Zhang, J.; Ning, Y.Q. Reliability Study of Grating Coupled Semiconductor Laser Based on Raman Spectra Technique. *Spectrosc. Spectr. Anal.* **2016**, *36*, 1745–1748.
72. Yamasaki, K.; Yoshida, M.; Sugano, T. Deep Level Transient Spectroscopy of Bulk Traps and Interface States in Si MOS Diodes. *Jpn. J. Appl. Phys.* **1979**, *18*, 113–122. [CrossRef]
73. Shin, Y.C.; Dong, H.; Kang, J.; Chang, Y.K.; Kim, T.G. Optimization of 660-nm-Band AlGaInP Laser Diodes by Using Deep-Level Transient Spectroscopy Analyses. *J. Korean Phys. Soc.* **2007**, *50*, 866–870. [CrossRef]
74. Evtikhiev, V.P.; Kotel'Nikov, E.; Kudryashov, I.V.; Tokranov, V.E.; Faleev, N.N. Correlation between the reliability of laser diodes and the crystal perfection of epitaxial layers estimated by high-resolution x-ray diffractometry. *Semiconductors* **1999**, *33*, 590–593. [CrossRef]
75. Qiao, Y.B.; Feng, S.W.; Xiong, C.; Wang, X. Spatial hole burning degradation of AlGaAs/GaAs laser diodes. *Appl. Phys. Lett.* **2011**, *99*, 103506. [CrossRef]
76. Fernández-Caballero, A. A Review on Machine and Deep Learning for Semiconductor Defect Classification in Scanning Electron Microscope Images. *Appl. Sci.* **2021**, *11*, 9508.
77. Shin, Y.C.; Kim, B.J.; Dong, H.K.; Kim, Y.M.; Kim, T.G. Investigation of Zn diffusion by SIMS and its effects on the performance of AlGaInP-based red lasers. *Semicond. Sci. Technol.* **2005**, *21*, 35. [CrossRef]
78. Zhou, Q.; Li, J.Y.; Liang, H.D.; Wu, C.P. Recent developments on secondary ion mass spectroscopy. *J. Chin. Mass Spectrom. Soc.* **2004**, *25*, 113–120.
79. Yamamoto, S.; Matsuda, I. Time-Resolved Soft X-ray Photoelectron Spectroscopy: Real-Time Observation of Photo-Excited Carriers at Semiconductor Surfaces. *Hyomen Kagaku* **2016**, *37*, 9–13. [CrossRef]
80. Xie, J.; Xu, C.; Huang, W.; Chen, G. An infrared thermal image processing framework based on superpixel algorithm to detect cracks on metal surface. *Infrared Phys. Technol.* **2014**, *67*, 266–272.
81. Kuo, W.; Yue, K. Facing the headaches of early failures: A state-of-the-art review of burn-in decisions. *Proc. IEEE* **1983**, *71*, 1257–1266. [CrossRef]
82. Guijun, H.; Jiawei, S.; Yinbing, G.; Xiaosong, Y.; Jing, L. Noise as reliability screening for semiconductor lasers. *Appl. Phys. B* **2003**, *76*, 359–363. [CrossRef]
83. Zhang, L.P. The Screening Method for Early Failure of Semiconductor Devices. *Manag. Technol. SME* **2021**, *9*, 137, 138, 141.
84. Lee, J.M.; Min, B.G.; Ju, C.W.; Ahn, H.K.; Lim, J.W. High temperature storage test and its effect on the thermal stability and electrical characteristics of AlGaN/GaN high electron mobility transistors. *Curr. Appl. Phys.* **2016**, *17*, 157–161. [CrossRef]

Review

Research on Silicon-Substrate-Integrated Widely Tunable, Narrow Linewidth External Cavity Lasers

Xuan Li, Junce Shi, Long Wei, Keke Ding, Yuhang Ma, Zaijin Li *, Lin Li, Yi Qu, Zhongliang Qiao, Guojun Liu and Lina Zeng

Key Laboratory of Laser Technology and Optoelectronic Functional Materials of Hainan Province, Academician Team Innovation Center of Hainan Province, College of Physics and Electronic Engineering, Hainan Normal University, Haikou 571158, China; lixuan13025794412@163.com (X.L.); jcs1075106215@163.com (J.S.); weilong0096@163.com (L.W.); dingkeke1205@163.com (K.D.); mayuhang327@163.com (Y.M.); licust@126.com (L.L.); quyihainan@126.com (Y.Q.); qzhl060910@hainnu.edu.cn (Z.Q.); gjliu626@126.com (G.L.); zenglinahainan@126.com (L.Z.)
* Correspondence: lizaijin@126.com

Abstract: Widely tunable, narrow linewidth external cavity lasers on silicon substrates have many important applications, such as white-light interferometry, wavelength division multiplexing systems, coherent optical communication, and optical fiber sensor technology. Wide tuning range, high laser output power, single mode, stable spectral output, and high side-mode suppression ratio external cavity lasers have attracted much attention for their merits. In this paper, two main device-integrated structures for achieving widely tunable, narrow linewidth external cavity lasers on silicon substrates are reviewed and compared in detail, such as MRR-integrated structure and MRR-and-MZI-integrated structure of external cavity semiconductor lasers. Then, the chip-integrated structures are briefly introduced from the integration mode, such as monolithic integrated, heterogeneous integrated, and hybrid integrated. Results show that the silicon-substrate-integrated external cavity lasers are a potential way to realize a wide tuning range, high power, single mode, stable spectral output, and high side-mode suppression ratio laser output.

Keywords: silicon substrate; narrow linewidth; widely tunable; external cavity

1. Introduction

Silicon-substrate-integrated narrow linewidth tunable external cavity semiconductor lasers (SINLT-ECSLs) are devices composed of the substrates Si, SiO_2, Si_3N_4, or other containing Si materials and external optical feedback elements (low-loss waveguide, waveguide filter, or other elements). By adjusting the external cavity elements, such as polarizer, prism, gratings, etc., narrow linewidth and wide tuning range can be achieved. SINLT-ECSLs have the characteristics of tunable [1], narrow, or even ultra-narrow linewidth [2,3], low noise [4,5], wide application, and so on. In this paper, the device-integrated structures, chip-integrated structures of silicon-based external cavity semiconductor lasers are introduced, and, especially, the integration technology and development are introduced. Silicon-based external cavity semiconductor lasers have significant advantages over Littow and Littman configurations of the external cavity semiconductor lasers in terms of structure design, function types, and application range. In recent years, with the continuous development of optical fiber communication, coherence technology, and other fields, the silicon substrate external cavity semiconductor laser will be applied in more and more fields, with its unique characteristics, and will become the ideal light source.

2. Principle of SINLT-ECSLs

SINLT-ECSLs mainly include a semiconductor optical amplifier (SOA) and a silicon photonic chip, which are integrated through a spot size converter (SSC). An SOA is an optoelectronic device that, under suitable operating conditions, can amplify an input light signal. A schematic diagram of a basic SOA is shown in Figure 1. The active region in the device imparts gain to an input signal. An external electric current provides the energy source that enables gain to take place. An embedded waveguide is used to confine the propagating signal wave to the active region. When the light signal passes through the active region, it will cause these electrons to lose energy in the form of photons and return to the ground state. The excited photon has the same wavelength as the optical signal; thereby, the optical signal is amplified.

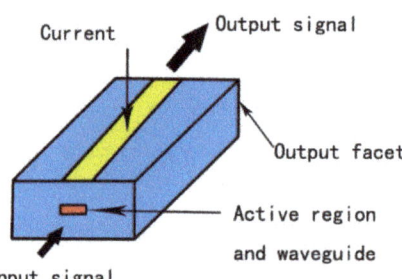

Figure 1. Schematic diagram of an SOA.

The role of the SOA is to provide gain amplification, while the silicon photonic chip is mainly for wavelength selection. Figure 2 shows the typical structure of a silicon-based external cavity semiconductor laser. The light wave of the SOA coupled to the silicon wire waveguide is filtered through two microring resonators (MRR). Two microring resonators with different radii are designed. According to Formula (1), the free spectral range (FSR) is also different due to the different radii of the microring resonators.

$$FSR = \lambda^2 / 2\pi r n_{eff} \tag{1}$$

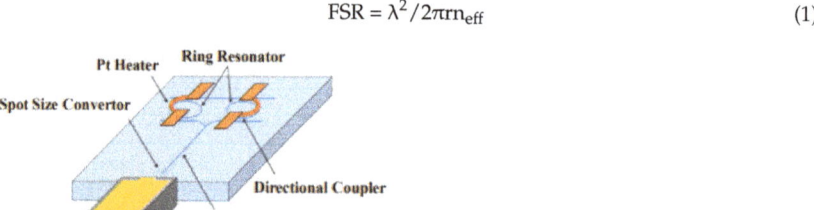

Figure 2. Typical structure of silicon substrate external cavity semiconductor laser.

In Formula (1), the wavelength of the light wave is λ, the radius of the microring resonator is r, and the effective refractive index of the waveguide is n_{eff}. The transmission spectra of the two microring resonators are superimposed on each other, and the wavelength of the mutually matched peak is determined by mode competition. The free spectral range of the microring resonator is changed by adjusting the heater through thermo-optic effect, and the transmission peak moves. The wavelength is tuned through the Vernier effect.

Figure 3 shows the working principle of wavelength tuning of a silicon-based external cavity semiconductor laser. When passing through two annular resonators with different radii, the wavelength difference of resonance results in a Vernier effect. The tuning wavelength range is determined by the radius difference between the two ring resonators. A small radius difference provides a wide wavelength-tuning range, although the transmittance difference between the main peak and the side peak adjacent to the main peak may be small. However, by heating one of the two ring resonators, the peak wavelength of the transmission spectrum of the dual-ring resonator filter changes discretely according to the resonant wavelength of the other ring resonator filter [6].

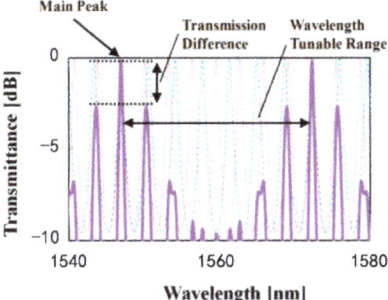

Figure 3. Working principle of wavelength tuning of silicon substrate external cavity semiconductor laser.

3. Research Progress of SINLT-ECSLs

SINLT-ECSLs can be mainly divided into the external cavity semiconductor lasers integrated with a microring resonator (MRR) [7–9], the external cavity semiconductor lasers integrated with an MRR and Mach–Zehnder interferometer (MZI), and the external cavity semiconductor lasers integrated with an MRR and others [10,11]. In order to subdivide it, MRR-integrated external cavity semiconductor lasers can be divided into double-MRR, three-MRR, and multiple-MRR integration. This paper mainly discusses the research progress of double-MRR integration and MRR-and-MZI-integrated external cavity semiconductor lasers.

3.1. MRR-Integrated External Cavity Semiconductor Laser

In 2006, Masahige Ishizaka et al. [12] reported an external cavity semiconductor laser integrated with SiO$_2$ dual MRR and SOA. The structure is shown in Figure 4, and the wavelength-tuning range is 45 nm, which can completely cover the C-band and L-band in a wavelength division multiplexing (WDM) optical communication system.

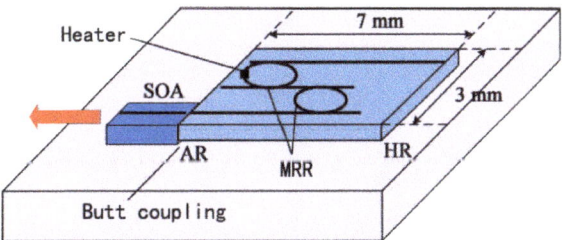

Figure 4. Structure diagram of double-MRR silicon substrate external cavity semiconductor laser.

In 2009, Takeshi Takeuchi et al. [13] reported the use of silicon waveguide (core material is SiON) three-MRR-and-SOA-integrated external cavity semiconductor laser, coupling SOA with silicon substrate through passive alignment technology, where the coupling mode is direct coupling. A waveguide reflector is used instead of high-reflection (HR) mirror to reduce the manufacturing cost. In the design of the microring structure, the threshold difference is fully considered, and the three-MRR structure is adopted. Compared with the two-MRR structure, the three-MRR structure has a larger threshold gain difference and can provide a more stable laser in a larger tuning range. The silicon-based external cavity semiconductor laser has a simple structure and is suitable for mass production. It has a high fiber output power of more than 15 dBm, is able to tune wavelengths in the 60 nm range in the L-band, and contains 147 ITU-T channels with channel spacing of 50 GHz.

In the same year, Tao Chu et al. [14] proposed a silica-based external cavity tunable laser, which is mainly integrated through dual MRR and SOA. The structure is compact, and the size of the external cavity is only 0.7×0.45 mm^2, about 1/25 of that of the traditional tunable laser. There is a wide tuning range, covering the optical communication C-band (1530–1565 nm) or L-band (1565–1610 nm), at the power of 26 mW, obtaining the maximum wavelength-tuning range of 38 nm.

Its structure is shown in Figure 5, consisting of an SOA and an external resonator. The resonator is made of silicon photonic line waveguide, and it is a double MRR. It is the first external cavity semiconductor laser made by silicon photonic technology. The ring resonator has a wide FSR due to its short cavity length. In addition, compared with the ring resonators made of SiON material, the ring resonators made of silicon photonic line waveguides have wider FSR due to their smaller bending radius of several microns. Therefore, the larger gain difference and wavelength-tuning range required for single-mode laser oscillation can be obtained more easily using a silicon ring resonator.

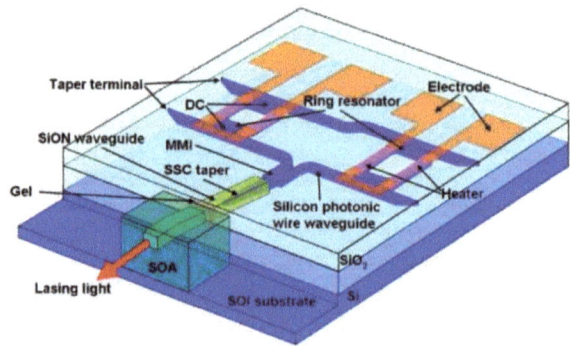

Figure 5. Structure of an external cavity tunable laser integrated with a silicon photonic line waveguide dual MRR and SOA.

In 2012, Keita Nemoto et al. [15] optimized the design of silicon substrate outer cavity semiconductor laser, using the ring resonator of silicon optical wire as the outer cavity, and produced a semiconductor laser with adjustable wavelength. The size of the ring resonator wavelength filter with outer cavity length of 6.0 mm is 1.78×0.52 mm^2, which is about 1/8 of that of silicon (SiON) material. The maximum laser output power is 18.9 MW, using heating power of 115.7 mW and tuning operation of wavelength above 45.1 nm. The spectral linewidth of the whole L-band is less than 100 kHz, which is suitable for being used as the light source of the digital coherent light transmission system. The structure is shown in Figure 6.

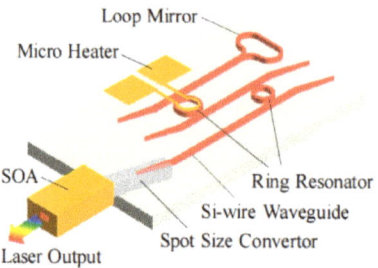

Figure 6. Schematic diagram of silicon photonic line waveguide ring cavity laser structure.

In 2013, Tomohiro Kita et al. [16] fabricated a tunable semiconductor laser with a maximum output power of 25.1 mW using a silicon photonic line waveguide ring resonator as an external optical cavity. The micro-heater can be continuously tuned to wavelengths above 50 nm, with a linewidth less than 100 kHz and a smaller size. While improving laser stability, it can be used in the actual digital coherent transmission system, as shown in Figure 7.

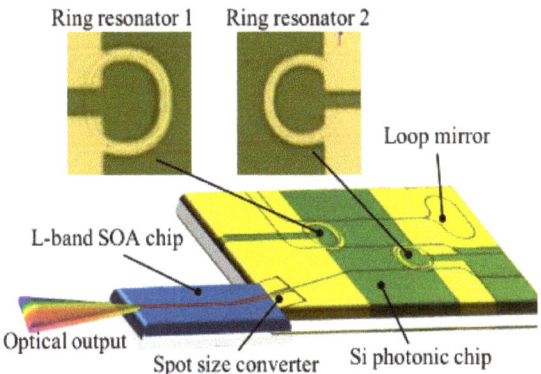

Figure 7. Schematic diagram of silicon-substrate tunable semiconductor laser.

In 2014, Sato et al. [17] integrated a silicon-based tunable filter, gain chip, and boost SOA, as shown in Figure 8. The tunable filter consists of two ring resonators, and the waveguide core of the gain part is composed of InGaAsP/InGaAsP base multiple quantum well. The laser side of the gain chip is coated with low-reflection (LR) coating and the output side of the boost SOA is coated with anti-reflection (AR) coating. The optical fiber coupling output power is greater than 100 mW, the linewidth is less than 15 kHz, the side-mode suppression ratio (SMSR) is greater than 45 dB, and the wavelength-tunable range is about 65 nm, enough to cover the entire C-band.

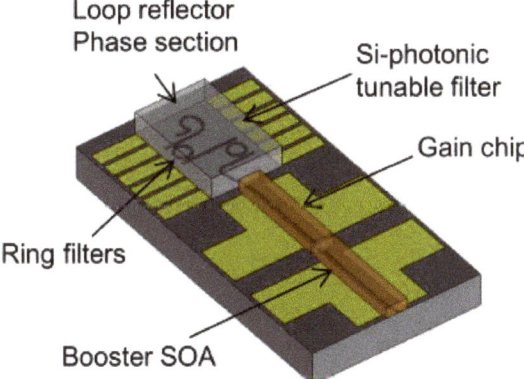

Figure 8. Schematic diagram of a silicon–photon mixed-ring external cavity tunable laser.

In 2015, Tin Komljenovic et al. [18] demonstrated a widely tunable external cavity semiconductor laser with an external cavity length of 4 cm through monolithic integration. The laser works in O-band and can be tuned in the range of 1237.7–1292.4 nm, with a tuning range of about 54 nm. Over the entire tuning range, SMSR is greater than 45 dB, output power is more than 10 mW, linewidth is less than 100 kHz, and the best single mode linewidth is 50 kHz.

In 2016, Zhao et al. [19] fabricated a low loss (0.1 dB/cm), high Q factor microring resonator based on double fringe SiN/SiO and developed a tunable InP/SiN mixed external cavity semiconductor laser, and the waveguide has good performance. The wavelength-tuning range of the laser is about 1530–1580 nm, the output power is 16 mW, the SMSR is more than 45 dB, and the linewidth is 65 kHz. It has broad application prospects in coherent transmission systems.

The schematic diagram of the laser is shown in Figure 9. It consists of a high-power InP/n-GaAsP SOA gain chip and two microring resonators. The front and back of the SOA are coated with highly reflective and AR coatings. Using the cursor effect, two MRR with slightly different radii are used to increase the wavelength-tuning range. Phase and power tuning sections are for fine-tuning longitudinal mode and output power, respectively.

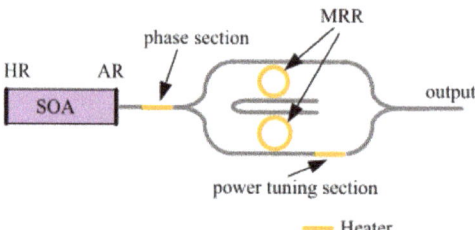

Figure 9. Schematic diagram of silicon-substrate MRR external cavity laser.

In 2017, Jing Zhang et al. [20] proposed a non-uniformly integrated, wide-tuned, unidirectional III–V ring laser on silicon. The wavelength-tuning range of the laser is 1560–1600 nm, and the ring radius of the ring resonator structure is 25 μm and 27 μm, respectively. The FSR in the wavelength range of 1550 nm is 4.1 nm and 3.7 nm, respectively. Using the Vernier effect of two ring resonators, a wide tuning range of 40 nm is obtained. Unidirectional operation is achieved throughout the tuning range, with a clockwise SMSR of about 10 dB. The linewidth is less than 1 MHz throughout the tuning range and can be reduced to 550 kHz at the optimal operating point.

In 2018, Hang Guan et al. [21] demonstrated a III–V/Si mixed external cavity laser with a tuning range greater than 60 nm, a maximum output power of 11 mW, a minimum linewidth of 37 kHz, a C-band that is always less than 80 kHz, a maximum SMSR of 55 dB, and a C-band that is always greater than 46 dB. It consists of a reflective external cavity constructed by RSOA, SSC, and a ring resonator.

In the same year, Minha Tran et al. [22] designed and manufactured a narrow-linewidth tuned laser with multi-ring mirrors. The structure is shown in Figure 10, including two-ring mirrors and three-ring mirrors. In heterogeneous silicon photons, a laser using a three-ring mirror was implemented, with an average SMSR of 55 dB in the 30 nm tuning range and a linewidth reduced to 17.5 kHz.

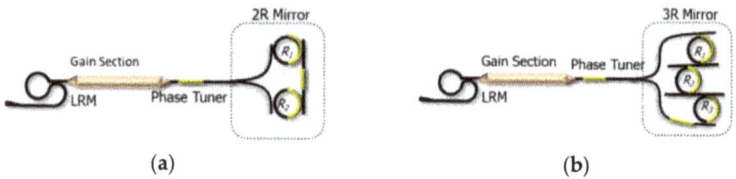

Figure 10. (a) Schematic diagram of a double-MRR tunable laser; (b) schematic diagram of a triple-MRR tunable laser.

In 2019, Yongkang Gao et al. [23] demonstrated a miniaturized packaged hybrid integrated silicon–photon (SiPh) tunable laser for coherent modules of small size. By integrating an internally designed high-power SOA, the SiPh laser developed achieved a record 21.5 dBm C-band output power with a linewidth of 60 kHz, an SMSR greater than 50 dB, a relative intensity noise less than 150 dB/Hz, and a tuning range of 65 nm. In addition, frequency stability of the SiPh tunable laser at 1 GHz was achieved over the package temperature range from 10 °C to 80 °C and over SOA current variations of more than 200 mA, as shown in Figure 11.

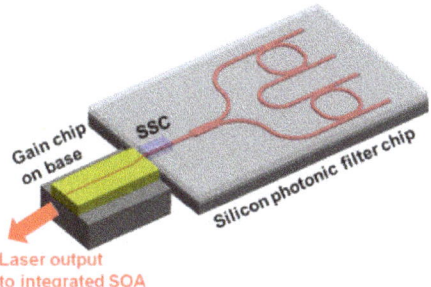

Figure 11. Schematic diagram of a hybrid integrated SiPh tunable laser.

In 2020, Jia Xu Brian Sia et al. [24] studied a cursors-based hybrid silicon–photonic tunable laser. By tuning the wavelength by MRR, the slope efficiency of the laser is 0.232 W/A, the output power is 28 mW, and the SMSR is 42 dB. When the thermal power of a single MRR reaches 47.2 mW, it can be tuned in the wavelength range of 1881–1947 nm, and the tuning range is 66 nm. In the same year, Jia Xu Brian Sia et al. [25] also reported a III–V/Si mixed wavelength-tunable laser with a working wavelength of 1647–1690 nm and a tuning range of 53 nm. Room-temperature continuous wave operation is realized, with output power up to 31.1 mW and corresponding maximum SMSR of 46.01 dB. The laser is hypercoherent with an estimated linewidth of 0.7 kHz, extending the coverage of the III–V/Si hybrid laser with a sub-kHz linewidth to the 1650 nm wavelength region beyond the L-band.

In 2021, Ruiling Zhao et al. [26] demonstrated a dual-gain InP–Si_3N_4 mixed external cavity laser, whose structure is shown in Figure 12. The working wavelength of the laser is 1550 nm, the tuning range is 44 nm of the working wavelength, the linewidth is 6.6 kHz, SMSR is greater than 67 dB, and the two gain parts work in parallel, thus providing high output power; the maximum power is about 23.5 mW.

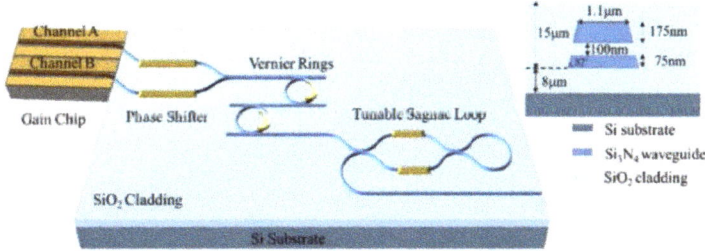

Figure 12. Schematic structure of tunable INP Si_3N_4 mixed external cavity laser with double parallel gain (the inset shows a cross-section of the Si_3N_4 waveguide).

In the same year, Yuyao Guo et al. [27] introduced a III–V/Si_3N_4 hybrid integrated laser with faster switching time, and its structure is shown in Figure 13. The working wavelength of the laser is 1516.5–1575 nm, the tuning range is 58.5 nm, the linewidth is 2.5 kHz, and the SMSR is greater than 70 dB. The maximum output power is 34 mW at 500 mA injection current.

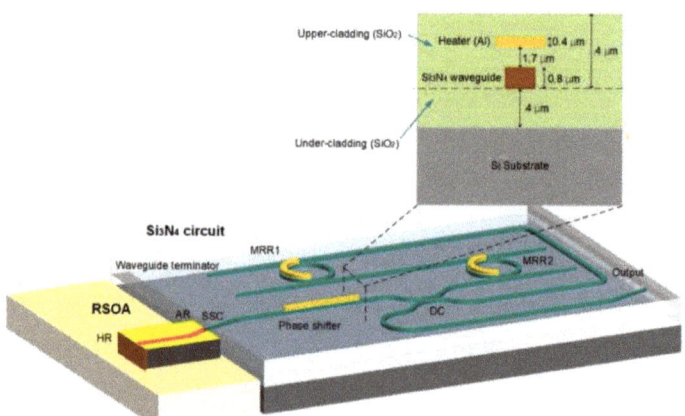

Figure 13. Schematic structure of the III–V/Si$_3$N$_4$ hybrid laser (illustration shows a cross section of a thermally tunable Si$_3$N$_4$ waveguide).

3.2. MRR-and-MZI-Integrated External Cavity Semiconductor Lasers

The principle of how MZI works is shown in Figure 14. D1 and D2 are detectors, BS1 and BS2 are beam splitters, and M1 and M2 are mirrors. A beam splitter, BS1, splits an incoming monochromatic light beam from source S into two beams, which, after reflection by mirrors M1 and M2, recombine and interfere at BS2 to result in two outgoing beams (collected by detectors D1 and D2). When the phase along one of the paths varies, signals in both D1 and D2 oscillate out of phase, and as no photons are being lost, the sum of both signals stays always equal to the input, S.

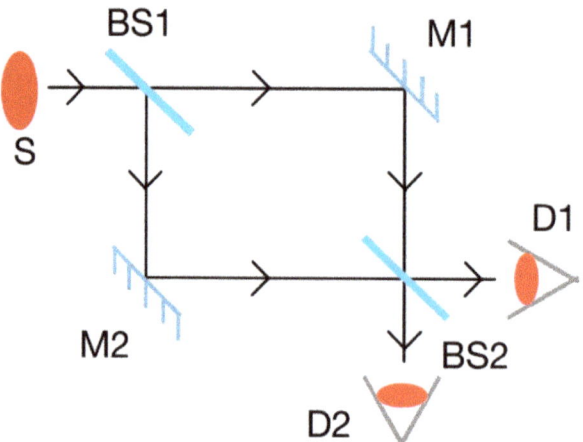

Figure 14. Principle of how MZI works.

In 2014, Debregeas et al. [28] proposed an integrated tunable laser that combines a reflective SOA (RSOA) with a silicon ring resonator-based outer cavity and MZI. The structure is shown in Figure 15. The external cavity of the laser is composed of two MRRs. The first ring is set to 25 GHz and the second ring is integrated with MZI.

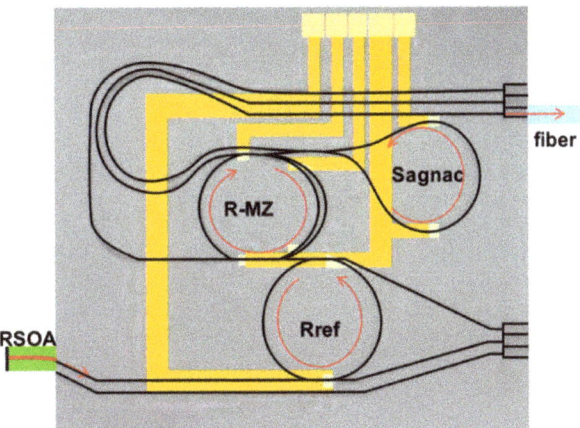

Figure 15. Structure diagram of RSOA and silicon substrate external cavity hybrid integrated laser.

In the same year, Tomohiro Kita et al. [29] fabricated a wavelength-tunable laser using a silicon photonic wavelength filter consisting of a ring resonator and an asymmetric MZI (A-MZI), as shown in Figure 16. The size, including the SOA, is very small, only 2.6 × 0.5 mm², about 1/9 of the size of the silicon nitrous tunable laser. The wavelength-tuning range is more than 61.7 ± 0.2 nm, covering the whole optical communication L-band, and SMSR is more than 38 dB. When the SOA injection current is 300 mA, the maximum optical output power is 42.2 mW, achieving a stable single-mode laser output. By optimizing the outer cavity design, a spectral linewidth of less than 100 kHz is obtained.

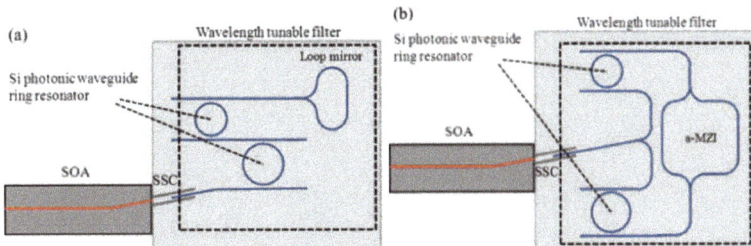

Figure 16. Schematic diagram of a wavelength-tunable silicon photon laser: (**a**) series configuration of ring resonators; (**b**) ring structure of ring resonators with A-MZI.

In 2015, Tomohiro Kita et al. [30] proposed a wavelength-tunable laser using silicon photonics to create a compact wavelength-tunable filter with high wavelength selectivity. Two ring resonators and A-MZI are used to realize a silicon photonic wavelength-tunable filter with a wide wavelength-tuning range. A wavelength-tunable laser made by docking a silicon photonic filter and an SOA achieved stable single-mode operation in a wide wavelength range. The size of the chip is 2.5 × 0.6 mm², the laser threshold is 25 mA, the maximum fiber coupling output power is 8.9 mW, and the maximum output power is estimated to be 35 mW. The tuning wavelength range is 99.2 nm (1527.9–1627.1 nm), covering both C-band and L-band. Through the fine control of heating power, the side mode rejection ratio is greater than 29 dB.

Figure 17 is a schematic diagram of a silicon photonic tunable laser. The light from the SOA is filtered using two MRRs with slightly different FSRs and an A-MZI, and the FSR is about twice as large as the ring resonator FSRs. Laser wavelength is determined by the Vernier effect between two ring resonators. Wavelength selectivity is defined as the transmittance difference between the main mode and the nearest mode. A large transmittance difference can achieve a stable single-mode laser.

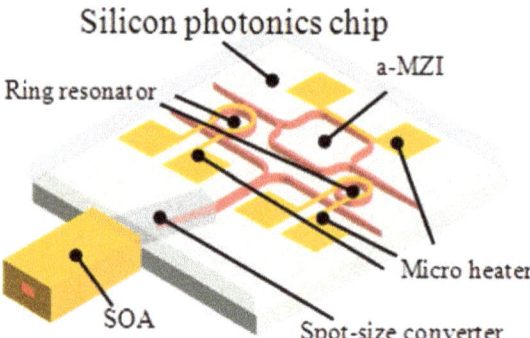

Figure 17. Schematic diagram of a silicon photonic tunable laser.

In the same year, Rui Tang et al. [31] proposed a narrow line-width silicon photonic tunable laser with a high A-MZI. The laser consists of two silicon ring resonators with different perimeters and a highly asymmetric MZI, with significantly different optical path lengths. The calculation and experimental results show that the high A-MZI increases the gain difference between the longitudinal modes. The result is a stable single-mode oscillation with a narrow band width of 12 kHz, which can be tuned in the wavelength range of 42.7 nm. The surface structure can also be applied to other ring resonator filters, regardless of the waveguide type.

The basic structure of the laser is shown in Figure 18. Both structures consist of an SOA and an external wavelength-tunable filter. SOA is a gain medium in C-band, and the filter consists of two ring resonators with different perimeters. The Vernier effect of two ring resonators is used to roughly select the oscillation wavelength.

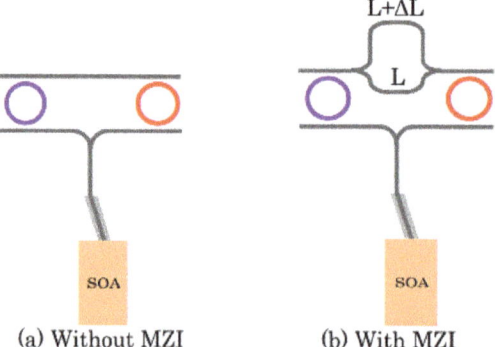

Figure 18. (a) Silicon-based laser without MZI. (b) Silicon-based laser with MZI.

In 2020, Aditya Malik et al. [32] proposed a widely tunable quantum dot laser heterointegrated on a silicon insulator substrate, and its structure is shown in Figure 19. The tuning mechanism is based on the Vernier double-ring geometry, and the tuning range is 47 nm at 52 dB SMSR. When the wavelength filter in the form of MZI is added to the cavity, the SMSR is increased to 58 dB, the tuning range is increased to 52 nm, and the linewidth is as low as 5.3 kHz.

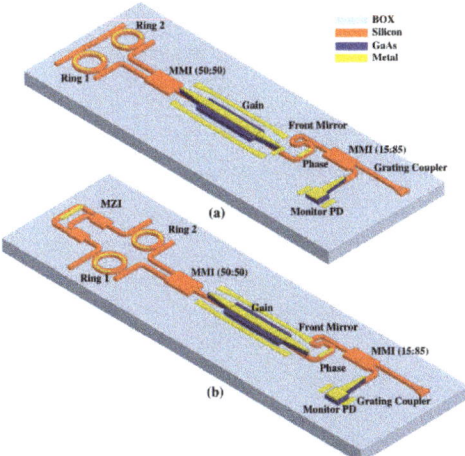

Figure 19. Schematic diagram of (**a**) and (**b**) double-ring cursors with MZI tunable lasers.

For Vernier ring lasers, the linewidth is in the range of 10–20 kHz, but due to the poor SMSR at the edge of the gain spectrum, the linewidth is as high as 50 kHz when the output wavelength is close to 1290 nm. When MZI is used, better SMSR can be obtained, so the linewidth is always less than 10 kHz in the total tuning range.

The research development on performance of SINLT-ECSLs in recent years is listed in Table 1. Compared with MRR-integrated external cavity semiconductor lasers, MRR-and-MZI-integrated external cavity semiconductor lasers have the characteristics of narrow-band filtering, which can minimize the transmittance of adjacent wavelengths at the maximum transmittance and improve the wavelength selectivity in the waveguide. By changing the temperature of the material through the micro-heater, the refractive index of the waveguide is changed, and wider wavelength-tuning range is realized. Therefore, MRR-and-MZI-integrated external cavity semiconductor lasers can make the lasers obtain narrower linewidth, wider wavelength-tuning range, and higher SMSR.

Table 1. A survey displaying developments in the field of the SINLT-ECSLs.

Laser Type	Band	λ (nm)	Tuning Range (nm)	Linewidth (kHz)	SMSR (dB)	Output Power mW	Output Power dBm	Year
MRR	C, L	-	45.2	-	>45	>2.9 *	>4.7	2006 [12]
MRR	C, L	1540–1636	96	-	>50	>20 *	>13	2009 [13]
MRR	C, L	1530–1610	38	-	>30	26	14.1 *	2009 [14]
MRR	L	-	45.1	<100	>40	18.9	12.8 *	2012 [15]
MRR	L	-	53	<100	>25	25.1	14 *	2013 [16]
MRR	C, L	1510–1575	65	<15	>45	>100 *	>20	2014 [17]
MRR and MZI	C, L	-	35	2	>60	3.2 *	5	2014 [28]
MRR and MZI	L	-	61.7±0.2	<100	>38	42.2	16.3 *	2014 [29]
MRR	O	1237.7–1292.4	54.7	<100	>45	10	10 *	2015 [18]
MRR and MZI	C, L	1527.9–1627.1	99.2	-	>29	35	15.4 *	2015 [30]
MRR and MZI	C, L	-	42.7	12	-	30	14.8 *	2015 [31]
MRR	C, L	1530–1580	50	65	>45	16	12 *	2016 [19]
MRR	C, L	1560–1600	40	<1000	10	>1.4 *	>1.5	2017 [20]

Table 1. Cont.

Laser Type	Band	λ (nm)	Tuning Range (nm)	Linewidth (kHz)	SMSR (dB)	Output Power		Year
						mW	dBm	
MRR	C	-	60	<80	>46	11	10.4 *	2018 [21]
MRR	C, L	1557–1587	30	17.5	>55	6.9	8.4 *	2018 [22]
MRR	C	-	65	60	>50	141.3 *	21.5	2019 [23]
MRR	-	1881–1947	66	-	42	28	14.5 *	2020 [24]
MRR	-	1647–1690	43	0.7	46	31.1 *	14.9	2020 [25]
MRR and MZI	O	-	52	5.3	58	10	10 *	2020 [32]
MRR	C, L	1524–1568	44	6.6	>67	23.5 *	13.7	2021 [26]
MRR	C, L	1516.5–1575	58.5	2.5	>70	34 *	15.3	2021 [27]

Legend: *—Calculated.

4. Integration of SINLT-ECSLs

4.1. Monolithic Integrated

Monolithic integration mainly refers to the direct epitaxial growth of group III–V compound semiconductor materials on the silicon substrate and synchronous device fabrication process. Due to the high-density thread dislocation in heteroepitaxy, the laser device performance and reliability will be poor due to the direct growth on silicon [33]. However, the gain characteristics can be fine-tuned by changing the growth conditions, so that the device has a long life even when it is epitaxial grown on silicon with high dislocation density [34]. For example, Chen et al. realized high-performance quantum dot lasers on silicon by combining the nucleation layer and dislocation filter layer with in situ thermal annealing and adopting the molecular beam epitaxy (MBE) epitaxial growth method to achieve high-quality GaAS-on-Si layer with low defects. The large lattice mismatch between III–V materials and silicon is no longer an obstacle to the single epitaxial growth of III–V photonic devices on silicon substrates, demonstrating the ability to grow uniformly high-quality III–V materials on the entire Si substrate, which is a significant advance in silicon-based photonics and optoelectronics integration [35].

In 2020, Bahawal Haq et al. [36] produced a C-band monolithic integration laser; its structure is shown in Figure 20. At 20 °C, the threshold current of 80 mA and the maximum single-waveguide coupled output power exceeding 6.9 mW are obtained, with the slope efficiency of 0.27 W/A and the SMSR greater than 33 dB.

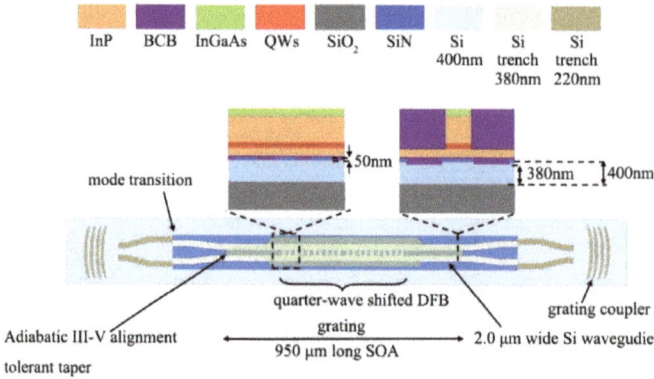

Figure 20. Schematic structure of monolithic integration laser.

4.2. Heterogeneous Integration

Herointegration [37] refers to the integration of an epitaxial growth group III–V compound semiconductor with silicon substrate through bonding technology, and then the device manufacturing process. Bonding can be divided into direct bonding, adhesive bonding, and metal bonding. Similar to hybrid integration, heterogeneous integration has the advantage of selecting the best materials for each function (i.e., lasers, low-loss waveguides, detectors), resulting in highly complex photonic integrated circuits (PIC). Thus, heterogeneous integration has all the scaling advantages of monolithic integration, while gaining greater flexibility in material selection, resulting in superior performance. However, because the output power of heterogeneous integration is relatively low, it cannot be well applied in multi-channel communication, and the whole processing process is complicated, the bonding repetition rate is not high, and it is difficult to carry out large-scale mass production.

In 2018, Sulakshna Kumari et al. [38] designed a heterogeneously integrated continuous-wave electrically-pumped vertical-cavity Si-integrated laser (VCSIL). Its structure is shown in Figure 21. The VCSIL structure consists of two distinct parts, called the upper half and the lower half. The upper part of the structure is a semi-vertical cavity surface-emitting laser based on GaAs, oxide layer, DBR, and gold film. The lower part of the structure is a SiN waveguide/dielectric DBR combination on a silicon substrate. SiO_2 cladding at the top and bottom prevents the waveguide mode from leaking into the high-refractive-index GaAs semi-vertical cavity surface-emitting laser and the high-refractive-index dielectric DBR and Si substrates. A VCSIL with a 5 μm oxide aperture diameter has a threshold current of 1.13 mA and produces a maximum single-sided waveguide coupled output power of 73 μW at 856 nm. The slope efficiency and the thermal impedance of the corresponding device is 0.085 W/A and 11.8 K/mW, respectively. The SMSR is 29 dB at a bias current of 2.5 mA.

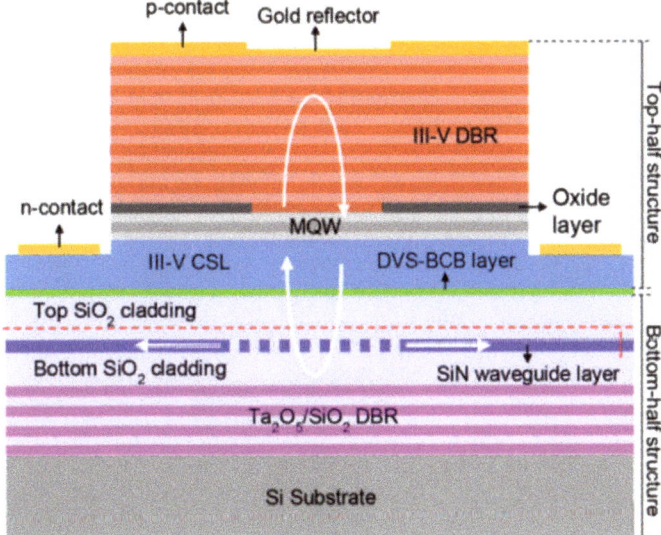

Figure 21. Schematic cross-section of heterogeneously integrated VCSIL.

In 2020, Chao Xiang et al. [39] designed a multilayer heterointegrated III–V/Si/Si_3N_4 laser structure, which can achieve a high-efficiency electro-pumped laser in a fully integrated Si_3N_4-based outer cavity. The structure is shown in Figure 22. The linewidth of the laser is 6 kHz, and it has good temperature stability and low phase noise. The Si_3N_4 spiral grating provides a narrowband filter together with high extinction ratio. This results in a large lasing SMSR of over 58 dB.

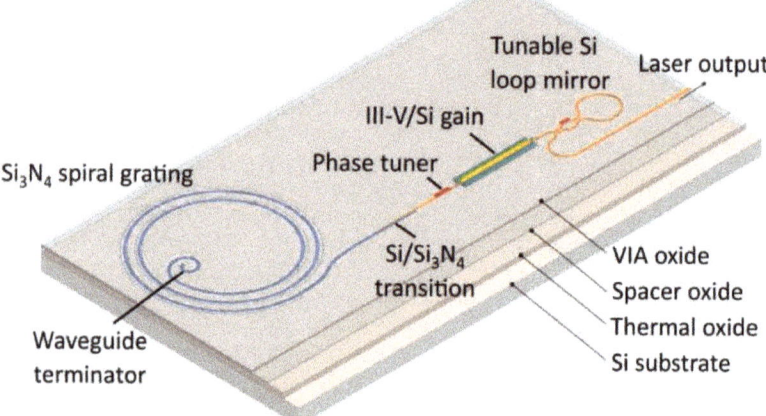

Figure 22. Schematic structure of hetero integrated laser.

4.3. Hybrid Integrated

Hybrid integration refers to the integrated assembly of the III–V laser chip and silicon substrate. These can be achieved with prior technology, but there is a drawback that two devices of different sizes or materials must be aligned to sub-micron accuracy for effective coupling [40]. However, it can also be improved in certain ways to avoid defects as much as possible. For example, Alexander W. Fang et al. [41] designed a special structure that can be performed at the wafer, partial wafer, or wafer level so that multiple lasers do not require any critical alignment of the silicon waveguide with III–V materials. In addition, this highly scalable structure can be extended to other active devices on silicon, such as optical amplifiers, modulators, and optical detectors, by selectively changing the III–V structure through processes such as quantum well mixing or non-planar wafer bonding.

In 2020, Yeyu Zhu et al. [42] designed the hybrid integration of low-loss passive Si_3N_4 outer cavity and 1.3 μm quantum dot RSOA, and its structure is shown in Figure 23. Chip-scale, tunable, narrow-linewidth hybrid-integrated diode lasers based on quantum-dot RSOAs at 1.3 μm are demonstrated through butt-coupling to a silicon–nitride photonic integrated circuit. The hybrid laser linewidth is around 85 kHz, and the tuning range is around 47 nm.

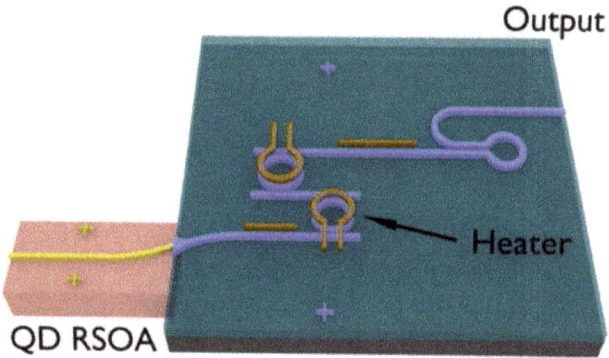

Figure 23. Schematic structure of hybrid-integrated laser.

In 2021, Yilin Xu et al. [43] designed a new hybrid-integrated laser, with the structure shown in Figure 24. The device consists of an InP-based RSOA that is connected to a thermally tunable feedback circuit on a silicon photonic (SiP) chip. A photonic wire bond connects the facet of the RSOA to the SiP external-cavity feedback circuit. The assembly is built on a metal submount that simultaneously acts as an efficient heat sink. The photonic wire bonding can be written in situ in a fully automated process and is shaped to fit the size of the mode field and the positions of the chips at both ends, thus providing low loss coupling even with limited placement accuracy. It demonstrates a tuning range from 1515 to 1565 nm along with side-mode suppression ratios above 40 dB and intrinsic linewidths down to 105 kHz. The approach combines the scalability advantages of monolithic integration with the performance and flexibility of hybrid multi-chip assemblies and may thus open a path towards integrated external cavity semiconductor lasers on a wide variety of integration platforms.

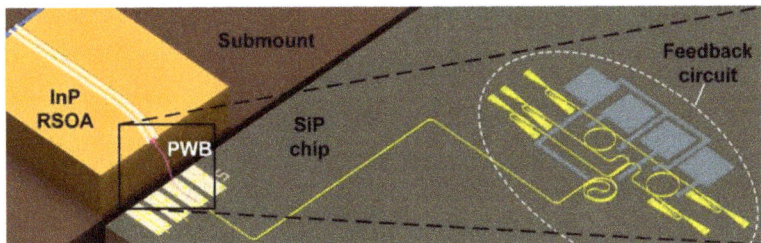

Figure 24. Schematic structure of hybrid-integrated laser using photonic line bonds as intracavity coupling elements.

5. Conclusions

SINLT-ECSLs are developing towards wider tuning range, narrower linewidth, and higher side-mode rejection ratio. Through the selection of gain media, materials, integrated devices, etc., and the design of new silicon-based outer cavity structure, the epitaxial design of SOA is improved, the loss of silicon waveguide is reduced [44], the coupling efficiency is increased, the reflectivity is reduced, and the stability is enhanced, so as to meet the application requirements in various fields. High-performance silicon-based external cavity semiconductor lasers with narrow linewidth or even ultra-narrow linewidth, wide tuning range, stable output, low noise, small volume, and low cost are realized. With the development of the information age, silicon-based external cavity semiconductor lasers will have a broader application market in optical communication, coherent detection, and other fields [45]. How to realize wide tuning range, high power, single mode, stable spectral output, and high SMSR laser output is a main research direction for the future development of external cavity semiconductor lasers.

In summary, the advantages of the SINLT-ECSLs over solitary diode lasers are so important that their future looks encouraging. This conclusion becomes still better supported if we consider multichannel external cavity lasers that are particularly interesting for optical interconnection applications. New materials and configurations appear that make these lasers still more attractive due to extension of the operation spectral range and high power. Last, but not least, the SINLT-ECSLs have the potential to be fabricated cheaply enough to promote more applications.

Author Contributions: Conceptualization, X.L. and J.S.; methodology, K.D. and Y.M.; writing—original draft preparation, X.L. and Z.L.; writing—review and editing, L.W., L.L. and L.Z.; visualization, Y.Q., Z.Q. and G.L.; supervision, Z.Q.; funding acquisition, G.L. All authors have read and agreed to the published version of the manuscript.

Funding: This work was supported in part by the Major Science and Technology Program of Hainan Province of China under Grant ZDKJ2019005; in part by Scientific Research Projects of Higher Education Institutions in Hainan Province under Grant hnky2020-24, Grant Hnjg2021ZD-22, Grant hnky2020ZD-12; in part by the Hainan Provincial Natural Science Foundation of China under Grant 120MS031, Grant 2019RC190, Grant 2019RC192; in part by the Special Research Project of Hainan Academician Innovation Platform under Grant SPTZX202034; in part by the National Natural Science Foundation of China under Grant 61774024, Grant 61864002, Grant 11764012, Grant 62174046, Grant 62064004 and Grant 61964007; in part by the Key Research and Development Projects in Hainan Province under Grant ZDYF2020020, Grant ZDYF2020036, and Grant ZDYF2020217; in part by specific research fund for Innovation Platform for Academicians of Hainan Province under Grant YSPTZX202034 and Grant YSPTZX202127; in part by Open Fund for Innovation and Entrepreneurship of college students under Grant 202111658021X, Grant 202111658022X, Grant 202111658023X, Grant 202111658013.

Institutional Review Board Statement: Not applicable.

Informed Consent Statement: Not applicable.

Data Availability Statement: Not applicable.

Acknowledgments: The authors thank Dongxin Xu, Hao Chen, and Yanbo Liang for helping with this article.

Conflicts of Interest: The authors declare no conflict of interest.

References

1. Verdier, A.; de Valicourt, G.; Brenot, R.; Debregeas, H.; Dong, P.; Earnshaw, M.; Carrère, H.; Chen, Y. Ultrawideband wavelength-tunable hybrid external-cavity lasers. *J. Lightwave Technol.* **2017**, *36*, 37–43. [CrossRef]
2. Tran, M.A.; Huang, D.; Guo, J.; Komljenovic, T.; Morton, P.A.; Bowers, J.E. Ring-resonator based widely-tunable narrow-linewidth Si/InP integrated lasers. *IEEE J. Sel. Top. Quantum Electron.* **2019**, *26*, 1500514. [CrossRef]
3. Zhu, Y.; Zhu, L. Narrow-linewidth, tunable external cavity dual-band diode lasers through InP/GaAs-Si$_3$N$_4$ hybrid integration. *Opt. Express* **2019**, *27*, 2354–2362. [CrossRef] [PubMed]
4. Malik, A.; Xiang, C.; Chang, L.; Jin, W.; Guo, J.; Tran, M.; Bowersa, J. Low noise, tunable silicon photonic lasers. *Appl. Phys. Rev.* **2021**, *8*, 031306. [CrossRef]
5. Liang, W.; Ilchenko, V.S.; Eliyahu, D.; Savchenkov, A.A.; Matsko, A.B.; Seidel, D.; Maleki, L. Ultralow noise miniature external cavity semiconductor laser. *Nat. Commun.* **2015**, *6*, 7371. [CrossRef]
6. Suzuki, K.; Kita, T.; Yamada, H. Wavelength tunable laser diodes with Si-wire waveguide ring resonator wavelength filters. In *Silicon Photonics VI*; International Society for Optics and Photonics: San Francisco, CA, USA, 2011; Volume 7943.
7. Bowers, J.E.; Komljenovic, T.; Srinivasan, S.; Hulme, J.; Davenport, M. A comparison of widely tunable, narrow linewidth ring cavity lasers on silicon substrates. In Proceedings of the 2015 IEEE Photonics Conference (IPC), Reston, VA, USA, 4–8 October 2015; pp. 533–534.
8. Lee, J.; Bovington, J.; Shubin, I.; Luo, Y.; Yao, J.; Lin, S.; Cunningham, J.E.; Raj, K.; Krishnamoorthy, A.V.; Zheng, X. 12.2% waveguide-coupled wall plug efficiency in single mode external-cavity tunable Si/III–V hybrid laser. In Proceedings of the 2015 IEEE Optical Interconnects Conference (OI), San Diego, CA, USA, 20–22 April 2015; pp. 142–143.
9. Srinivasan, S.; Davenport, M.; Komljenovic, T.; Hulme, J.; Spencer, D.T.; Bowers, J.E. Coupled-ring-resonator-mirror-based heterogeneous III–V silicon tunable laser. *IEEE Photonics J.* **2015**, *7*, 2700908. [CrossRef]
10. Wang, R.; Sprengel, S.; Vasiliev, A.; Boehm, G.; van Campenhout, J.; Lepage, G.; Verheyen, P.; Baets, R.; Amann, M.; Roelkens, G. Widely tunable 2.3 µm III-V-on-silicon Vernier lasers for broadband spectroscopic sensing. *Photonics Res.* **2018**, *6*, 858–866. [CrossRef]
11. Hulme, J.C.; Doylend, J.K.; Bowers, J.E. Widely tunable Vernier ring laser on hybrid silicon. *Opt. Express* **2013**, *21*, 19718–19722. [CrossRef]
12. Ishizaka, M.; Yamazaki, H. Wavelength tunable laser using silica double ring resonators. *Electron. Commun. Jpn.* **2006**, *89*, 34–41. [CrossRef]
13. Takeuchi, T.; Takahashi, M.; Suzuki, K.; Watanabe, S.; Yamazaki, H. Wavelength tunable laser with silica-waveguide ring resonators. *IEICE Trans. Electron.* **2009**, *92*, 198–204. [CrossRef]
14. Chu, T.; Fujioka, N.; Ishizaka, M. Compact, lower-power-consumption wavelength tunable laser fabricated with silicon photonic wire waveguide micro-ring resonators. *Opt. Express* **2009**, *17*, 14063–14068. [CrossRef] [PubMed]
15. Nemoto, K.; Kita, T.; Yamada, H. Narrow spectral linewidth wavelength tunablelaser with Si photonic-wire waveguide ring resonators. In Proceedings of the 9th International Conference on Group IV Photonics (GFP), San Diego, CA, USA, 29–31 August 2012; pp. 216–218.
16. Kita, T.; Nemoto, K.; Yamada, H. Narrow spectral linewidth and high output power Si photonic wavelength tunable laser diode. In Proceedings of the 10th International Conference on Group IV Photonics, Seoul, Korea, 28–30 August 2013; pp. 152–153.

17. Kobayashi, N.; Sato, K.; Namiwaka, M.; Yamamoto, K.; Watanabe, S.; Kita, T.; Yamada, H.; Yamazaki, H. Silicon photonic hybrid ring-filter external cavity wavelength tunable lasers. *J. Lightwave Technol.* **2015**, *33*, 1241–1246. [CrossRef]
18. Komljenovic, T.; Srinivasan, S.; Norberg, E.; Davenport, M.; Fish, G.; Bowers, J.E. Widely tunable narrow-linewidth monolithically integrated external-cavity semiconductor lasers. *IEEE J. Sel. Top. Quantum Electron.* **2015**, *21*, 214–222. [CrossRef]
19. Zhao, J.L.; Oldenbeuving, R.M.; Epping, J.P.; Hoekman, M.; Heideman, R.G.; Dekker, R.; Fan, Y.; Boller, K.-J.; Ji, R.Q.; Fu, S.M.; et al. Narrow-linewidth widely tunable hybrid external cavity laser using Si_3N_4/SiO_2 microring resonators. In Proceedings of the 2016 IEEE 13th International Conference on Group IV Photonics (GFP), Shanghai, China, 24–26 August 2016; pp. 24–25.
20. Zhang, J.; Li, Y.; Dhoore, S.; Morthier, G.; Roelkens, G. Unidirectional, widely-tunable and narrow-linewidth heterogeneously integrated III-V-on-silicon laser. *Opt. Express.* **2017**, *25*, 7092–7100. [CrossRef]
21. Guan, H.; Novack, A.; Galfsky, T.; Ma, Y.; Fathololoumi, S.; Horth, A.; Huynh, T.N.; Roman, J.; Shi, R.; Caverley, M.; et al. Widely-tunable, narrow-linewidth III-V/silicon hybrid external-cavity laser for coherent communication. *Opt. Express* **2018**, *26*, 7920–7933. [CrossRef]
22. Tran, M.A.; Huang, D.; Komljenovic, T.; Liu, S.; Liang, L.; Kennedy, M.J.; Bowers, J.E. Multi-ring mirror-based narrow-linewidth widely-tunable lasers in heterogeneous silicon photonics. In Proceedings of the 2018 European Conference on Optical Communication (ECOC), Rome, Italy, 23–27 September 2018.
23. Gao, Y.; Lo, J.; Lee, S.; Patel, R.; Zhu, L.; Nee, J.; Tsou, D.; Carney, R.; Sun, J. High-power, narrow-linewidth, miniaturized silicon photonic tunable laser with accurate frequency control. *J. Lightwave Technol.* **2020**, *38*, 265–271. [CrossRef]
24. Sia, J.X.B.; Wang, W.; Qiao, Z.; Li, X.; Guo, T.X.; Zhou, J.; Littlejohns, C.G.; Liu, C.; Reed, G.T.; Wang, H. Analysis of Compact Silicon Photonic Hybrid Ring External Cavity (SHREC) Wavelength-Tunable Laser Diodes Operating From 1881–1947 nm. *IEEE J. Quantum Electron.* **2020**, *56*, 2001311. [CrossRef]
25. Sia, J.X.B.; Li, X.; Wang, W.; Qiao, Z.; Guo, X.; Zhou, J.; Littlejohns, C.G.; Liu, C.; Reed, G.T.; Wang, H. Sub-kHz linewidth, hybrid III-V/silicon wavelength-tunable laser diode operating at the application-rich 1647–1690 nm. *Opt. Express* **2020**, *28*, 25215–25224.
26. Zhao, R.; Guo, Y.; Lu, L.; Nisar, M.S.; Chen, J.; Zhou, L. Hybrid dual-gain tunable integrated InP-Si_3N_4 external cavity laser. *Opt. Express* **2021**, *29*, 10958–10966. [CrossRef]
27. Guo, Y.; Zhao, R.; Zhou, G.; Lu, L.; Stroganov, A.; Nisar, M.S.; Chen, J.; Zhou, L. Thermally Tuned High-Performance III-V/Si_3N_4 External Cavity Laser. *IEEE Photonics J.* **2021**, *13*, 1500813. [CrossRef]
28. Debregeas, H.; Ferrari, C.; Cappuzzo, M.A.; Klemens, F.; Keller, R.; Pardo, F.; Bolle, C.; Xie, C.; Earnshaw, M.P. 2kHz linewidth C-band tunable laser by hybrid integration of reflective SOA and SiO_2 PLC external cavity. In Proceedings of the 2014 International Semiconductor Laser Conference, Palma de Mallorca, Spain, 7–10 September 2014; pp. 50–51.
29. Kita, T.; Nemoto, K.; Yamada, H. Silicon photonic wavelength-tunable laser diode with asymmetric Mach–Zehnder interferometer. *IEEE J. Sel. Top. Quantum Electron.* **2013**, *20*, 344–349. [CrossRef]
30. Kita, T.; Tang, R.; Yamada, H. Compact silicon photonic wavelength-tunable laser diode with ultra-wide wavelength tuning range. *Appl. Phys. Lett.* **2015**, *106*, 111104. [CrossRef]
31. Tang, R.; Kita, T.; Yamada, H. Narrow-spectral-linewidth silicon photonic wavelength-tunable laser with highly asymmetric Mach–Zehnder interferometer. *Opt. Lett.* **2015**, *40*, 1504–1507. [CrossRef] [PubMed]
32. Malik, A.; Guo, J.; Tran, M.A.; Kurczveil, G.; Liang, D.; Bowers, J.E. Widely tunable, heterogeneously integrated quantum-dot O-band lasers on silicon. *Photonics Res.* **2020**, *8*, 1551–1557. [CrossRef]
33. Liu, A.Y.; Zhang, C.; Norman, J.; Snyder, A.; Lubyshev, D.; Fastenau, J.M.; Liu, A.W.K.; Gossard, A.C.; Bowers, J.E. High performance continuous wave 1.3 μm quantum dot lasers on silicon. *Appl. Phys. Lett.* **2014**, *104*, 041104. [CrossRef]
34. Norman, J.C.; Jung, D.Z.Z.; Wan, Y.; Liu, S.; Shang, C.; Herrick, R.W.; Chow, W.W.; Gossard, A.C.; Bowers, J.E. A review of high-performance quantum dot lasers on silicon. *IEEE J. Quantum Electron.* **2019**, *55*, 2000511. [CrossRef]
35. Chen, S.; Li, W.; Wu, J.; Jiang, Q.; Tang, M.; Shutts, S.; Elliott, S.N.; Sobiesierski, A.; Seeds, A.J.; Ross, I.; et al. Electrically pumped continuous-wave III–V quantum dot lasers on silicon. *Nat. Photonics* **2016**, *10*, 307–311. [CrossRef]
36. Hay, D.; Vaskasi, J.R.; Zhang, J.; Gocalinska, A.; Pelucchi, E.; Corbett, B.; Roelkens, G. Micro-transfer-printed III-V-on-silicon C-band distributed feedback lasers. *Opt. Express* **2020**, *28*, 32793–32801. [CrossRef] [PubMed]
37. Grund, D.W.; Ejzak, G.A.; Schneider, G.J.; Murakowski, J.; Prather, D.W. Heterogenous integrated silicon-photonic module for producing widely tunable narrow linewidth RF. In Proceedings of the 2013 IEEE International Topical Meeting on Microwave Photonics (MWP), Alexandria, VA, USA, 28–31 October 2013; pp. 100–103.
38. Kumari, S.; Haglund, E.P.; Gustavsson, J.S.; Larsson, A.; Roelkens, G.; Baets, R.G. Vertical-Cavity Silicon-Integrated Laser with In-Plane Waveguide Emission at 850 nm. *Laser Photonics Rev.* **2018**, *12*, 1700206. [CrossRef]
39. Xiang, C.; Jin, W.; Guo, J.; Peters, J.D.; Kennedy, M.J.; Selvidge, J.; Morton, P.A.; Bowers, J.E. Narrow-linewidth III-V/Si/Si_3N_4 laser using multilayer heterogeneous integration. *Optica* **2020**, *7*, 20–21. [CrossRef]
40. Fan, Y.; Oldenbeuving, R.M.; Klein, E.J.; Lee, C.J.; Song, H.; Khan, M.R.H.; Offerhaus, H.L.; Van der Slot, P.J.M.; Boller, K.J. A hybrid semiconductor-glass waveguide laser. In *Laser Sources and Applications II*; International Society for Optics and Photonics: Brussels, Belgium, 2004; Volume 9135, pp. 231–236.
41. Fang, A.W.; Park, H.; Cohen, O.; Jones, R.; Paniccia, M.J.; Bowers, J.E. Electrically pumped hybrid AlGaInAs-silicon evanescent laser. *Opt. Express* **2006**, *14*, 9203–9210. [CrossRef] [PubMed]
42. Zhu, Y.; Zeng, S.; Zhu, L. Optical beam steering by using tunable, narrow-linewidth butt-coupled hybrid lasers in a silicon nitride photonics platform. *Photonics Res.* **2020**, *8*, 375–380. [CrossRef]

43. Xu, Y.; Maier, P.; Blaicher, M.; Dietrich, P.; Marin-Palomo, P.; Hartmann, W.; Bao, Y.; Peng, H.; Billah, M.R.; Singer, S.; et al. Hybrid external-cavity lasers (ECL) using photonic wire bonds as coupling elements. *Sci. Rep.* **2021**, *11*, 16426. [CrossRef] [PubMed]
44. Stern, B.; Ji, X.; Dutt, A.; Lipson, M. Compact narrow-linewidth integrated laser based on a low-loss silicon nitride ring resonator. *Opt. Lett.* **2017**, *42*, 4541–4544. [CrossRef]
45. Kita, T.; Tang, R.; Yamada, H. Narrow spectral linewidth silicon photonic wavelength tunable laser diode for digital coherent communication system. *IEEE J. Sel. Top. Quantum Electron.* **2016**, *22*, 23–34. [CrossRef]

Editorial

Editorial for Special Issue "Frontiers of Semiconductor Lasers"

Yongyi Chen [1,2,*] and Li Qin [1]

[1] State Key Laboratory of Luminescence and Applications, Changchun Institute of Optics, Fine Mechanics and Physics, Chinese Academy of Sciences, Changchun 130033, China
[2] Jlight Semiconductor Technology Co., Ltd., Changchun 130033, China
[*] Correspondence: chenyy@ciomp.ac.cn

Since the end of the last century, in which semiconductor lasers were fast developing, this kind of laser and its applications have greatly improved our world. Semiconductor lasers are carriers of both energy for industry and information for optical communications. The recent progress in research on semiconductor lasers has offered new perspective regarding both their material growth and structural design. In this Special Issue of *Crystals*, we have gathered twelve peer-reviewed papers that shed light on recent advances in the field of semiconductor lasers and their applications.

Shunhua Wu et al. reported on the theoretical and experimental lasing performance of a 2 mm laser bar. [1]. An experimental electro-optical conversion efficiency of 71% was demonstrated, with a slope efficiency of 1.34 W/A and an injection current of 600 A at a heatsink temperature of 223 K. Qiaoxia Gong et al. established a thermal–structural coupling model and analyzed the influences of the pump power, cavity structure, and crystal size [2]. The highest temperature rise was also analyzed. Meanwhile, increasing the curvature radius of the cavity mirror and the length and width of the crystal, or decreasing the thickness of the crystal, was also found to be beneficial. Jinliang Han et al. reported a method of compressing the spectral linewidth and tuning the central wavelength of multiple high-power diode laser arrays in an external cavity feedback structure based on one volume Bragg grating (VBG) [3]. A diode laser source producing 102.1 W at an operating current of 40 A was achieved using a combination of beam collimation and spatial beam technologies. Moreover, a tuning central wavelength ranging from 776 to 766.231 nm was realized by precisely controlling the temperature of the VBG, and the locked central wavelength, as a function of temperature, shifted at a rate of approximately 0.0076 nm/°C. Zhuo Zhang et al. designed an ultra-long stable oscillating laser cavity with a transmission distance of 10 m [4]. The proposed wireless energy transmission scheme based on a VECSEL laser is the first of its kind to yield a 1.5 m transmission distance output power that exceeds 2.5 W. Yuhang Ma et al. reviewed the progress in the development and application of external cavity quantum cascade lasers [5]. Nan Zhang et al. discussed the merits of solution-processed perovskite semiconductors as lasing gain materials and summarized the characteristics of a variety of perovskite lasers [6]. Yanxin Shen et al. reviewed the 1.3 μm laser crystals and the progress made in their research, as well as some new optical crystals and novel materials [7]. Bin Wang et al. reviewed the principles of selective area epitaxy, including growth rate enhancement, composition variation, the vapor phase diffusion model, and bandgap engineering, as well as its applications, such as BH lasers, QD lasers, heteroepitaxial lasers on Si, EML, and MWLA, which are introduced in detail [8]. Shen Niu et al. briefly introduced the properties and working principles of the DFB laser array [9]. Keke Ding et al. reviewed the narrow linewidth external cavity semiconductor laser [10]. Yue Song et al. reviewed the reliability issues affecting semiconductor lasers throughout the whole supply chain, including the failure modes and causes of failure for high-power semiconductor lasers, the principles and application status of accelerated aging experiments and lifetime evaluation, and common techniques used for high-power semiconductor laser failure analysis [11]. Xuan Li reviewed two main device-integrated

Citation: Chen, Y.; Qin, L. Editorial for Special Issue "Frontiers of Semiconductor Lasers". *Crystals* **2023**, *13*, 349. https://doi.org/10.3390/cryst13020349

Received: 8 February 2023
Accepted: 15 February 2023
Published: 18 February 2023

Copyright: © 2023 by the authors. Licensee MDPI, Basel, Switzerland. This article is an open access article distributed under the terms and conditions of the Creative Commons Attribution (CC BY) license (https://creativecommons.org/licenses/by/4.0/).

structures for achieving widely tunable, narrow-linewidth external cavity lasers on silicon substrates, such as the MRR-integrated structure and MRR–MZI-integrated structure of external cavity semiconductor lasers. The results show that the silicon-substrate-integrated external cavity lasers offer a potential way to realize a wide tuning range, high power, single mode, stable spectral output, and high side mode suppression ratio laser output. [12].

As shown in this Special Issue of *Crystals*, the study of semiconductor lasers and their applications continues to grow and expand as we, as a community, strive to acquire further understanding of the underlying potential of these lasers. The goal of this Special Issue is to bring these and other new concepts closer to application in the field of semiconductor lasers and beyond.

Data Availability Statement: The authors declare no data availability for this editorial.

Conflicts of Interest: The authors declare no conflict of interest.

References

1. Wu, S.; Li, T.; Wang, Z.; Chen, L.; Zhang, J.; Zhang, J.; Liu, J.; Zhang, Y.; Deng, L. Study of Temperature Effects on the Design of Active Region for 808 nm High-Power Semiconductor Laser. *Crystals* **2023**, *13*, 85. [CrossRef]
2. Gong, Q.; Zhang, M.; Lin, C.; Yang, X.; Fu, X.; Ma, F.; Hu, Y.; Dong, L.; Shan, C. Analysis of Thermal Effects in Kilowatt High Power Diamond Raman Lasers. *Crystals* **2022**, *12*, 1824. [CrossRef]
3. Han, J.; Zhang, J.; Shan, X.; Zhang, Y.; Peng, H.; Qin, L.; Wang, L. Tunable, High-Power, Narrow-Linewidth Diode Laser for Potassium Alkali Metal Vapor Laser Pumping. *Crystals* **2022**, *12*, 1675. [CrossRef]
4. Zhang, Z.; Zhang, J.; Gong, Y.; Zhou, Y.; Zhang, X.; Chen, C.; Wu, H.; Chen, Y.; Qin, L.; Ning, Y. Long-Distance High-Power Wireless Optical Energy Transmission Based on VECSELs. *Crystals* **2022**, *12*, 1475. [CrossRef]
5. Ma, Y.; Ding, K.; Wei, L.; Li, X.; Shi, J.; Li, Z.; Qu, Y.; Li, L.; Qiao, Z.; Liu, G.; et al. Research on Mid-Infrared External Cavity Quantum Cascade Lasers and Applications. *Crystals* **2022**, *12*, 1564. [CrossRef]
6. Zhang, N.; Na, Q.; Xie, Q.; Jia, S. Development of Solution-Processed Perovskite Semiconductors Lasers. *Crystals* **2022**, *12*, 1274. [CrossRef]
7. Shen, Y.; Fu, X.; Yao, C.; Li, W.; Wang, Y.; Zhao, X.; Fu, X.; Ning, Y. Optical Crystals for 1.3 μm All-Solid-State Passively Q-Switched Laser. *Crystals* **2022**, *12*, 1060. [CrossRef]
8. Wang, B.; Zeng, Y.; Song, Y.; Wang, Y.; Liang, L.; Qin, L.; Zhang, J.; Jia, P.; Lei, Y.; Qiu, C.; et al. Principles of Selective Area Epitaxy and Applications in III–V Semiconductor Lasers Using MOCVD: A Review. *Crystals* **2022**, *12*, 1011. [CrossRef]
9. Niu, S.; Song, Y.; Zhang, L.; Chen, Y.; Liang, L.; Wang, Y.; Qin, L.; Jia, P.; Qiu, C.; Lei, Y.; et al. Research Progress of Monolithic Integrated DFB Laser Arrays for Optical Communication. *Crystals* **2022**, *12*, 1006. [CrossRef]
10. Ding, K.; Ma, Y.; Wei, L.; Li, X.; Shi, J.; Li, Z.; Qu, Y.; Li, L.; Qiao, Z.; Liu, G.; et al. Research on Narrow Linewidth External Cavity Semiconductor Lasers. *Crystals* **2022**, *12*, 956. [CrossRef]
11. Song, Y.; Lv, Z.; Bai, J.; Niu, S.; Wu, Z.; Qin, L.; Chen, Y.; Liang, L.; Lei, Y.; Jia, P.; et al. Processes of the Reliability and Degradation Mechanism of High-Power Semiconductor Lasers. *Crystals* **2022**, *12*, 765. [CrossRef]
12. Li, X.; Shi, J.; Wei, L.; Ding, K.; Ma, Y.; Li, Z.; Li, L.; Qu, Y.; Qiao, Z.; Liu, G.; et al. Research on Silicon-Substrate-Integrated Widely Tunable, Narrow Linewidth External Cavity Lasers. *Crystals* **2022**, *12*, 674. [CrossRef]

Disclaimer/Publisher's Note: The statements, opinions and data contained in all publications are solely those of the individual author(s) and contributor(s) and not of MDPI and/or the editor(s). MDPI and/or the editor(s) disclaim responsibility for any injury to people or property resulting from any ideas, methods, instructions or products referred to in the content.

MDPI
St. Alban-Anlage 66
4052 Basel
Switzerland
Tel. +41 61 683 77 34
Fax +41 61 302 89 18
www.mdpi.com

Crystals Editorial Office
E-mail: crystals@mdpi.com
www.mdpi.com/journal/crystals

www.ingramcontent.com/pod-product-compliance
Lightning Source LLC
LaVergne TN
LVHW070415100526
838202LV00014B/1460